High Power Lasers

Pergamon Titles of Related Interest

INTERNATIONAL INSTITUTE OF WELDING
Electron and Laser Beam Welding
Physics of Welding, 2nd Edition

KING
Surface Treatment and Finishing of Aluminium

NIKU-LARI
Advances in Surface Treatments, Volumes 1–5

RYKALIN et al
Laser Machining and Welding

WATERHOUSE & NIKU-LARI
Metal Treatments Against Wear, Corrosion, Fretting and Fatigue

Pergamon Related Journals *(free specimen copy gladly sent on request)*

Acta Metallurgica

Corrosion Science

Engineering Fracture Mechanics

Fatigue and Fracture of Engineering Materials and Structures

Journal of Physics and Chemistry of Solids

Journal of the Mechanics and Physics of Solids

Materials and Society

Materials Research Bulletin

Progress in Materials Science

Scripta Metallurgica

Solid State Communications

Welding in the World

High Power Lasers

Edited by
A. NIKU-LARI
*Institute for Industrial Technology Transfer, 40 Promenade Marx Dormoy
Gournay-sur-Marne, F-93460, France*

and

B. L. MORDIKE
*Technische Universität Clausthal, Agricolastrasse 2
D-3392 Clausthal-Zellerfeld, Federal Republic of Germany*

PERGAMON PRESS
OXFORD · NEW YORK · BEIJING · FRANKFURT
SÃO PAULO · SYDNEY · TOKYO · TORONTO

U.K.	Pergamon Press plc, Headington Hill Hall, Oxford OX3 0BW, England
U.S.A.	Pergamon Press, Inc., Maxwell House, Fairview Park, Elmsford, New York 10523, U.S.A.
PEOPLE'S REPUBLIC OF CHINA	Pergamon Press, Room 4037, Qianmen Hotel, Beijing, People's Republic of China
FEDERAL REPUBLIC OF GERMANY	Pergamon Press GmbH, Hammerweg 6, D-6242 Kronberg, Federal Republic of Germany
BRAZIL	Pergamon Editora Ltda, Rua Eça de Queiros, 346, CEP 04011, Paraiso, São Paulo, Brazil
AUSTRALIA	Pergamon Press Australia Pty Ltd., P.O. Box 544, Potts Point, N.S.W. 2011, Australia
JAPAN	Pergamon Press, 5th Floor, Matsuoka Central Building, 1-7-1 Nishishinjuku, Shinjuku-ku, Tokyo 160, Japan
CANADA	Pergamon Press Canada Ltd., Suite No. 271, 253 College Street, Toronto, Ontario, Canada M5T 1R5

Copyright © 1989 Pergamon Press plc

All Rights Reserved. No part of this publication may be reproduced, stored in a retrieval system or transmitted in any form or by any means: electronic, electrostatic, magnetic tape, mechanical, photocopying, recording or otherwise, without permission in writing from the publishers.

First edition 1989

Library of Congress Cataloging in Publication Data

High power lasers/edited by A. Niku-Lari and
B. L. Mordike.—1st ed.
p. cm.— (Technology transfer handbook series)
Includes index.
1. High power lasers. 2. Lasers—Industrial applications.
I. Niku-Lari, A.   II. Mordike, Barry L.   III. Series.
TA1677.H535   1989   621.36'6—dc19   89–31104

British Library Cataloguing in Publication Data

High power lasers.
1. Lasers
I. Niku-Lari, A.   II. Mordike, B.L.   III. Series
621.36'6
ISBN 0-08-035918-3

Printed in Great Britain by BPCC Wheatons Ltd, Exeter

Technology Transfer Series

INTRODUCTION

The main objective of the Technology Transfer Series is to increase the calibre of engineers and scientists by updating and widening their ability.

It is well known that the international competitiveness of companies is dependent on a whole range of skills and knowledge. Engineers and technicians will therefore need to make a vital contribution to ensure that companies sharpen their competitive edge in world markets. This is particularly true in the case of new technologies (represented by such buzzwords as CAD, CAM, FMS, CAPM, CAI, CAQA, robotics and control, simulation, processing by laser, plasma, water-jet techniques, surface treatment, BEM, FEM, composites ... These are all computer related, and are part of the advanced manufacturing technology).

The increasing demand worldwide for skill and expertise in the above sectors of industry cannot be satisfied simply by new graduates in these disciplines. New and practising engineers should be trained to meet this demand.

The Technology Transfer Series, which will consist of conference proceedings on the above-mentioned and related disciplines, will help industrial and educational institutions to achieve many of these objectives. In addition, materials from seminars and short courses will also be published in this series. Two Pergamon guide book series, 'Advances in Surface Treatments' and 'Structural Analysis Systems', will presently become subgroups of the overall Technology Transfer Series. Several additional subseries are being planned (High Power Lasers, Fatigue and Stress).

Technology transfer is about people. IITT-International events and Pergamon publications will therefore bring together people for the successful interaction of their needs.

Dr A. Niku-Lari
Head of IITT-International
Editor

Institute for Industrial Technology Transfer
40 Promenade Marx Dormoy
F-93460 Gournay-sur-Marne
France

Tel: 33 (1) 45 92 17 71
Telex: 250303 (ATTN IITT)

PREFACE

The application of lasers to material working and treatment is becoming more widespread. It is commonplace for stress-free cutting and drilling. Welding applications are being developed where the advantages of the laser, namely very localized heating and low distortion, are fully exploited. Surface treatment is a relative newcomer to this list of laser applications. This comprises surface melting, surface alloying and cladding. Potentially these are very important developments in laser technology.

The papers in this Guide Book were presented at the Paris and Zurich meetings of the Institute for Industrial Technology Transfer. They show not only the developments in the various laser processes, but also the properties of laser-treated materials and the effect of the laser treatment on other properties. Although laser technology is still in its infancy, progress is rapid. It will not be long before these developments are exploited industrially.

Professor B. L. Mordike
Clausthal University

INTERNATIONAL SCIENTIFIC COMMITTEE

Dr. S. AMARJIT, *Canada*
Dr. L. P. BORUC, *Poland*
Prof. CHING PIAO HU, *Taiwan*
Dr. R. DEKUMBIS, *Switzerland*
Prof. T. ERICSSON, *Sweden*
Dr. H. E. FRANKS, *USA*
Prof. L. M. GALANTUCCI, *Italy*
Prof. S. K. GHOSH, *UK*
Prof. J. T. M. DE HOSSEN, *Netherlands*
Prof. P. MEYRUEIS, *France*
Prof. T. MIKIO, *Japan*
Prof. B. L. MORDIKE, *FRG*
Dr. A. NIKU-LARI, *France*
Prof. C. PALMA, *Italy*
Prof. E. RAMOUS, *Italy*
Dr. M. ROTH, *Switzerland*
Dr. M. C. SEEGERS, *Netherlands*
Prof. A. VANNE, *France*
Dr. C. VERPOORT, *Switzerland*
Dr. J. ZAHAVI, *Israel*

CONTENTS

Technology Transfer Series: Introduction   v
Preface   vii
International Scientific Committee   ix

1 Laser Surface Treatment   1

Properties of Laser-nitrided Surface Layers on Titanium
B.L. Mordike   3

Surface Treatment by Melting with Laser
E. Ramous   13

Wear Performance of Laser-melted Steels
H. de Beurs and J.T.M. de Hosson   27

2 Laser Surface Coating and Alloying   39

Plasma-deposited Laser-remelted Wear-resistant Layers
B.L. Mordike and W.N. Kahrmann   41

Laser Applications for Steel Surface Alloying with Carbon
E. Ramous, A. Tiziani and L. Giordano   49

An Experimental Study on CO_2 500 W Laser Surface Alloying of Chromium on Copper
L.M. Galantucci, G. Ruta and S. Magnanelli   57

Laser Surface Modification of Plasma-sprayed Oxide Ceramic Coat Layer for Anti-corrosion Performance
M. Takemoto   75

Effect of Laser Melting on Eutectium of Boriding Layer on Steel
Zhang Luting, Shao Huimeng and Zhu Jingpu   89

3 The Mechanism of Surface Modifications by Laser   99

The Effect of Laser Irradiation on Structure and
Microhardness of AISI 1045 Steel
    M. Riabkina-Fishman and J. Zahavi   101

Microstructure and Microcracking of Laser-glazed
Fe-B-C Surfaces
    S. Staniek and E. Hornbogen   117

Effect of Laser Heating on the Substructure of
0.4% Carbon Steel
    M.C. Seegers, S. Mandziej and J. Godijk   131

Laser Surface Remelting and Alloying of
Aluminium Alloys
    P.L. Antona, S. Appiano and R. Moschini   143

Coating Design for Laser Transformation Hardening
    E.W. Kreutz and K. Wissenbach   159

4 Laser Machining   193

Laser Cutting
    M. Querry   195

Laser Cutting Using CNC Techniques
    J.C. Beitialarrangoitia, G.E. Garcia de Vicuna and
    S.K. Ghosh   213

Defects Arising from Laser Machining of Materials
    G.E. Garcia de Vicuna, J.C. Beitialarrangoitia and
    S.K. Ghosh   227

SL 25, a New High Performance 2.5 kW
Industrial CO_2 Laser
    V. Fantini and G. Incerti   251

5 Research and Development in Laser Technology and Application   257

Measuring of Surface Roughness by Holographic Phase
Shifting Interferometry
    Ming Chang and Ching-Piao Hu   259

Superresolution in Microscopy Through Holographic
Synthetic Aperture
    P. De Santis, F. Gori, G. Guattari and C. Palma   271

Holographic Optical Elements on Dichromated Gelatin,
from Refractive to Diffractive Optics
    P. Meyrueis and R. Piel   279

A Fast Experimental Method of Measuring Laser Beam
Absorption as a Function of Temperature in Solids
    R. Dekumbis, H. Mayer and P. Fernandez   289

Optical Fiber Temperature Sensing by Mode Filtering
   A. Cahkari and P. Meyrueis   297

Surface Microanalysis with Laser Fourier Transform and Synthetic Hologram Processing
   E. Soubari, P. Meyrueis and M. Torzynski   307

Flexible Metal Waveguide for CO_2 Laser Processing
   M. Torjinsky, P. Meyrueis and C. Liegeois   325

Author Index   335

Chapter 1

LASER SURFACE TREATMENT

PROPERTIES OF LASER NITRIDED SURFACE LAYERS ON TITANIUM

B. L. Mordike

Institut für Werkstoffkunde und Werkstofftechnik
Technische Universität Clausthal Agricolastr. 2
D-3392 Clausthal-Zellerfeld, FRG

ABSTRACT

Laser gas nitriding of titanium is shown to be a technologically feasible method of greatly improving the surface properties.

KEYWORDS

Surface Melting, Gas Alloying, Titanium Nitride.

INTRODUCTION

Over the last few years studies have been made into the possible uses of lasers in the surface treatment of metals (1,2,3). Although lasers are not employed on a wide scale in industry for surface treatment many possible applications have been identified. The advantages of laser treatment is that it can be applied locally. The total heat input is low and hence distortion can be avoided as can microstructural changes in the bulk of the material. The high rate of heating and cooling of the surface layer can result in microstructural refinement and associated with that an improvement in mechanical and chemical properties.

Initially much attention was devoted to surface melting of alloys to obtain the benefit of microstructural refinement. Now the emphasis is on producing the alloy composition on the surface which gives the best results; in other words laser surface alloying. In this paper techniques of laser surface alloying are discussed with particular emphasis on production and properties of laser produced layers on titanium.

ALLOYING TECHNIQUES

Traditionally laser surface alloying has been a two step process in which an alloying element has been deposited on the surface

and subsequently melted into the substrate by passing a laser
beam over the surface. The thickness of the deposited layer and
the depth of the substrate melted are used to control total alloy
layer depth and its composition. Long range convection in the
molten layer ensures good mixing. Although titanium transforms
martensitically on quenching and although this can be influenced
by alloying, no improvement in properties is observed as is the
case for many iron based alloys. An alternative means of improving the strength must be sought e.g. by a dispersion of a second
hard phase. Elements can be found which produce hard second
phases on surface melting. It is known genrally from surface
treatment of metals that nitrides possess good properties.
Titanium is often plasma nitrided to good effect.

Plasma nitriding is a slow process due essentially to the low
rate of diffusion of nitrogen. By laser melting in the presence
of nitrogen, nitrides are produced and distributed throughout
the molten layer. The layer, which is a mixture of TiN dendrites
and Ti is much thicker then the solid TiN layer produced by
other techniques.

The process of gas alloying can be employed for several systems
e.g. Fe-C, Ti-N, Ti-C (4). The precise mechanism of absorption/
desorption depends in each particular case on the nature of the
gas-metal system.

The behaviour in general can be described by the law of mass
action,

e.g. $uA + vB \rightleftarrows wC + xD + \Delta H$

$$Kc = \frac{C^w \cdot D^x}{A^u \cdot B^v}$$

In gaseous systems the concentrations can be replaced by the
partial pressures. In the case of nitriding of titanium

$Ti + 1/2\ N_2 \rightleftarrows TiN + \Delta H$

the partial pressure of N will together with the temperature
determine the rate and the direction of the reaction. The
nitrogen is diluted in practice by an inert carrier gas to
facilitate control.

The phase diagrams for Ti-N and Ti-C are shown in Fig. 1. The
two diagrams are similar. TiN forms at temperatures in excess
of 2600 °C and precipitates dendritically. At sufficiently
long nitriding times a completely TiN melt can be obtained.

EXPERIMENTAL DETAILS

CO_2 laser radiation is absorbed well by titanium. Very high
power lasers are not required. The present work was undertaken
using a Coherent Everlase CO_2 laser with a continuous power
output of 600 - 650 watts with a pulsed mode with up to 4 kW.
For experimental purposes a simple arrangement as shown in Fig.
2 was used to provide the reactant gas. This is perfectly

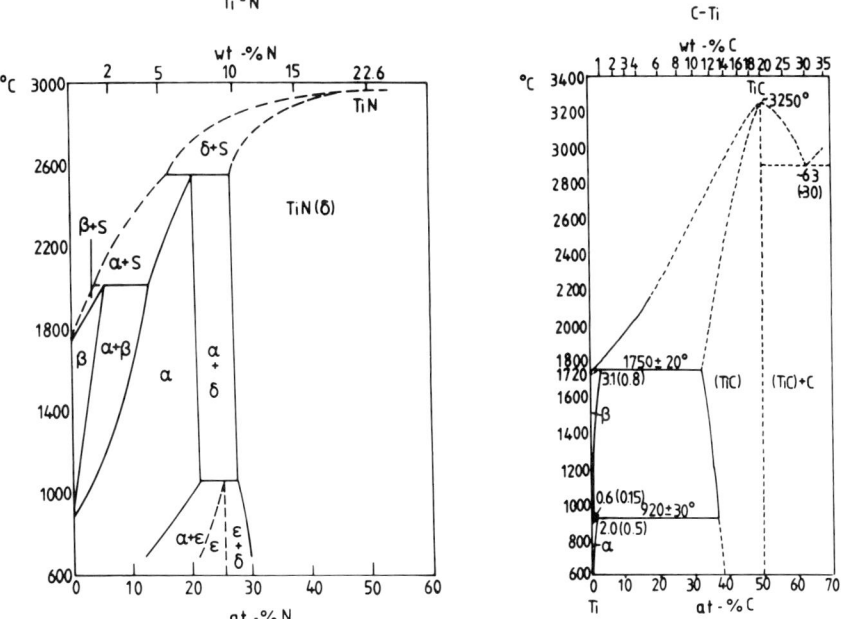

Fig. 1. Phase diagrams for the systems Ti-N and Ti-C.

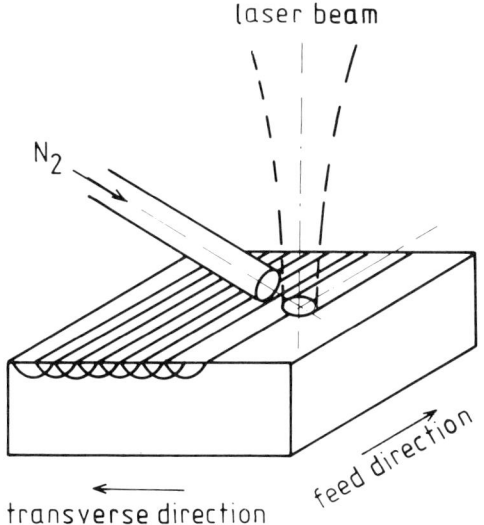

Fig. 2. Simple arrangement for providing gas atmosphere.

satisfactory for flat massive specimens which remain essentially cold. For more complicated shapes or where the whole sample becomes hot more elaborate arrangements are necessary to prevent oxidation and distortion. Typical parameters for specimens about

10 mm thick are: focussed beam 0.2 mm ∅, specimen feed rate
0.1 → 100 m/min, pulsed mode 2 ms 100 Hz or 0.4 ms 500 Hz to
produce a wide range of melted depths. The degree of nitride
formation depends on the interaction time, power, volume of
molten metal and partial pressure of nitrogen. The optimum
parameters must be determined empirically.

Figure 3 shows the effect of feed rate on melt depth for various
atmospheres and specimen temperatures for constant laser power.
The maximum depth for a Coherent Everlase 650 Watt is about 0.5
mm. Increasing the laser power to 1.2 kW laser enables a thicker
layer to be melted. The gas atmosphere influences the plasma
formation and this can aid energy absorption. Consequently the
melted depth in different atmospheres is different.

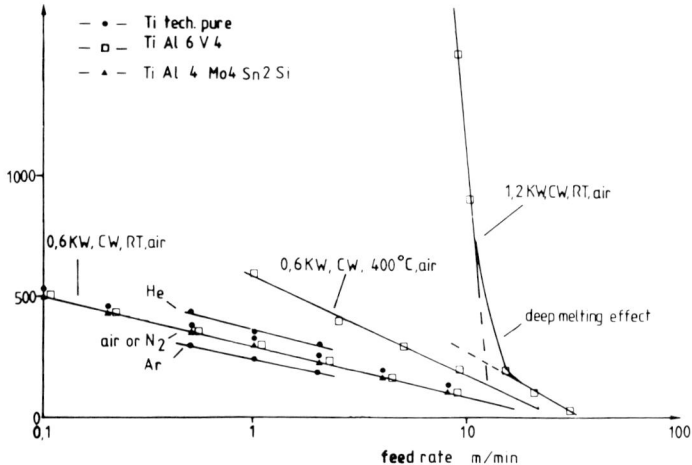

Fig. 3. Melted depth as a function of feed rate for various
atmospheres and laser power.

If, as in nitriding, we require a reaction to take place then at
high feed rates the interaction time is too low. Several passes
may be required to obtain an acceptable degree of nitride formation. There is thus a practical lower limit. At low feed rates
the tendency to cracking increases. This is essentially a function of the thickness of the massive layer and the result of
differential expansion.

The very rapid increase in depth obtainable with lasers > 1 kW
is due to the deep welding or keyhole effect. Sometimes suitable
fixing and support devices must be developed to prevent distortion. This is particularly the case for small parts. The heat
removal must be improved so that the temperature of the part does
not become too high.

PROPERTIES OF THE LAYER

PROPERTIES OF THE LAYER

TiN forms dendritically in the melt. The fineness of the dendrites depends on the feed rate or rate of cooling. The surface consists of a mixture of TiN and α-titanium. The proportion of TiN increases with decreasing feed rate and number of passes. The hardness of the layer increases with the proportion of TiN. The maximum hardness of the layer is obviously that for pure TiN but the thickness is limited by cracking problems. A thicker, mixed, less hard layer may even be more suitable for the application. A pure TiN covering layer can always be produced if necessary.

Figure 4 shows a series of micrographs of laser-nitrided surfaces demonstrating how the density and quantity of dendrites increase on multiple treatment. The extent of the heat affected zone is also apparent.

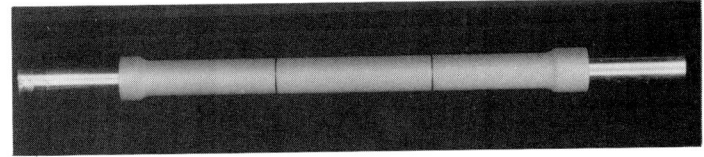

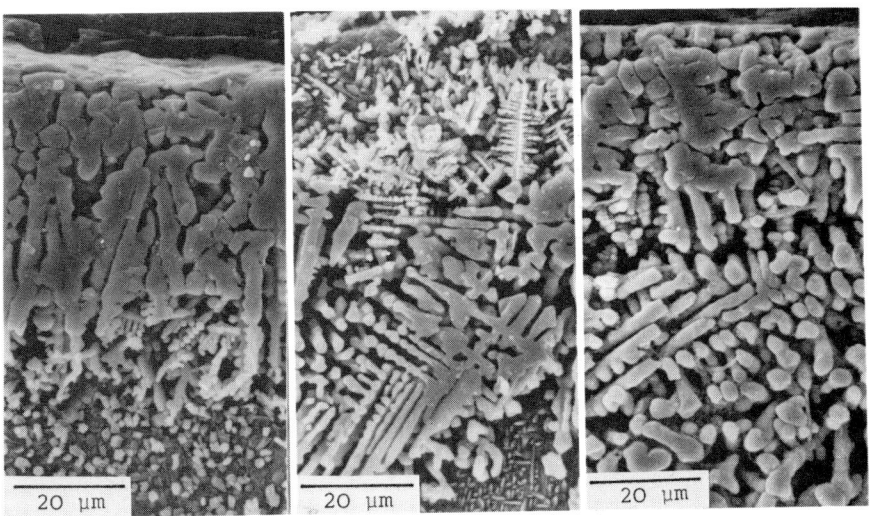

Fig. 4. TiAl6V4 (IMI 318) specimen gas nitrided for one, two and three passes, together with the corresponding micrographs.

The hardness of a laser nitrided surface of IMI 318 is shown in Fig. 5. The phases produced in the various regions are also indicated. Under optimum conditions the melted layer is about

400 μm thick. The dense mixed TiN-α'Ti-layer is about 250 μm thick.

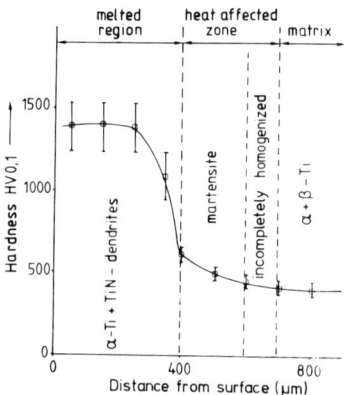

Fig. 5. Hardness profile for a typical laser nitrided surface.

The roughness of the laser melted surface depends on the thickness of the layer melted and degree of overlap of successive passes. Nitrided layers are somewhat rougher than simple melted layers. Figure 6 shows the roughness measured for 3 conditions, as received (ground), melted in air, melted in nitrogen. The as ground condition was in fact the roughest.

The improvement in wear properties due to TiN layers is well established. The question which must be answered is whether other properties are affected e.g. fatigue, strength or chemical properties.

Melting a surface layer will produce tensile stresses in the surface and these may assist crack propagation and hence reduce the fatigue life. A simple test of fatigue is rotating bending. Figure 7 shows Wöhler curves for TiAl6V4 after various treatment: Titanium as received in the solution treated and aged state has the best fatigue behaviour. Annealed titanium has relatively poor fatigue properties. Laser surface melting or nitriding reduces the fatigue life as is shown in Fig. 7. This is due mainly to the presence of tensile stresses in the surface layer. This deterioration may not be important as not all applications involve the danger of fatigue.

The tensile behaviour is also influenced by laser treatment. There is a reduction in ultimate tensile strength of about 5 %. The reduction in the ductility to fracture is more pronounced (Table I). Cracking of the nitride layer occurs before the metal macroscopically yields. The lasered layer does not flake off apart from isolated sites in the heavily necked region.

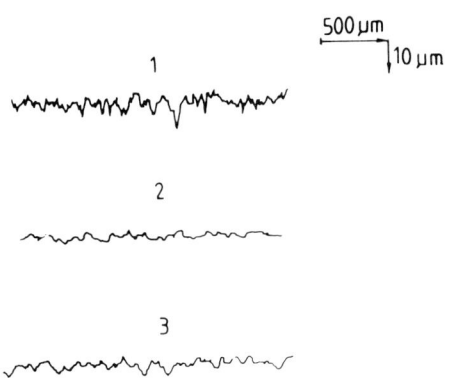

Fig. 6. Surface roughness for as received (ground) specimens, and specimens after surface melting and surface nitriding.

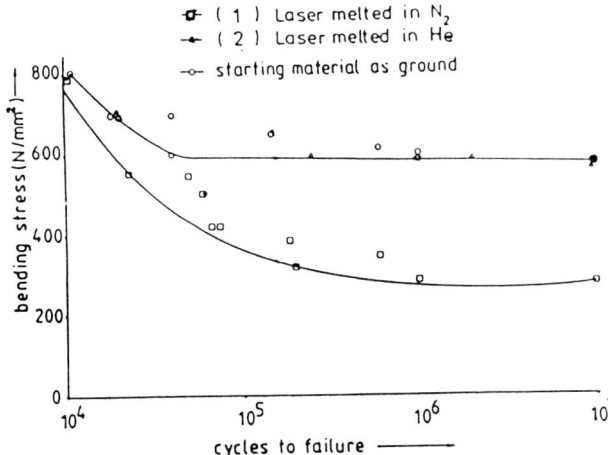

Fig. 7. Wöhler curves for titanium IMI 318 in the states as received, melted and nitrided.

Table I: Tensile Test Data for IMI 318

Condition	U.T.S. (N/mm²)	% Elongation
As received	1010	19
Laser melted in He (c.w.)	955	6
Laser melted in N_2 (c.w.)	975	4
Laser melted in N_2 (pulsed)	985	6,5

The corrosion resistance of titanium is relatively good. Surface gas alloying produces in general further improvement. TiN and TiC are both nobler than Ti which leads to a displacement of the corrosion potential. They passivate at lower current densities (10 - 100 times less) as is shown in Fig. 8.

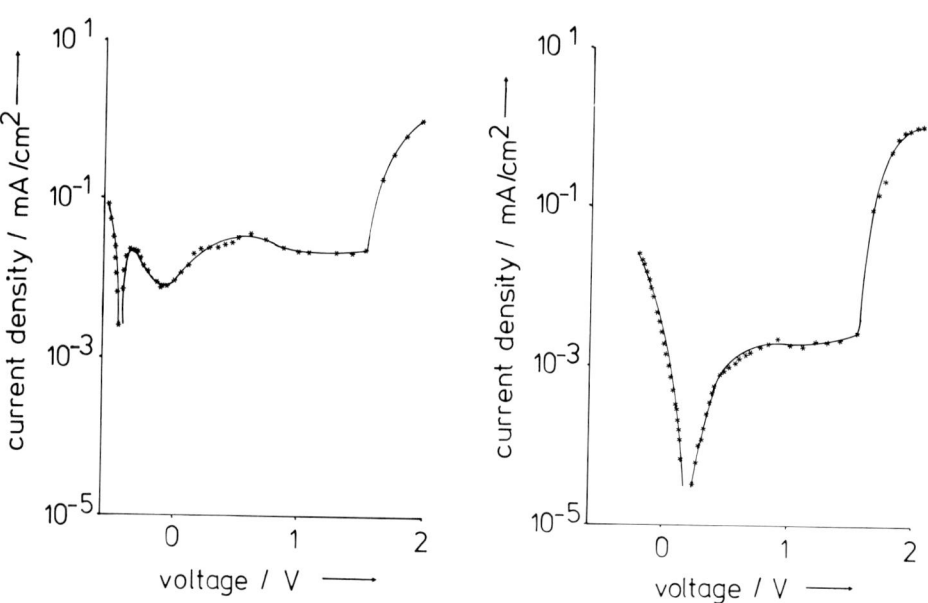

Fig. 8. Effect of laser treatment on corrosion behaviour.

CONCLUSION

It has been shown that laser surface treatment of titanium is possible. Much thicker layers can be produced than with alternative techniques. An extension in the application of titanium components is thus possible.

REFERENCES

1) C. Draper, Int. Metals Review 1985.

2) B.L. Mordike, H.W. Bergmann, N. Groß, Z. Werkstofftechnik $\underline{14}$ (1983) 253.
3) B.L. Mordike, H.W. Bergmann, Mat. Res. Sci. Symp. Proc. Vol. 8 North Holland (1982) 463.
4) B.L. Mordike, Nato School San Miniato 1985 in press.

SURFACE TREATMENT BY MELTING WITH LASER

E. Ramous

University — Istituto Chimica Industriale
Via Marzolo, 9–35128 Padova, Italy

INTRODUCTION

Rapid solidification processing (RSP) commonly results in departure from conventional microstructures, giving an improvement of grain refinement and chemical homogeneity (1). Fine grained structures generally have improved strength and reduced anisotropy. Chemical homogeneity deriving from non-equilibrium extended solutions offers opportunities for subsequent heat treatment to produce final microstructure of optimized properties.

High-density heating devices, such as laser beams, by rapid melting and subsequent solidification of surface layers, may produce typical effects of RSP in surface treatments. Range of surface modifications is enlarged by possibility of introducing other elements by surface alloying during melting (2).

Microstructures resulting from laser surface melting (LSM) are influenced by thermal transients induced into surface layers. Therefore the background of laser surface treatment is a suitable knowledge of relationships between treatment parameters, like power density of the beam and interaction time, the induced thermal transient during solidification and its influence on resulting microstructures (3).

In this paper some peculiar features and problems arising in metal surface treatment by laser melting will be examined.

MICROSTRUCTURAL EFFECTS

The main effect of rapid solidification after surface melting by laser is the possibility to achieve relatively large supercooling of the melt. This may result both in constitutional changes of phases and in marked modifications in morphology of final microstructures.

Of course, the type of phase produced from solidification is determined both by thermodynamic and kinetic characteristics. Limits and possibilities of metastable, or eventually amorphous, **phase productions can be illustrated by simple thermodynamics** considerations, starting from ΔG diagrams. For example, extension of solid solution range, or suppression of intermetallic compounds may be examined. To this end a minimum thermodynamic condition is that the alloy must be cooled below its T_0 temperature (where free energy values of liquid and first solid phase are equal), without formation of any other phase, Fig. 1. Generally **temperature** against composition curves extend over the range of equilibrium solid solutions, and, in some cases (i.e. Ag-Cu system, Fig. 1) are continuous between the component, in the entire range of alloy composition. Undercooling below T_0 then allows the possibility of extended solid solubility under conditions of sufficiently rapid solidification, as easily attainable by laser surface melting. In systems with discontinuous T curves, in the intermediate composition range, where partitionless solidification based on either of the terminal phases is undesirable, formation is favoured of different non-equilibrium crystalline or glassy phases (4).

Quenching rate $\dot{T}$ during solidification, together with other parameters like thermal gradient G and solidification front velocity R ($\dot{T}$ = GxR) determine the morphology of solid phases. Normally the undercooling degree before solidification is increasing with $\dot{T}$, favouring formation of finer microstructure for any morphology, coming from increasing nucleation frequency. Indeed morphology may be planar, cellular or dendritic, being determined by the level of "constitutional supercooling". This effect is the result of the liquidus temperature varying with composition and composition varying at solidification front, due to solute partitioning there. This means that liquidus temperature rises over the actual temperature of liquid just ahead the melt front, encouraging growth of liquid/solid interface in the form of cells or dendrites. It may be defined a critical value Gcr (varying with R), above which flat interface becomes unstable. Decreasing G under Gcr, solid morphology turns from cellular to dendritic. As summarized in Fig. 2, microstructure of solid will depend upon G/R ratio (5).

Dendrite shape is also influenced by cooling rate $\dot{T}$, as it can be easily demonstrated assuming no convection in the interdendritic liquid. Relation between dendrite arm spacing 1 and GxR : $1\alpha(GxR)^{-\frac{1}{2}}$ have been verified for a number of different alloys composition. Often actual quenching rate during solidification is derived from metallographic measurement of dendritic arm spacing (6).

Surface finishing is actually one of limiting factor for current application of LSM, being strongly affected by flow of material in the interaction zone. Surface rippling is the result of several mechanisms, and was examined by only a few studies. In any case there is a backward flow of material around the depression where beam strikes. This can result in a molten thread flowing down the middle of the track. Of course the effect of surface tension is also important, strongly depending on this variation with temperature. However modelling of complex flow phenomena has been only attempted, and factors controlling final

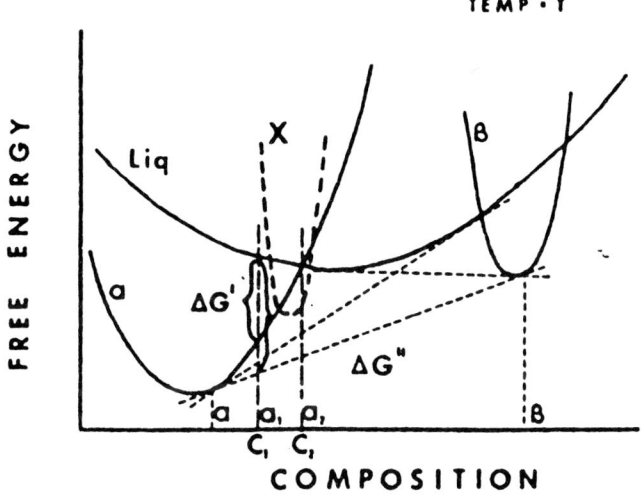

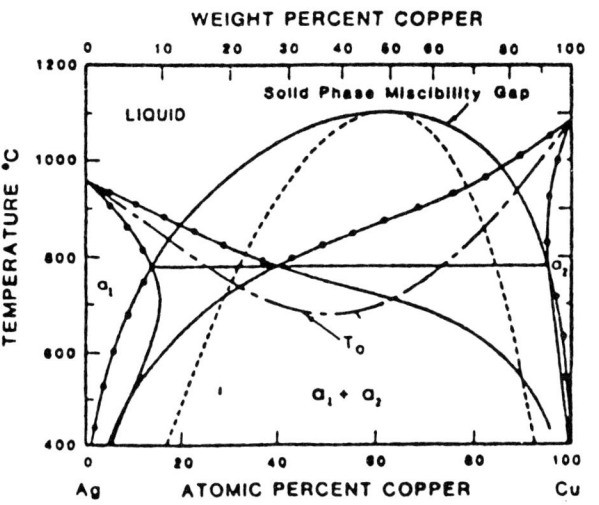

Fig. 1. Effect of rapid quenching on extending solid solutions limits. Example of Ag-Cu system.

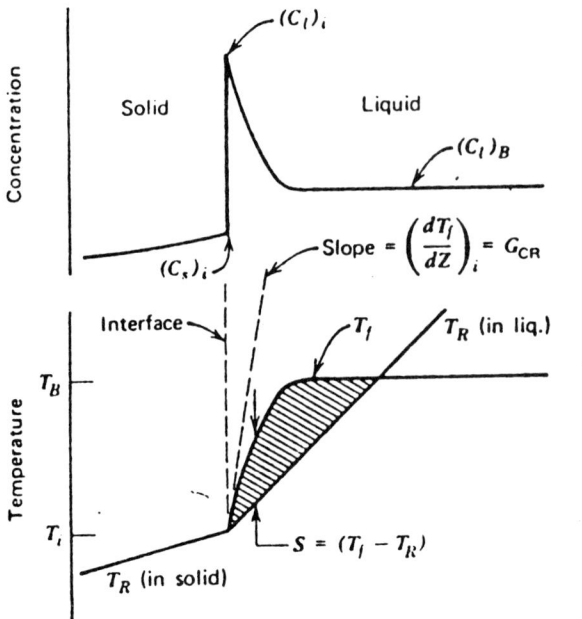
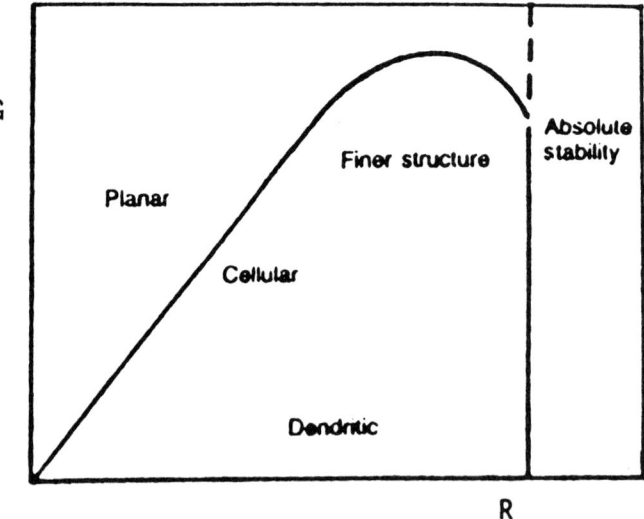

Fig. 2. Influence of G and R variations on solidification microstructures (schematic).

surface finishing are not yet completely understood, owing to this very important influence to determining feasibility of LSM (7).

Technological behaviour of treated surfaces is highly affected by final stress field generated by surface treatment. A schematic illustration of stress/temperature relationship during a laser melt run, is given in Fig. 3. Starting from point 1 (no stress), compressive stress first developed during heat fall due to plastic flow, down to zero at melting point [3], so remaining during melting [4]. Tensile stress starts to build up at solidification [5] and increases up to final value.

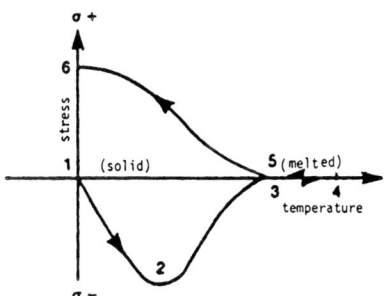

Fig. 3. Stress variation during a laser surface melting run (schematic).

This figure is completely modified in the case of transformation in solid phase. In steels the austenite to martensite transition with an approximate 4% volume increase, can produce compressive stress in treated surface, so increasing fatigue life and reducing the tendency to crack.

More common checks in laser melted tracks arise as **liquefaction** cracking in the dendritic structures before complete solidification, or, in iron alloys, in the heat affected zone under melted layers. In any case, troublesome residual stress may be removed by subsequent heat treatment.

THERMAL MODELS

The heat flow in a semi-infinite body beneath high intensity heat sources was looked at in a number of studies (8-10). However, most mathematical formulations of surface melting aimed at the determination of the position and shape of liquid-solid interface, in a pure substrate material, or, recently, in alloys solidifying in a temperature range.

These models behave quite satisfactorily to describe thermal transients in solid state surface treatments, like transformation hardening. However, application of some models to transformation involving surface melting and resolidification often results in a lack of agreement between calculated and experimental data. Often mathematical models of heat flow suffer from a number of limitations, such as the use of medium constant values of physical properties in a large range of temperatures, neglect of latent heat of fusion, etc.

Moreover, in addition to laser surface melting other difficulties arise from the complex phenomena of interaction between the laser beam and processed surface. Following surface temperature variation during interaction with laser, large differences may occur between actual and calculated thermal transients, even during heating period. This results from variation of absorption coefficient with temperature, which also makes variable heat to be actually absorbed. Indeed absorption coefficient presents a sharp discontinuity during transition of treated surface from solid to liquid phase: after surface melting absorbed power density decreases abruptly. Other particular features are striking agitation effects induced in melted pool. These phenomena are probably very effective in reducing thermal and composition gradients in liquid phase.

Finally, from all these considerations, it seems that up to now no reliable relationships have been established among absorbed heat flux, temperature gradients, and kinetic driving process parameters, like superheating and undercooling at liquid/solid interface, which however are very important in determining solidification microstructures.

Mathematical models, based on integration of Jaeger's equation or on application of finite elements, may supply only qualitative information on melted zone profile and on heating and cooling rates.

APPLICATION OF MODEL WITH VARIABLE HEAT FLUX

It is generally accepted that the absorbed energy varies during heating of metal surfaces by laser beam, particularly over the melting point. A tentative assessment of a model using variable thermal fluxes has been performed, using a finite difference model and comparing calculated and experimental data on a group of steel samples treated for surface melting by laser. On each steel sample, maximum melt depths and transformation hardening depths were determined. Therefore we assumed that temperature of about 900 °C was reached at transformation hardening depth and about 1150 °C at maximum melting depth (graphite dag as coating). Consequently we modified the profile of absorbed heat flux by a test and trend procedure to deduce absorbed flux profile better by fitting positions of aforementioned isotherms in treated samples.

First attempt was made using the profile shown in Fig. 4a, but finally better results were obtained using a slightly different profile, as in Fig. 4c. An example of corresponding thermal transients induced in a sample is shown in Fig. 5. It is interesting to note the plateau of temperature values near melting and subsequent slower temperature rise to achieve final maximum temperature on the surface. This type of temperature/time profile appears almost necessary to fit satisfactorily positions of both experimental isotherms (melting and austenitizing) in treated samples. However purely latent heat cannot entirely justify the plateau of surface temperature. This effect is possibly related to other phenomena influencing absorption behaviour during melting. In such a particular situation clearly a large modification of absorption occurs. This effect may be approximately simulated by an absorbed flux profile as indicated However a complete justification of discrepancy between calculated and experimental data up to now is not available to date.

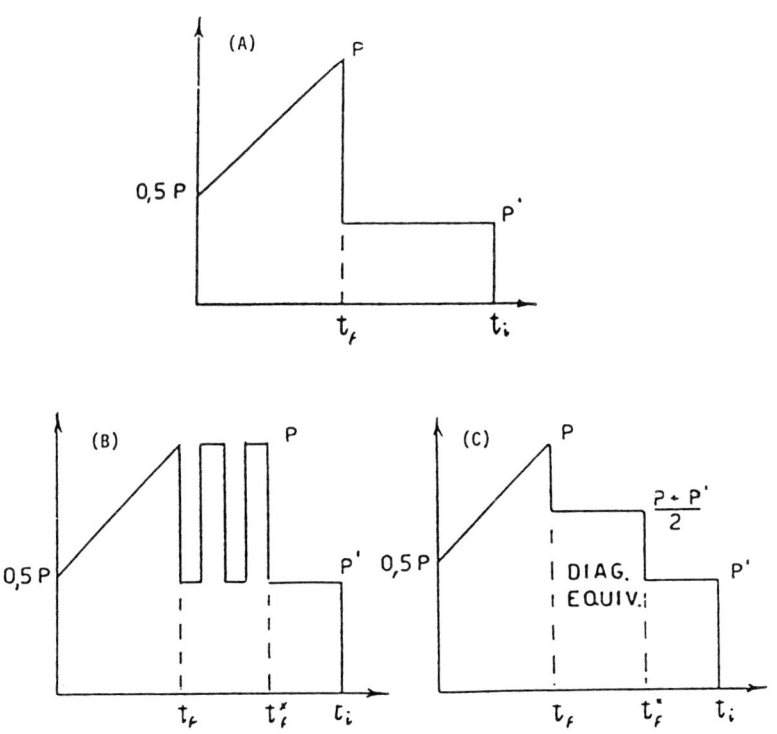

Fig. 4. Diagrams of absorbed heat fluxes used for tests of optimization of thermal models for laser surface melting processes.

EXPERIMENTAL

In this section, for example, some results are summarized concerning two typical applications of melting by laser for surface treatment of cast iron and tool steels.

Laser surface melting runs were performed using a cw CO_2 AVCO laser of 15 kW maximum power. Beam was focussed by an optical device "beam integrator" supplying a rectangular spot of adjustable dimensions (typically 10*10 mm). A moveable table allowed the sample to be moved at constant velocity beneath fixed laser beam. Incident power densities were in the range between 5 and 15 kW cm², and interaction time between 0.2 and 10 s. Actual

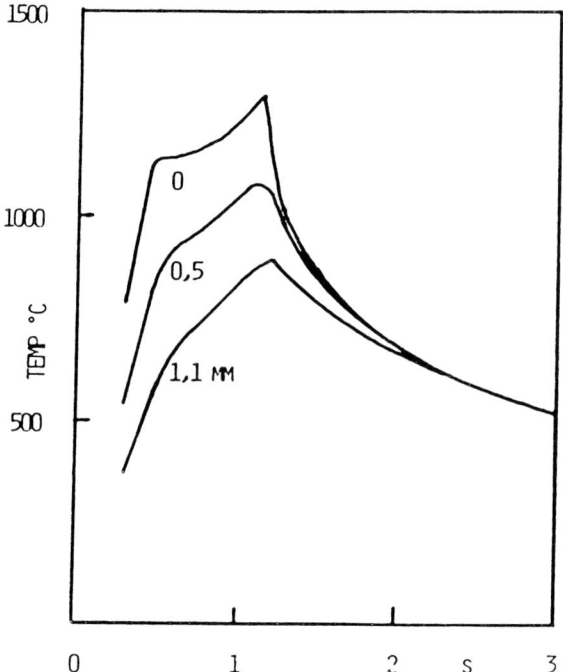

Fig. 5. Example of thermal transients calculated using a varying absorbed heat fluxe.

power arriving on the sample surface was measured by a cone calorimeter. Radiation absorption was improved by covering sample surfaces by a graphite dag in alcohol. Treated samples were examined by light and SEM microscopy, X-ray diffraction, and for some samples, by Mossbauer spectroscopy (CEMS).

CAST IRON

Treated samples were grey iron (3,1 % carbon, 2,16 % silicon, 0,5 % manganese) with flake graphite, between 3 and 5 ASTM grade, in pearlitic matrix (11). Best results were obtained using about 5 kW cm^2 and from 1 to 7 s interaction time. Typical examples of hardness profiles and microstructures are shown in Fig. 6. Figure 6a illustrates influence of sample thickness, and consequently of different thermal transients, on final hardness of melted layers. In thin samples, hardness improvement of surface layer is greatly reduced. Only in samples of sufficient thickness, over about 10 mm, did the laser melting and subsequent self-quenching induce a noticeable surface hardening, though more discontinuous, in samples over 20 mm thick.

These different patterns correspond to very different microstructures in samples of various thickness, coming from different

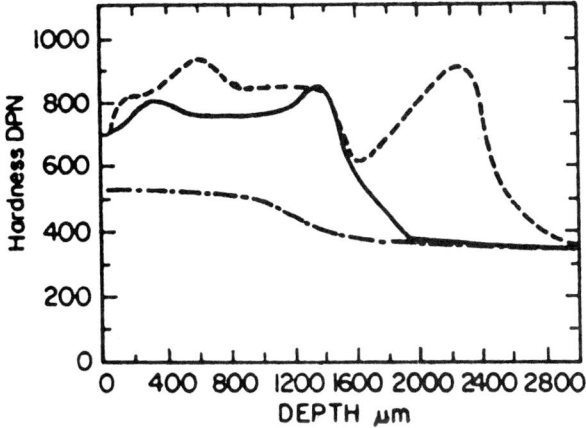

Fig. 6(a). Hardness profiles of laser treated cast iron samples of different thickness.

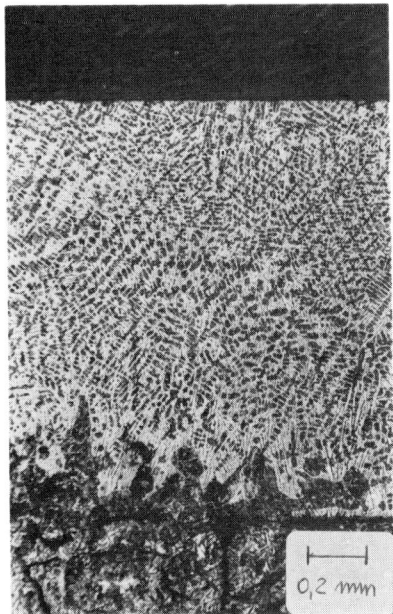

Fig. 6(b). Microstructure of treated layer.

thermal transients and self-quenching rates in heat affected layers. Indeed laser parameters directly determine only depth of melted layer, but final microstructure and hardness profile are determined by self-quenching rate, which depend on the workpiece thickness (or on preheating temperature of sample to be treated). So in thinner samples final temperature remains too high to attain a sufficient quenching rate for both martensitic and from grey to white iron transformation. Then final structure is only slightly modified resulting in small quantities of cementite in a matrix of fine pearlite. In the thicker samples, over 20 mm, structure obtained in treated layers is typical of white cast iron: primary cementite and dendrites of transformed austenite (Fig. 6b).

However in these samples unmelted transition layer between melted zone and the base material appears transformed into martensite and some retained austenite, by solid state transformation. This latter zone was found to favour initiation of microcracks going towards the surface of sample. But cracks formation, being related with martensitic transformation, may be avoided by choosing treatment parameters in order to control and reduce cooling rate in these zones, below critical rate for martensitic transformation. This is confirmed by the results we obtained on workpieces of intermediate thickness, between 10 and 20 mm. In these samples the martensitic transformation was confined to the melted layer, transition zone having a more complex microstructure ranging from bainite to fine pearlite. This structure pattern corresponds to the best hardness profile, without any crack or other defect. Some structural results may also be obtained in samples of higher thickness, by pre-heating the workpiece to modify thermal transients and reduce cooling rate after the heating by laser beam.

By our results, only workpieces of thickness ranging between about 10 and 20 mm may be surface hardened directly by laser melting. Corresponding values of melted layers vs. interaction time, are shown in Fig. 7. Shorter interaction times led to some microcracks while with longer interaction times a decrease of hardness was observed.

Wear tests, in order to verify tribological behaviour of laser treated surfaces were carried out at low specific load but high number of cycles, typical of the slide-bed couple, coupling material being cast iron G32. In our tests conditions wear of laser remelted cast iron was reduced from 2 to 4 times, in comparison with other materials examined, like induction hardened steel or with other surface treatment procedures.

HIGH CARBON STEELS

High carbon steels are commonly used for tools production. High hardness is obtained, after a suitable heat treatment, from a microstructure composed of primary carbides and low tempered martensite. In this case, the aim of rapid solidification after surface melting by laser is the refinement of primary carbides together with the attainment of carbon supersaturated austenites (12). These, being metastable, would be transformed by **subse-**

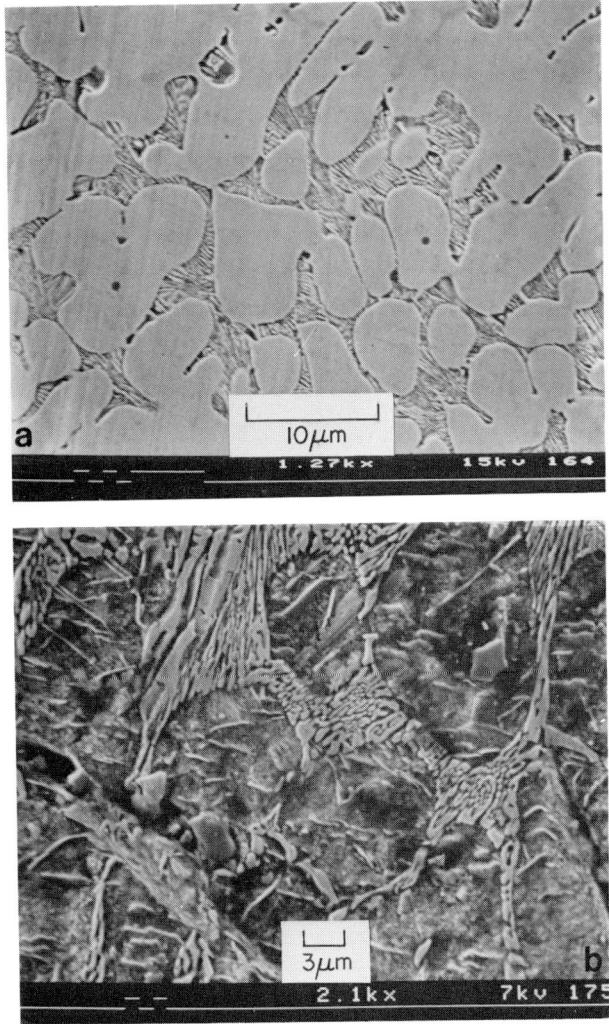

Fig. 7. (a) Microstructure of AISI A2 steel after laser melting and (b) idem heat treated 550 °C.

quent heat treatment, to other structures of improved both hardness and toughness.

Treatment on some high carbon (1-2 %) and chromium alloyed (5-8 %) steels was examined. High quenching rates, necessary for retention of metastable austenites, may be obtained using shorter interaction times, and, consequently higher power densities, typically about 10 kW/cm² and times below 1 s.

Melted layers exhibit a variety of different microstructural morphology: needle-shaped, dendritic and cellular. In samples treated using intermediate values of interaction time, 0.2-0.3 s, three zones may be identified: the first one, near the surface, of austenite dendrites surrounded by eutectic, a second one of austenite only and the third, inner unmelted zone but heat affected by solid state transformation to martensite or bainite. However prevailing morphology is the dendritic one. Indeed samples treated using longer interaction times, besides presenting a more deep melted zone show a more homogeneous cellular morphology of austenite together with a small area of eutectic. Only these samples of more homogeneous microstructures show also a regular microhardness profile. Examinations by X-ray diffraction and Mossbauer spectroscopy confirmed the austenitic structure of the melted zone as well as large carbon supersaturation of that phase, up to about 2 %. Carbon content indicates, being over the value of nominal steel composition, some carbon alloying effect from graphite coating used to enhance absorption capability of treated surfaces.

Surface hardness, of laser samples, as treated, is quite poor, but may be easily improved by subsequent destabilizing treatment Figure 8, at temperatures between 500 and 550 °C, in the range occurring solid state transformation and austenite to bainite type microstructures, as shown in Fig. 7. To this end, best results have been obtained for samples having cellular austenitic morphology, corresponding both to higher supersaturation level and to the better refinement of primary eutectic carbides. Tentative treatments below 500 °C proved insufficient to obtain significant transformation of metastable austenite. Treatments during many hours produced only partial transformation of cellular austenite to bainite.

Destabilization treatment produce also tempering of quenched layer under melted zone, however avoiding any modification of bulk structure and properties. Therefore by coupling laser surface melting and a subsequent heat treatment, surface hardness

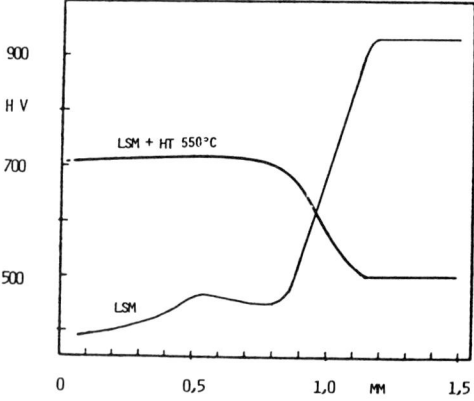

Fig. 8. Hardness profiles of AISI A2 samples laser melted and after subsequent heat treatment.

may be increased maintaining and sometimes improving toughness of the bulk. This situation would give best technological performance for this type of steel, typically when used for tools production. Another significant advantage appears to be the possibility to complete finishing operations after laser melting, on austenitic surface layer, of lower hardness, before final heat treatment is done to achieve maximum hardness.

REFERENCES

(1) J. F. Wallace, J. Metals, 15 (1963) 372.
(2) B. H. Kear, E. M. Breinan and L. E. Greenwald, Metals Techn. 6 (1979) 121.
(3) W. M. Steen, 159 in Industrial Laser Annual Handbook, D. Belforte and M. Levitt Editors, Ed. PennWell Books, Tulsa (1986).
(4) H. Jones, J. Mat. Sci., 19 (1984) 1043.
(5) M. C. Flemings, Solidification Processing, Ed. McGraw Hill Book. Co., New York (1974).
(6) M. Lamb, W. M. Steen and D. R. F. West, 322 in Proc. Stainless Steel '84, Gothenburg, Ed Metals Soc., London (1984).
(7) I. C. Hawkes, M. Lamb, W. M. Steen and D. R. F. West, 125 Vol. 1 in Proc. CISFFEL, Lyon, Ed. Commissariat Energie Atomique, Paris (1983).
(8) H. S. Carslaw and J. C. Jaeger, Conduction of Heat in Solids, Ed. Oxford University Press, London (1978).
(9) J. Mazumder and W. M. Steen, J. Appl. Phys., 51 (1981) 941.
(10) S. Kou, S. C. Hsou and R. Mehrabian, Metall. Trans., 128 (1981) 33.
(11) A. Tiziani, P. Meneghello, V. Tagliaferri, P. Munari and M. Magrini, 847 in Proc. 2nd Int. Conf. Heat Treatment Materials, Ed. AIM, Milano (1983).
(12) A. Tiziani, L. Giordano and E. Ramous, 108 in Proc. ASM Conf. Laser in Materials Processing, Ed. ASM, Ohio (1983).

WEAR PERFORMANCE OF LASER MELTED STEELS

H. de Beurs and J. Th. M. de Hosson

Department of Applied Physics, Materials Science Centre, University of Groningen, Nijenborgh 18, NL 9747 AG Groningen, The Netherlands

ABSTRACT

Wear experiments show strong correlations between different laser parameters and the wear performance. The presence of metastable austenite in a laser melted high carbon chromium steel results in a hardening during the wear process. This hardening process during running-in-wear is well measurable. An increase in hardness from 600 HV to 900 HV is observed. TEM observations reveal deformation induced martensite. The alloying of extra carbon result in high hardnesses up to 1400 HV, but at the same time extensive cracking occurs. Layers free of cracks are made by preheating the samples up to 500 °C. These layers show a different wear behaviour.

INTRODUCTION

Laser surface melting is a powerful technique to produce wear resistant layers. It combines the advantages of local hardening, alloying and high quench rates. The latter may result in new metastable phases with novel tribological properties. Clearly, not only the hardness but also quantities like ductility, friction and internal stresses play an important role in the wear process. These material properties are also largely influenced by a laser treatment.

Most research on the relation between wear performance and laser processing is carried out on cast iron. It comprises studies on the wear behaviour of melted zones [1, 2] and not melted, i.e. only hardened melted [3]. As far as research on tool steels is concerned laser melting followed by conventional hardening is reported as well [4].

In this paper aspects of laser melt treatments, are linked to wear features. We will focus in particular on highly alloyed steels, which are hardened by laser melting resulting in a very fine homogeneous dispersion of carbides in a matrix that mainly consists of austenite. Further hardening can be generated by increasing the carbon content. This is achieved by coating the surface with DAG graphite prior to laser melting. Subsequently these laser treated steel surfaces are subjected to wear tests. The microstructure is investigated by means of light microscopy, hardness tests and (transmission) electron microscopy. Also a comparison between normally hardened steel and laser melted steel has been made.

EXPERIMENTAL PROCEDURE

In the investigation two commercial highly alloyed steels are used, a chromium and a tungsten steel. The chemical compositions are listed in Table 1.

Table 1

Chemical compositions (wt %)

Elements:	C	Cr	W	V	Fe
Chromium steel: DIN: X210CrW12	2.05	11.05	0.62	-	bal
Tungsten steel: DIN: S 18-0-1	0.75	4.25	18.0	1.10	bal

The surfaces are laser melted with or without alloying of extra carbon.

For the laser treatment a Spectra Physics 820 CO_2 laser is used with a maximum continuous power of 1500 W. The samples are mounted on a moveable XY-table. The optical system consists of one mirror and a ZnSe lens. The power losses are measured to be 10 %. As a consequence the power on the surface amounts 1350 W. Before the laser treatment, the samples are ground with SiC paper of 500 grit and thereafter coated with a black writing marker or with DAG graphite to alloy extra carbon. The thickness of the DAG graphite layer is about 20 g/m^2. The laser melted layers which are alloyed with extra carbon show extensive cracking. Therefore some samples are preheated up to 500 °C and kept at this temperature during the treatment. This results in samples free of cracks. It is aimed keeping the laser parameters constant for all experiments. However, this appears to be not always possible due to differences in thermal conductivity, absorption of coatings, and the resulting roughness. The laser processing parameters are given in Table 2.

Before the wear experiments, the hardened surfaces are slightly polished with SiC paper and diamond paste. This does not introduce extra hardening. Wear experiments are carried out with a pin-on-disk apparatus. A ruby crystal ball with a diameter of 5 mm is pressed upon a rotating surface at a load of 2.3 N. This load is kept to be constant over all tests. The normalized force of $\tilde{F}$ this configuration is about 0.03-0.06. For example at $\tilde{F} = 1$ seizure will occur and for low values of $\tilde{F}$ (10^{-4}) ultra mild wear occurs. For our value of $\tilde{F}$ mild oxidational wear or delaminational wear might be expected [5]. The sliding velocity is either 0.5 or 5.0 cm/sec. To reduce the effect of humidity and to create some lubrication pure ethanol is supplied. In addition the experiments are carried out under dry nitrogen atmosphere. The amount of wear is determined with an interference microscope. Figure 6 shows an example of such a wear track.

Table 2

Laser processing parameters

Material		Chromium steel		Tungsten steel	
Coating		ink	graphite	ink	graphite
Effective power	W	1350	1350	1350	1350
Focal length	mm	127	127	127	127
Position focus above surface	mm	15	15	5	5
Scan speed	cm/s	4	25	5/25	5
Distance between melt paths	mm	1.0	1.0	0.3	0.3
Shieldgas		argon	argon	argon	argon
Sample temperature	°C	20	20/500	20/250	500

Hardness measurements are carried out with a Vickers hardness tester. **The load used is 100 g, unless otherwise stated.** Laser melted samples as well as worn samples are tested.

The microstructure of laser melted steel is investigated using transmission electron microscopy (JEM 200 CX at 200 KV) both on worn and unworn samples. Disk type samples of 3 mm in diameter are cut out of laser melted steel. These samples are thinned electrochemically with a Struers Tenupol.

It is hardly possible to make TEM samples of the small worn tracks produced by the pin on disk tester. Therefore 3 mm TEM disks are thoroughly ground on SiC-paper, producing abrasive wear. After this treatment the surface is smoothed with diamond paste. Thinning these samples from the bulk side results in an electron transparent area just underneath the surface layer. To obtain information from deeper regions the worn surface is further thinned in an argon ion milling machine.

RESULTS

CHARACTERIZATION

The surface of a piece of chromium steel is covered with overlapping laser tracks. The resulting hardness for laser melted chromium steel wihtout extra carbon is 550-600 HV. This is in line with measurements reported in Ref. [6]. Alloying with extra carbon results in hardnesses up to 1400 HV, but extensive cracking occurs. A reduction in cracking is achieved if the material is kept at 500 °C during laser melting. In case of only melted chromium steel the hardness profile across the surface is constant. In cross section also a constant hardness over the melt is measured. In contrast when laser alloying with carbon a strongly varying hardness profile is observed. The highest hardnesses are found in the middle of the laser tracks where as in overlapping regions a decrease in hardness to 450 HV is determined. Figure 1a shows a TEM photo of only laser melted chromium steel. An austenite cell structure is observed with segregated M_3C carbides. The cell diameter varies from 0.5 to 8 μm. The defect structure is mainly characterized by a high dislocation density and pile-ups of dislocations. Figure 2 shows a pile-up bound by a cell wall. Coating the surface with DAG graphite gives a higher absorption **coefficient** and a higher scan speed

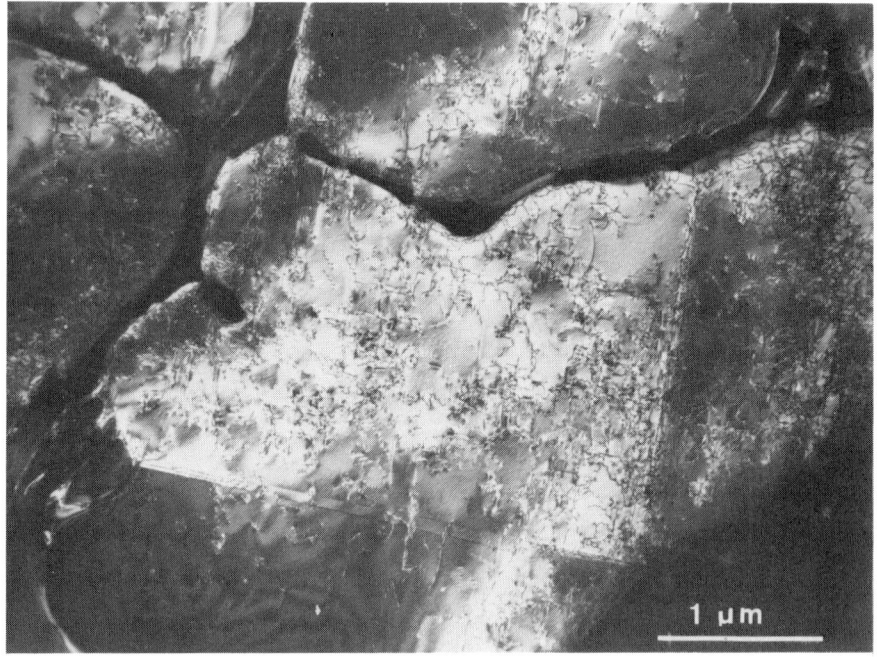

Fig. 1a. Austenite cells in X210CrW12 after laser melting.

Fig. 1b. Austenite cells in X210CrW12 after alloying with carbon.

is used. The resulting higher quench rate results in smaller cell sizes (0.5-2 μm). We are not able to determine the carbon concentration in the melted layer. Walker et al. (7) claim to have alloyed pure iron up to 6 wt% C after repeated treatments. Burning away and splashing of the graphite coating during laser melting makes it impossible to relate the final carbon content to the coating thickness. The higher carbon concentration gives a stronger carbide segregation as can be concluded from Fig. 1b. The resulting depth for laser melted chromium steel is about 60-100 μm. For alloyed surfaces it varies between 30-60 μm.

A higher thermal conductivity of tungsten steel forces us to use a lower focal point to increase the energy density. More variation in hardness on the surface is observed. Especially, the overlapping regions of laser tracks show a strong decrease in hardness. It lies between 500 and 800 HV. For different scan speeds in the range from 5 to 25 cm/sec no change of this hardness range is found. Tempering the samples after laser melting for one hour at 570 °C shifts the hardness range upwards to 650-1000 HV. This is also reported for splat quenched high alloy steels (8). By annealing the austenite phase transforms into a stable structure consisting of ferrite and carbide. For laser melted steel TEM observations reveal a mixed microstructure of austenite, martensite and segregated carbides. Tempered samples show segregation of carbides. Tungsten steel has a severe tendency to crack. The generation of cracks is successfully suppresse

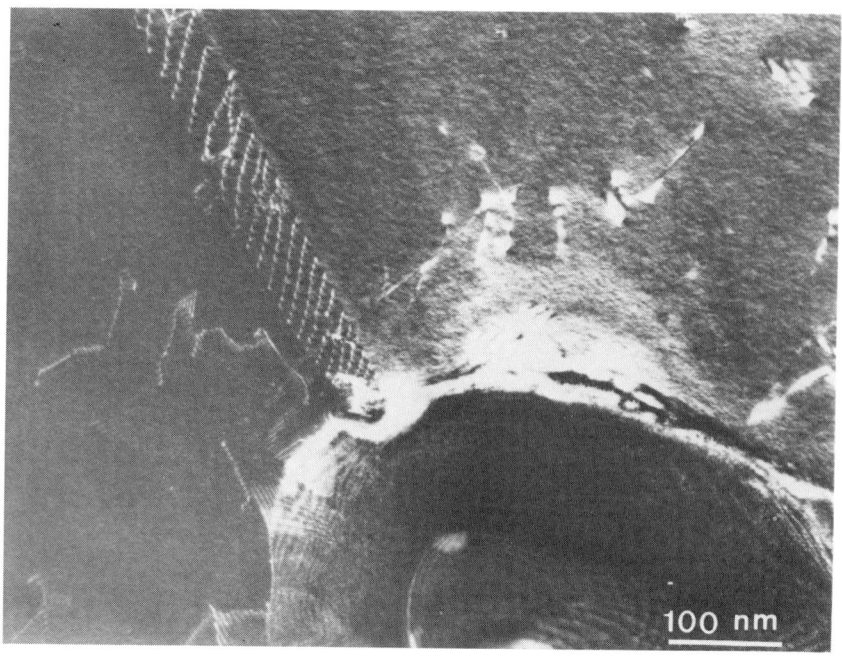

Fig. 2. Pile-up of dislocations at a cell boundary.

by keeping the material at 250 °C during the treatment. After laser melting the steel is cooled down in air which takes about 5 minutes. Alloying of graphite results in further hardening accompanied by severe cracking. Holding the temperature of the material now at 500 °C suppresses cracking completely. The hardness of these layers strongly increases and varies between 700-1100 HV.

WEAR PERFORMANCE

All wear tests are done under the same circumstances. For chromium steel the sliding speed is 5 cm/sec. However this speed caused oxidational wear in some tungsten steel samples. To prevent oxidation all tungsten steel samples are worn at a lower speed of 0.5 cm/sec. By absence of metal to metal contact no adhesion occurs and the wear process can be explained in terms of forces, stresses and deformation processes. Measured wear volumes for chromium steel are given as function of number of turns in Fig. 3. We can see a strong increase in wear resistance for the laser melted steel with respect to the untreated sample. In Table 3 the wear rates are given, i.e. the slopes of the fitted lines.

In some treatments there exists a kind of running in effect. This can be caused by the large pressure at the start of the experiment which will cause seizure. The effect will be strongest

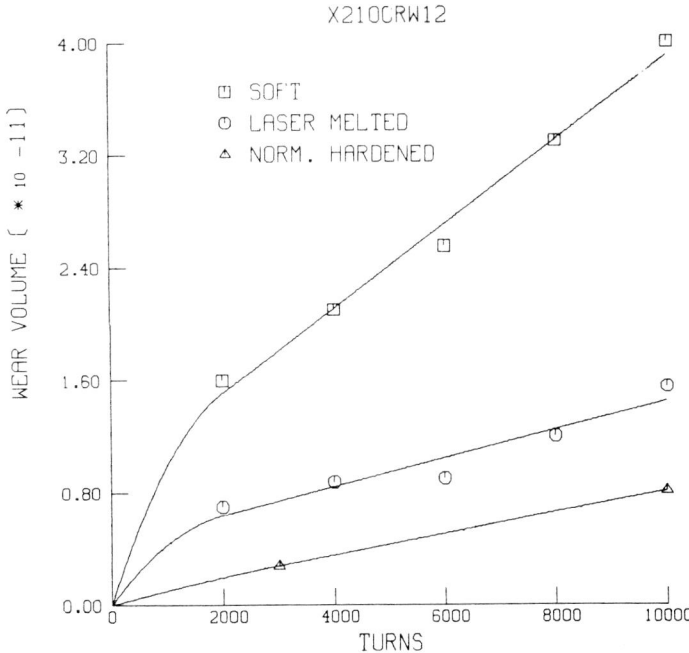

Fig. 3. Wear volume in m³/m as a function of the number of turns.

Table 3

Material	Treatment	Wear rates [10^{-15} m³/m]	hardness [HV]
X210CrW12	not treated	3.0	350
	laser-melted	1.0	550-600
	norm. hardened	0.8	900
S 18-0-1	laser-melted scan speed 5 cm/s	0.4	500-800
	other treatments	0.58-0.65	650-1100

for the soft (not hardened) steel and which wears more heavily. Not surprisingly, the normally hardened steel shows hardly any running in effect. The laser-melted layer with a somewhat lower hardness exhibits also a running in effect. However, although it is softer than the normally hardened steel, it shows nearly the same wear rate after 2000 turns. The wear rate can be described by Archard's relation:

$$V = K \cdot \frac{S \cdot L}{3 \cdot H}$$

where

V = Wear volume
S = Sliding distance

H = Hardness
K = Wear coefficient
L = Load

If we consider K as a material constant the hardness H should increase during the wear process. In fact, hardness measurements reveal a hardening of the wear track. Figure 4 shows the hardness profile of a wear track after 10,000 turns. The hardening behaviour during the wear process is shown in Fig. 5. This hardening behaviour can be explained **by a friction induced** martensitic transformation. TEM observations do confirm this (Fig. 8) and reveal α-martensite (9). This transformation into α-martensite is explained by a dislocation mechanism. Pile-ups of dislocations are easily generated during solidification and deformation, because the carbides on the cell walls are effective barriers to dislocation movements. If during wear the induced stresses are high enough to produce a shear of 1/6<112>, a pseudo bcc stacking is obtained. hcp ε-martensite is not found. Nucleation centres like stacking faults are absent in the laser melted chromium steel.

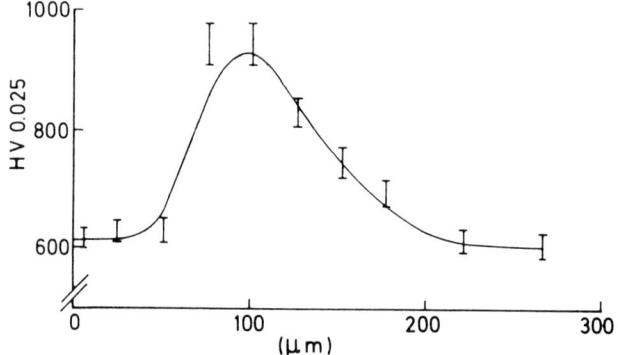

Fig. 4. Hardness profile of a wear track on laser melted chromium steel after 10 000 turns.

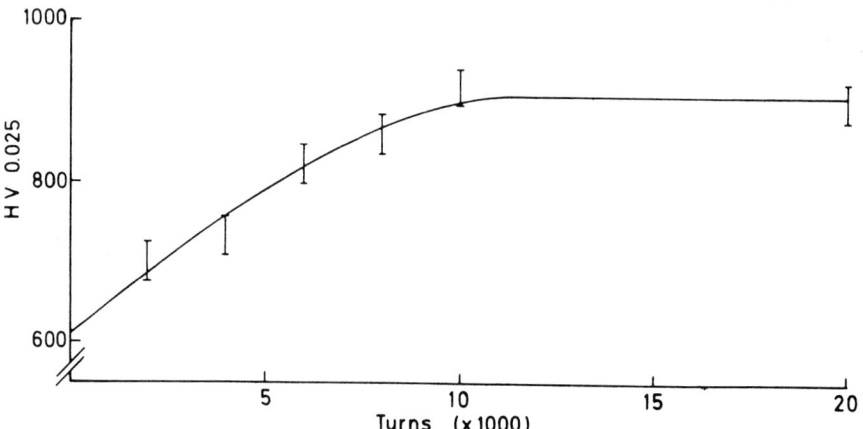

Fig. 5. Wear induced hardening for laser melted chromium steel.

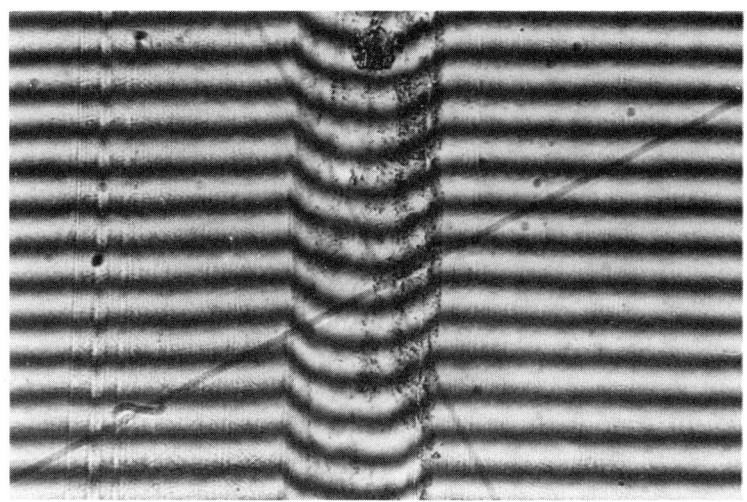

Fig. 6. Interference microscope image of worn laser melted chromium steel after 10 000 turns (sodium light source).

Besides the quantitative wear results the laser-melted material shows to be more ductile than the normally hardened steel. Figure 7 shows the more rough surface of this hardened chromium steel. It is explained by the absence of individual large carbides in laser-melted steel which take part in the wearing process of the normally hardened steel. This applies also to extra carbon alloyed steels.

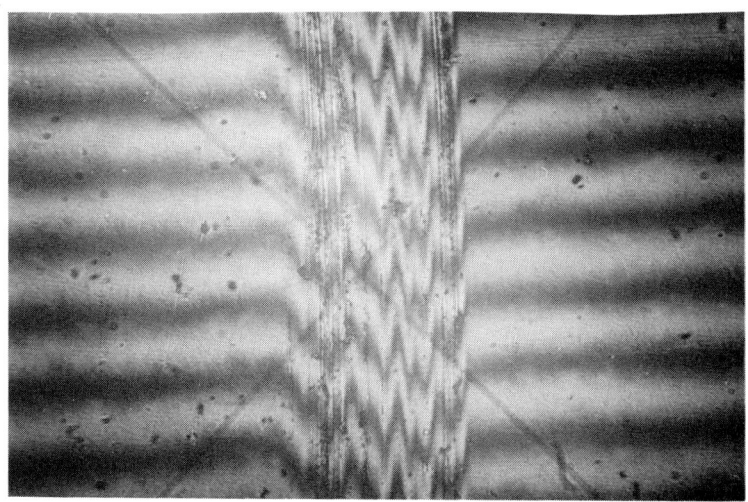

Fig. 7. Image of wear track on normally hardened chromium steel.

Fig. 8. Wear induced martensite near surface.

A lower wear rate is measured for laser-melted tungsten steel. In Fig. 9 the wear volume is plotted against the number of turns. The variations in hardness across the surface cause only little difference in wear volumes. The plotted wear volumes are averaged values. Although hardnesses for different laser treatments vary, no large differences in wear rates are measured. In Table 3 the wear rates are summarized. Only the material that is melted at a scan speed of 5 cm/s results in a lower wear rate, although it has a worse running in performance. The difference at the onset of wear process is quite obvious. The harder materials exhibit a much lower wear volume.

CONCLUSIONS

Laser melting of X210CrW12 results in a better wear performance. The microstructure consists of austenite with segregated M_3C carbides. During wear this transforms into α-martensite. The hardness increases from 550 to 900 HV in the worn material. The wear rate of the laser-melted material is comparable with conventionally hardened material. Nevertheless the laser-melted chromium steel shows a much more ductile behaviour.

Laser-melted S 18-0-1 steel shows a better wear performance than the chromium steel. The microstructure consists of a mixture of austenite, martensite and carbides.

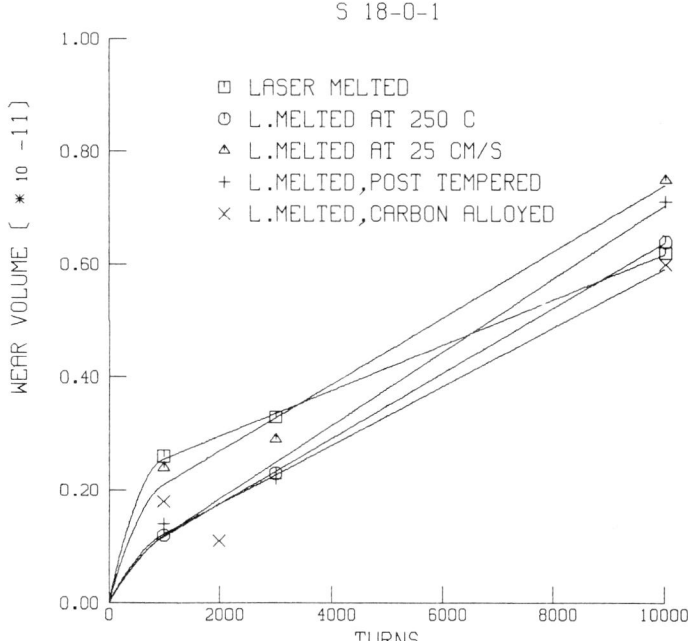

Fig. 9. Wear volume in m³/m as a function of the number of turns.

Alloying of carbon increases the hardness of S 18-0-1 steel. However extensive cracking occurs. This is completely suppressed by carrying out the laser-melting at 500 °C. The resulting hardness is 700-1100 HV. Alloying with extra carbon also results in a lower wear rate during running in wear. However it does not result in an overall better wear performance.

ACKNOWLEDGEMENTS

We would like to thank J. C. Kleuver, O. P. van Ravensteyn and J. A. Valk for their contributions. The work is part of the research program of the Foundation for Fundamental Research on Matter (F.O.M. - Utrecht) and has been made possible by financial support from the Netherlands Organization for the Advancement of Pure Research (Z.W.O. - The Hague).

REFERENCES

(1) A. Blarasin, S. Corcuto, A. Belmondo, D. Bacci. Wear, vol. 86, pp. 315-325, (1983).
(2) C. H. Chen, C. J. Altstetter and J. M. Rigsbee. Met. Trans., vol. 15a, pp. 719-728, (1984).
(3) P. A. Molian, M. Baldwin. Journ. of Tribology, vol. 108, pp. 326-333, (1986).

(4) L. Ahman, Met. Trans., vol. 15a, pp. 1829-1835, (1984).
(5) S. C. Lim, M. F. Ashby. Acta Metall., vol. 35, pp. 1-24, (1987).
(6) H. W. Bergmann and B. L. Mordike. Z. Metallkde., Bd. 71, pp. 658-668, (1980).
(7) A. Walker, H. M. Flower, D. R. F. West. Journ. of Materials Sci., vol. 20, pp. 989-995, (1985).
(8) T. Minemura, A. Inoue, Y. Kojima, T. Masumoto. Met. Trans., vol. 11a, pp. 671-673, (1980).
(9) H. de Beurs, J.Th.M. De Hosson. Scripta Metall., vol. 21, pp 627-632, (1987)

Chapter 2

LASER SURFACE COATING AND ALLOYING

PLASMA DEPOSITED LASER REMELTED WEAR RESISTANT LAYERS
B. L. Mordike and W. N. Kahrmann

Institut für Werkstoffkunde und Werkstofftechnik, Technische Universität Clausthal, Clausthal-Zellerfeld, FRG
Presented at "European Conference on Laser Treatment of Materials" 1986 in Bad Nauheim, FRG

INTRODUCTION

The use of surface coating has been common practice for many years for a wide range of metallurgical applications. It has been used when the properties of the bulk material are insufficient to withstand the mechanical loading or chemical attack imposed. Typical methods of coating are PVD, CVD, electroplating, thermal, plasma spraying and explosive spraying or coating. Thermal spray coating is the method most widely used. It can be used for a wide range of substrate and coating materials [1]. The main problem with thermal spraying is the poor adherence to the substrate. Porosity also presents for some applications a problem. Laser surface melting provides means of improving the adhesion to the substrate and at the same time eliminating porosity. The laser beam can be passed in a raster over the surface and can melt or partially melt, depending on the power [2], any surface layer. Ideally the power should be set to melt the surface onto the substrate producing a good bond without dilution of the layer [3]. In this work results are presented on the laser surface melting of plasma deposited hard metal layers.

EXPERIMENTAL DETAILS

The aim of the work is to produce a wear resistant layer on X10CrNiNb18-8 plates for use at high temperatures under abrasive conditions and thermal shock. The requirements are: - as thick a wear resistant layer as possible - smooth surface - good adhesive bond between layer and substrate - low degree of porosity in the surface layer - homogeneous microstructure.

A WC-Co (83/17 %) coating was considered most likely to meet the requirements. The initial layer can be applied either by flame spraying or plasma spraying. Plasma spraying was considered better as the temperature is higher, the kinetic energy of the powder particles is higher and consequently there will be less porosity and better bonding. The laser parameters power density,

focus to workpiece distance, degree of overlapping of beads and preheating all influence to some extent the final microstructure. The following parameters proved to be the best:
- 4,5 kW laser power
- beam size 2 x 1 mm
- working distance from focal point (5 to 65 mm) distance between passes 1.0 to 4.5 mm (degree of overlap)
- feedrate 0.3 - 4.0 m/min
- protective atmosphere argon
- protective gas flow perpendicular to direction of feed and opposite to raster direction 30 x 20 cm x 12 mm plates were coated by plasma and laser melted.

RESULTS

The first stage in coating is the plasma spraying. The conditions of spraying can markedly influence the subsequent laser treatment. Plasma produced layers tend to be brittle. Thick layers, due to the build up of internal stresses and the poor bond to the substrate may crack or peel off. Coatings thicker than 1.2 mm showed cracking.

The maximum coating thickness is determined by the tendency to cracking during plasma deposition (< 1.2 mm). If cracks are acceptable the maximum thickness depends on the laser power available. With coatings less than 0,5 mm there is a tendency for dilution of the layer by substrate material (Figs. 2 and 3). Strong intermixing leads to an inhomogeneous layer and contributes to an increased surface roughness. The best results were achieved for layers initially 1.0 mm thick. During laser melting the layer contracts by about 30 %. The pore and inclusion content in plasma sprayed layers is about 10 - 15 % and this is reduced by laser melting to less than 1 %. Evaporation and other losses accounts for the difference.

When large surfaces have to be remelted the beam is moved as shown in Fig. 1 over the workpiece in a raster like manner. The displacement set for adjoining passes depends on several factors, primarily on the beam diameter, which itself depends on the laser employed and the distance of the workpiece to the focal point.

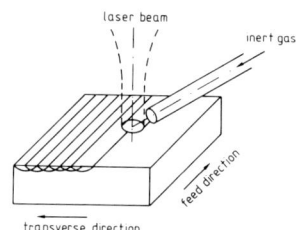

Fig. 1. Protecting the surface of the workpiece by a stream of inert gas. The movement of the workpiece during laser treatment is shown.

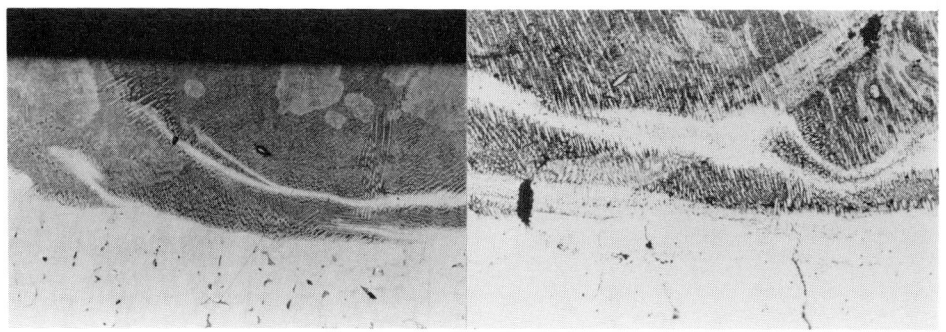

Fig. 2. Intermixing of base material with sprayed coating.

Fig. 3. Intermixing of base material with sprayed coating.

The beam diameter determines the size of **the molten pool**. It is obviously affected by other factors as well e.g. preheating temperature, absorbivity and thermal conductivity. The shape of the molten pool is determined by these factors and the translation speed of the workpiece. Experience has shown that even with optimum bead shapes about 1/3 of the volume has not been melted to a sufficient depth. This is corrected by setting a certain degree of overlap for the next pass (75 %) (Fig. 4).

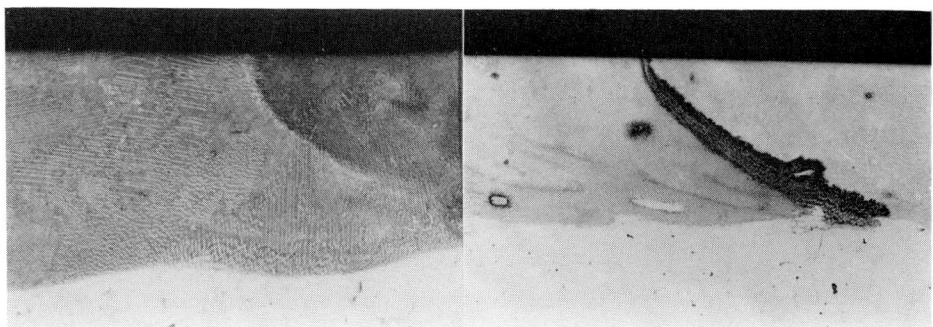

Fig. 4. Section of three overlapping beads.

Fig. 5. Interface between coating and substrate showing little roughness.

The distance of the workpiece to the focal point is chosen on the basis of preliminary tests and only after this has been fixed can parameters be chosen. The spot size depends on this distance. Then the feed rate and degree of overlap can be fixed to obtain the best degree of homogenisation and smoothness of the interface (Fig. 5). If the distance from the focus is too large, the spot size increases and the energy density is too low for complete melting. Similarly if the feed rate is too high the interaction time is too low and insufficient energy is transferred to the specimen, see Figs. 6, 7 and 8.

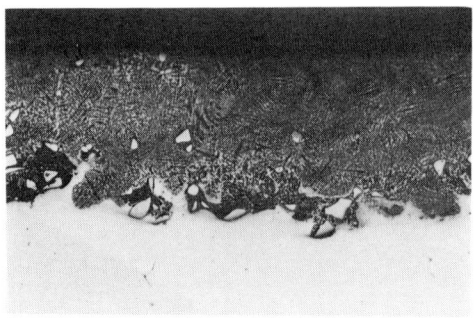

Fig. 6. Incompletely laser melted sprayed layer, interface.

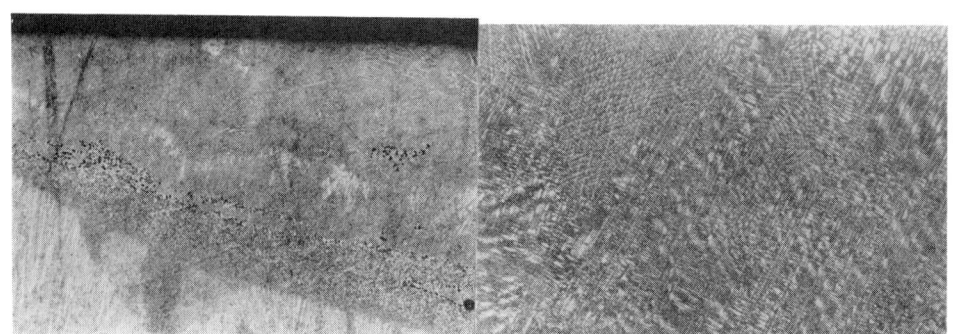

Fig. 7. Entrapped unmelted particles in a laser melted layer.

Fig. 8. Magnification of Fig. 7.

On melting sprayed powders inclusions or pores agglomerate to form spherical defects of various sizes. If the feed rate was too high the powder is molten for too short a time and inclusions and gas cannot escape via the surface (Fig. 9 - 11) and hence remain trapped. High feed rates imply high quenching rates and fine microstructures with high hardness (> 700 Hv 0.1). At lower speeds or working nearly in focus, the carbides in the layer dissolve completely in the matrix resulting in a low hardness about 400 Hv 0.1.

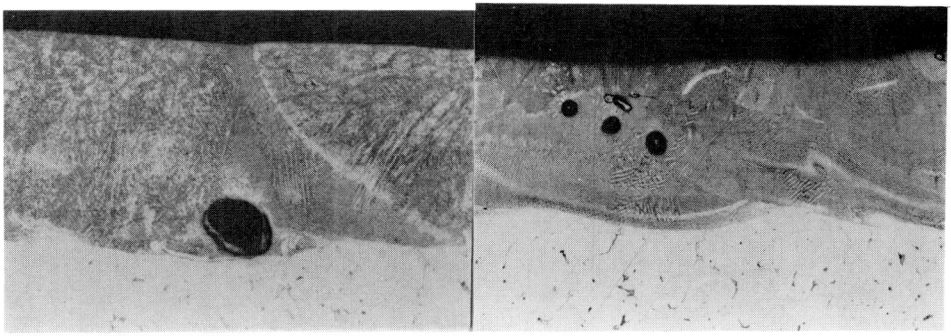

Fig. 9. Inclusion at the interface layer - substrate.

Fig. 10. Included particles in the middle of the laser melted layer.

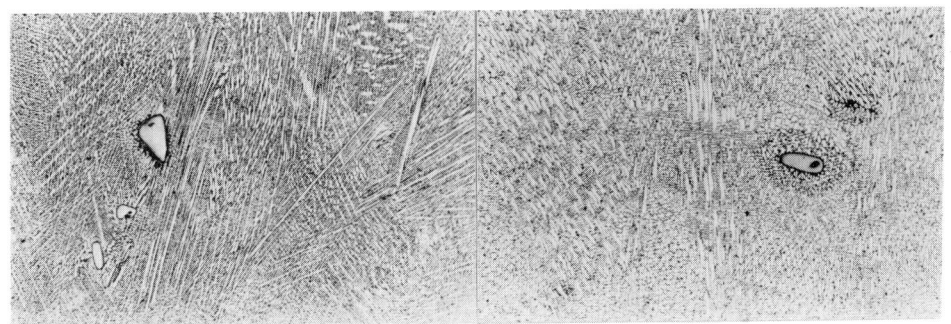

Figs. 11 and 12. Dendrically solidified structures with partially dissolved tungsten carbides.

The following microstuctures are typical for layers melted under optimised conditions (Figs. 11 and 12).

The final thickness of the layer was about 0.7 mm and as can be seen is essentially free of pores and cracks. The average hardness of the surface was about 650 Hv 0.1 or 500 Hv 30, Figs. 13 and 14.

The hardness falls sharply at the interface and then decreases further from 300 Hv 0.1 down to the hardness of the substrate material (240 Hv 0.1) throughout a transition zone. Some effect of the heat can be detected up to a depth of 0.7 mm. Subsequent annealing at the proposed operating temperature of 800 °C for 24 h did not modify the structure or hardness values. A limited number of thermal shock tests have been carried out. No spalling was observed. The coated layer was resistant to compressive, impact loading and could withstand some deformation together with the base material without spalling, although some cracking may occur.

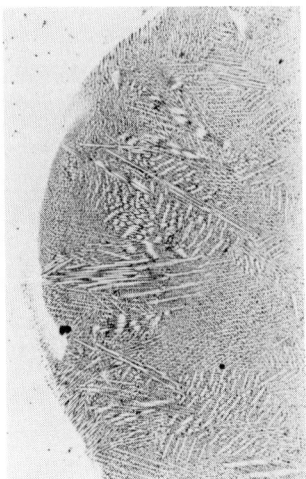

Fig. 13. Hardness across a laser bead.

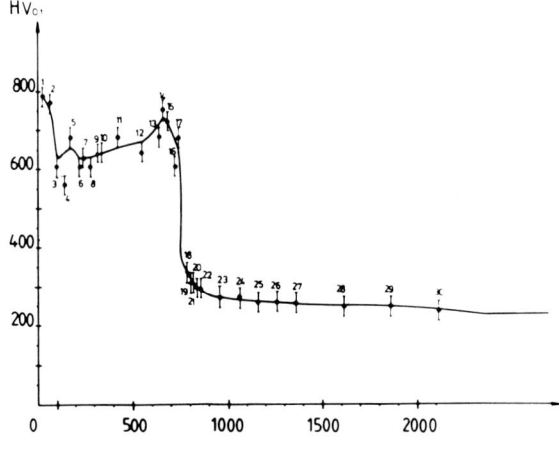

Fig. 14. Plot of hardness as a function of distance from the surface.

CONCLUSIONS

The attempt to produce wear and scale resistant layers on X10CrNiNb18-8 plates for operation at 800 °C was successful. The layer is characterised by good adhesion to the base material. It was possible to obtain a relatively gentle transition from the substrate material (240 Hv 0.1) and the antiwear layer (600 Hv 0.1).

Optimisation of the laser parameters and preheating the sheets up to 400 °C enables the formation of deep and uniform layers. Intermixing of substrate and layer material can be maintained low for sprayed layers 1.0 mm ± 10 % thick. The surface roughness is about 100 μm. Considerable warpage can occur and this can only be taken into account by prebending of the workpiece.

REFERENCES

(1) A. R. Nicoll, H. Cramer, R. Prince, G. West. Surface Engineering 1, (1985) 59.
(2) P. Loosen, L. Bakowsky, G. Herziger, F. Rühl. Laser-Optolektronik in der Technik, W. Waidelich, Springer Verlag, Berlin, (1983).
(3) H. W. Bergmann, B. L. Mordike, H. U. Fritsch. Z. Werkstofftechnik 14 (1983).

LASER APPLICATIONS FOR STEEL SURFACE ALLOYING WITH CARBON

E. Ramous, A. Tiziani and L. Giordano

Universita' — Instituto Chimica Industriale
Via Marzola, 9–35126 Padova, Italy

INTRODUCTION

Surface alloying by laser is a process which offers numerous potential applications. Almost any type of surface alloy can be prepared, and a substantial scientific activity has been devoted to this topic. Though the largest interest until now has been devoted to electronic devices area, however a large number of applications can be forecast for mechanical workpieces (1).

In this field, the aim of surface alloying may be to improve the wear and corrosion resistance of poor and cheaper base materials to reduce consumption of special alloying elements. In the case of steels, carbon surface enrichment is a very old and traditional process, to improve wear resistance and fatigue life. Using high-energy heating devices, like laser, the possibility of introducing similar alloying effects in a simpler way and shorter time may be explored (2).

Generally surface alloying by laser is performed by melting and adding another material to the melt pool. The same procedure may be used for carbon alloying of steels, though, owing to the high diffusion coefficient of carbon at temperatures over 1000 °C, also a solid state diffusional process may be considered. In this paper are some of the results obtained using the two different processes.

EXPERIMENTAL

Laser treatments were carried out using e cw CO_2 AVCO laser, at RTM Institute, Vico Canavese, Italy. The uniformity of the energy distribution on the spot was optimized by using a biaxial scanning system of the beam, with 10*10 mm mask to cut off lateral broadening. Argon was used as shielding gas, and all samples were coated by graphite dag in alcohol. Actual incident energy on treated surface was measured before and after any laser run by a cone calorimeter.

Base material for solid state diffusion tests was ARMCO iron, as 6 mm thick slabs. For carbon alloying by melting tests, both ARMCO iron and medium carbon steel (0,4 % C) were used, as 15 mm thick slabs.

ALLOYING BY SOLID STATE DIFFUSION

From solid state diffusion coefficient of carbon in austenite (3), it may be immediately deduced that surface temperature achieved by laser heating has to be maintained in the range between 1000 °C and 1200 °C at least for some tenths of seconds to obtain a significant carbon diffusion effect. To this end, laser treatment parameters have been chosen from calculations of thermal transients. The temperature time variation may be evaluated by the well known Carslaw and Jaeger equation (4), used to describe the thermal transient induced in the surface layers of a moving slab beneath a stationary heath source, like the laser beam. In our case it is sufficient to consider only the one dimensional heat flow, along the normal to the sample surface. Assuming the entering heat flux as a constant, one-dimensional equation can be integrated (4). However, for our samples, of limited thickness (6 mm), for increasing heating effect, solution must be modified according to La Rocca (5), taking into account the heating of the whole thin samples.

Examples of calculated thermal transients corresponding to selected treatment parameters are shown in Fig. 1. Interaction

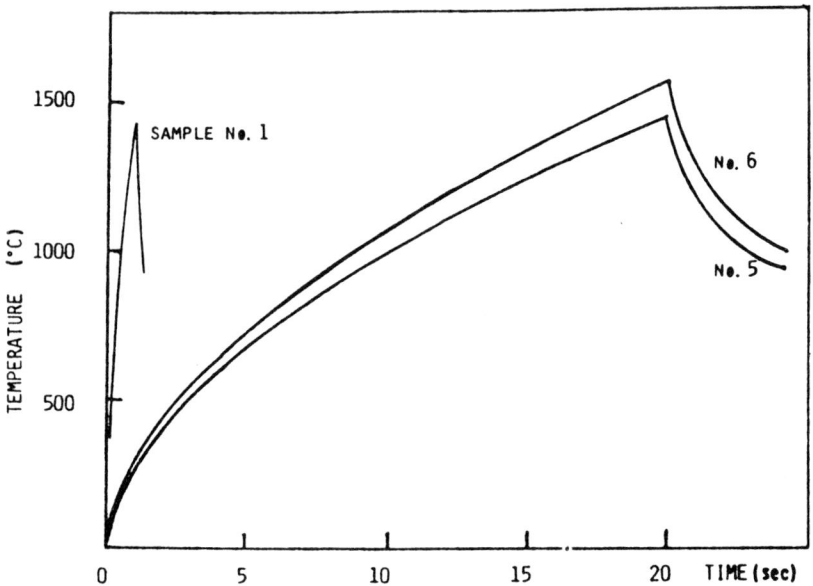

Fig. 1. Calculated thermal transients for solid state carbon diffusion by laser surface heating.

times are significantly longer than values used for surface
hardening and consequently power densities are reduced to avoid
surface overheating and incipient melting.

In Table 1 we summarize the treatment parameters of our tests
and corresponding values of carbon diffusion path obtained.

Table 1

Sample	F_0 (W/mm²) Intensity of Absorbed Radiation	Interaction Time (Sec)	T_{max} °C	Carbon Diffusion Path (mm)	
				Calculated	Experimental
1	20.00	1	1548	0.037	0.04
2	3.0	20	1051	0.049	0.055
3	3.4	20	1174	0.068	0.07
4	3.8	20	1330	0.08	0.09
5	4.1	20	1422	0.12	0.13
6	4.5	20	1546	0.17	0.17

Experimental values are in agreement with calculated data by a
simple one-dimensional diffusion model (6), integrating equation
of Fick's second diffusion law. Indicated values of carbon
diffusion path correspond to the depth where the carbon content
was reduced to 10 % of values reached in the surface.

It is interesting to note that the very short interaction time
used appears sufficient to achieve a significant diffusion effect
and a shallow but well defined carbon enriched layer on steel
surface. Typical microstructures obtained are shown in Fig. 2
and the carburized layer exhibits a bainitic structure. According to calculations from thermal model, interaction times with
laser beam over 5 seconds produce thermal cycles having cooling
times of several seconds (by direct heat transmission to the bulk
of the sample, and without any external cooling device). Therefore the corresponding cooling rates are lower than the critical
value for martensitic transformation, and the case layer has a
final bainitic structure showing maximum hardness in the range
of 550 to 600 HV.

It may be noted that optimization of experimental conditions,
treatment parameters, methods for carbon supplying on the surface,
and so on, may allow to obtain further improvement of carbon
surface enrichment, and particularly of the depth of alloyed
layer. However our results may be sufficient to confirm the
possibility of a noticeable carbon surface enrichment of iron
and steel surfaces by solid state diffusion and without melting

and by heating the surfaces with laser beam with large interaction times. The technological implications of this possibility are obvious.

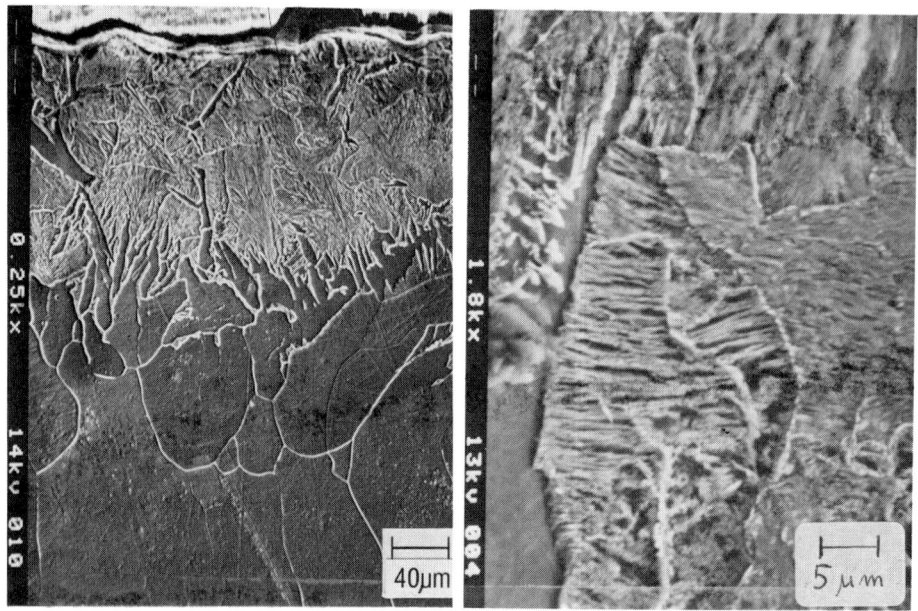

Fig. 2. Microstructures of carbon alloyed layers.

ALLOYING BY MELTING

Aims of carbon surface alloying by melting may be different:
- production of a ledeburitic structure in surface layer, like white cast iron;
- production of a martensitic structure, together with some retained austenite.

The main difference between the two situations is the presence of cementite type primary carbides, coming from a higher carbon content. Both situations may be obtained by laser treatment, and the most important parameter influencing final microstructure appear the carbon pick-up during melting.

In our experimental conditions carbon was supplied as predeposited graphite coating the surface to be alloyed. This is the simplest way to supply carbon, and should be easily used in industrial applications. However, at it is well known (7), part of predeposited graphite burns out during heating period. Therefore high power densities and long interaction times, though being favourable for deep surface melting, reduce actual possibility for carbon to be alloyed in the melt, carbon gasification being increased. This appears as the most limiting effect for carbon alloying in steel by laser, as observed in our experimental conditions. However this limitation would be overcome by using different, but perhaps more complex methods to put alloying material in the melted pool, like injection by an inert gas (and so on).

In conclusion, in our experimental conditions only a limited amount of predeposited graphite being actually dissolved during melting, it is possible to obtain alternatively deep melted layers with low carbon enrichment, i.e. of martensitic structure, or relatively thinner melted layers with higher carbon content and consequently showing the ledeburitic microstructure.

The situation previously discussed is illustrated by the results of our laser melting runs. Treatment parameters were 5 and 7 kW/cm² and 1 and 2 s interaction times. Melted depth obtained are shown in Fig. 3, together with some results for cast iron (3,4 % C) remelting for the purpose of comparison. Increasing melt depth with carbon content of base material may be explained by corresponding decrease of **liquidus temperature of surface being** treated.

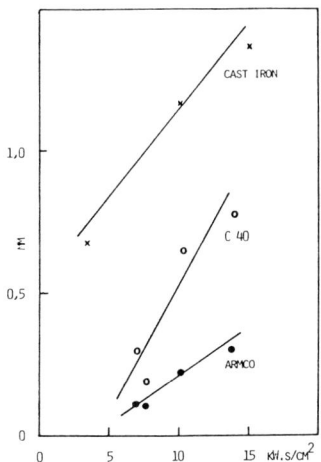

Fig. 3. Depths of melted layers for different base materials.

Microstructures of alloyed layers are clearly different, as shown in Fig. 4. At lowest kW*s conditions, corresponding to shallowest melted layer, microstructure is typically ledeburitic, with dendrites of transformed austenite embedded in the eutectic with primary cementite, Fig. 4b.

At both largest kW*s values and melted depths, eutectic disappears almost completely and the base structure is martensite with a noticeable content of retained austenite. This result appears the consequence of higher dilution of alloyed carbon, giving lower medium values, not sufficient to achieve primary carbides nucleation, owing to relatively high quenching rate from the melt.

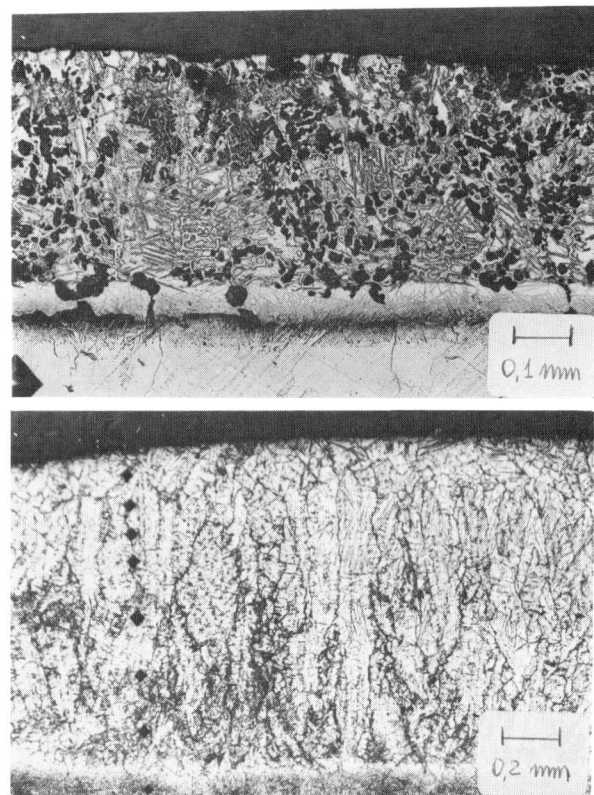

Fig. 4. Microstructures of layers melted and carbon alloyed.

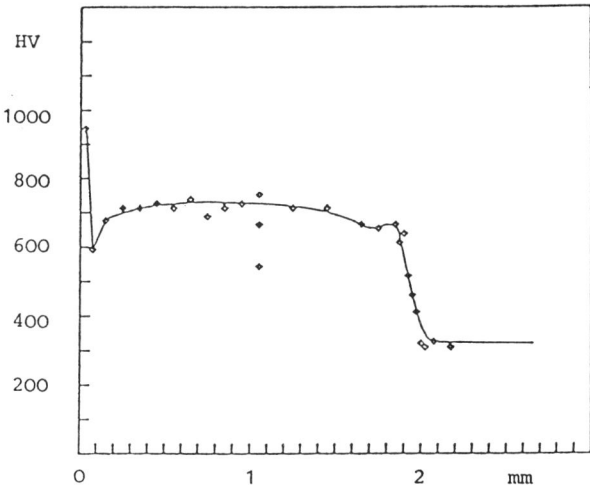

Fig. 5. Microhardness profile of carburized and melted layer.

These effects are confirmed by hardness values. Highest values, about 900 HV corresponding to white cast iron (3,4 % C) are never reached, and hardness obviously decreases, from 700 to 400 HV with increasing melt depth. Moreover it is interesting to point out that, for steel samples, hardness values may be further increased by a subsequent heat treatment to complete transformation of retained austenite to bainite.

CONCLUSIONS

Wear tests on alloyed layers are being carried out. First results indicate that ledeburitic layer behaves quite well, like similar layers obtained by laser melting on cast iron. Martensitic layers offer sometimes better performance after subsequent tempering at low temperature of both martensite and retained austenite.

From our results we may conclude that using laser, carbon surface alloying of steels may be performed in different ways corresponding to different final resulting microstructures:

1) solid state diffusion, with max surface temperature about 1200 °C, to obtain thin (0,10-0,25 mm) bainitic layers;
2) alloying during melting to obtain high carbon ledeburitic layers or martensitic layers, with lower carbon content.

REFERENCES

1. C. W. Draper and C. A. Ewing, J. Mat. Sci., 19 (1984) 3815.
2. W. M. Steen, p. 158 in Industrial Laser Annual Handbook, D. Belforte and M. Levitt Editors, Ed. PennWell Books, Tulsa (1986).
3. J. W. Christian, The Theory of Transformation in Metal and Alloys, Ed. Pergamon Press, Oxford (1975).
4. H. Carslaw and J. Jaeger, Conduction of heat in solids, Ed. Oxford University Press, London (1978).
5. A. V. La Rocca, p. 83 in Proc. 4th Int. Symp. Gas Flow and Chemical Lasers, A. S. Kaye and A. C. Walker Editors, Ed. Adam Hilger, Bristol (1982).
6. A. Canova and E. Ramous, J. Mat. Sci., 21 (1986) 2143.
7. E. Ramous, p. 475 in Laser Surface Treatment of Metals, C. W. Draper and P. Mazzoldi Editors, Ed. Martinus Nijhoff Pub., Boston (1986).

AN EXPERIMENTAL STUDY ON CO_2 500W LASER SURFACE ALLOYING OF CHROMIUM ON COPPER

L. M. Galantucci*, G. Ruta** and S. Magnanelli***

*Dipartimento di Progettazione e Produzione Industriale —
Università di Bari, Viale Japigia, 182-Bari, Italy
**Centro Laser, Via De Blasio, 1-Bari, Italy
***Centro Ricerche L.M.I., Fornaci di Barga (LU) — Italy

ABSTRACT

An experimental study has been carried out to investigate the technological possibility of obtaining metastable copper alloys with high chromium percentages.

These alloys, though having potential characteristics of extreme industrial interest, cannot be achieved with conventional methods owing to copper and chromium solid immiscibility.

In order to do this a 500 W - CO2 Laser source has been used and an experimental plan based on the analysis of the influence of various process parameters (coating thicknesses, Laser energy transfer, pulse or continuous, covering gases and speed rate) upon the treatments, has been set up.

Experimental results allow us to affirm that copper-chromium alloys with high chromium percentages are now possible by using low power Laser sources: pulse wave seems to be the most suitable, giving the greatest alloyed thicknesses and negligible heat treated zones in the base metal. Further investigations are in progress on the possibility of adopting higher power Laser sources and other coatings to improve the metallurgical characteristic of the alloyed zone and to find the optimal technological process.

INTRODUCTION

In the field of Laser applications, surface alloying is one of the processings which has led to the greatest scientific interest. This is because of its high potential in industry in the sector of wear and corrosion-proof coatings and heat-resistant materials (1,2,3).

The methods for obtaining these coatings and the advantages of this technique, which enables strongly anchored surface coatings to be obtained, have been described in depth in (4,5,6,7).

The high power density of the Laser beam gives such high heating and cooling speeds (from 10^6 to 10^{10} K/s) that one can obtain metastable metal alloys on the surface, due to the absence of diffusion processes in the solid state, while the structure of the base metal is left virtually unaltered (1,8,9).

The Laser Surface Alloying process is particularly interesting when one wants to alloy metals which are miscible in their liquid state though only slightly so in their solid state, obtaining solid solutions, which are otherwise impossible, and having particular chemical-physical characteristics (3).

The aim of this paper is to give the results obtained using a 500 W CO2 Laser, and it is part of a wider research programme into Laser Surface Alloying of chromium onto copper which is currently being completed and which has seen the collaboration of the 'Dipartimento di Progettazione e Produzione Industriale' of Bari University, 'Centro Laser' of Bari and the research centre 'L.M.I.' of Fornaci di Barga (LU) (10,11).

A preliminary description of this wider research work has already been given in (12).

The decision to work with chromium and copper depended on the need in the industry for surface alloyed coatings of copper alloys with a high chromium content. This is because the presence of Cr in alloys gives Cu excellent anticorrosive and mechanical characteristics (even at high temperatures), without appreciably reducing its thermal and electrical conductivity. These are qualities which can already be partially found in commercial alloys obtained with traditional processes (13,14,15), and containing from 0.4 % up to a maximum of just 1 % Cr (this is due to relative immiscibility of Cr and Cu in solid state - Fig. 1).

EXPERIMENTAL METHODS AND MATERIALS

The only previous experiment which to our knowledge can be found in literature on L.S.A. of high percentage of chromium on copper was done by Draper (16). Draper used a Q-switch Nd-Yag Laser source to work on samples with a copper substrate and a multi-layer coating made of thin, alternating layers of chromium and copper. These wafer multilayers were a few nanometres thick each. The technique and also the Nd-Yag Laser source described in this work seem to be based on the need to overcome the problems of the high reflectivity of both chromium and copper to the more common radiations, like CO2.

On the contrary, in this research we chose a CO2 Laser, this being the most common industrial Laser source. The samples used had relatively thick deposited layers and substrate dimensions so that the results would give the most relevant information for industrial applications.

A 500 W Valfivre Laser with slow axial flow was used working with continuous wave which could also work with pulsed one.

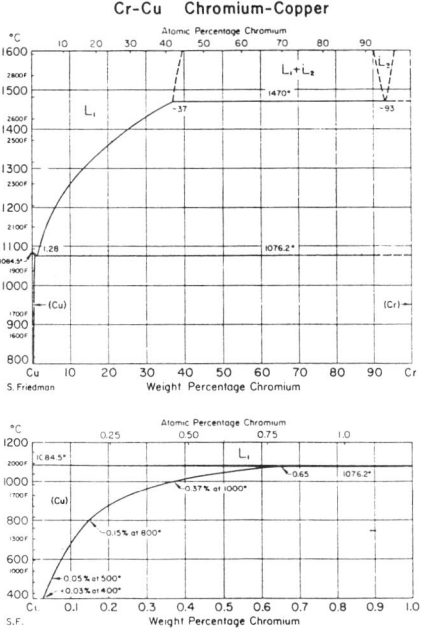

Fig. 1. Cr-Cu equilibrium diagram.

The focusing lens was made of KCl with 3.5" focal length and focus dimension Df 450-500 μm. The protection gas was aimed co-axially at the Laser beam and the nozzle used for this was in brass with a diameter of 12 mm.

The samples, in Cu ETP, were 50 x 25 mm and 3.2 mm thick.

The surface chromium deposit was obtained in an electrolytic bath at 30 °C, with various final thicknesses for the deposited layer (0.6 - 5 -11 μm).

Tables 1 and 2 show the chemical composition of the ETP copper and the main thermophysical characteristics of the Cu and Cr.

CHEMICAL COMPOSITION OF COPPER E.T.P

Cu MIN	O MAX	Bi MAX	Pb MAX	P
99.90%	0.04%	0.001%	0.005%	/

Table 1.

The chromium electrolytically deposited on the copper is very bright and appears particularly reflective to the CO2 Laser radiation (λ = 10.6 µm). The very poor absorption by the chromium of the infrared radiation coming from this source (5 % at ambient temperature) creates considerable difficulties for energy transfer to the samples. Furthermore, the high thermal diffusivity of the Cu substrate (1.16 cm^2/s for copper with respect to 0.26 cm^2/s for chromium) makes it very slow and difficult to raise the surface temperature up to the one at which chromium melts (1850 °C).

THERMO-PHISICAL CHARACTERISTICS OF COPPER AND CHROMIUM

		COPPER	CHROMIUM
MELTING POINT	T_F (°C)	1083	1850
BOILING POINT	T_V (°C)	2590	2620
DENSITY	AT 20°C (GR/CM3)	8.9	7.1
THERMAL DIFFUSIVITY	AT 20°C K_D (CM2/S)	1.16	0.26
THERMAL DIFFUSIVITY	AT T_F K_F (CM2/S)	0.75	0.07
ELECTRICAL RESISTIVITY	AT 20°C (Ω·MM2/M)	0.017	0.21
ELECTRICAL CONDUCTIVITY	AT 20°C (%IACS)	101	/
THERMAL CONDUCTIVITY	AT 20°C (CAL/CM·S·°C)	0.948	0.165
ABSORPTION AT 20°C AND	10.6 µM A_O%	1.45	5.2
ABSORPTION AT T_F AND	10.6 µM A_F%	3.6	12.4
REFLECTIVITY	AT 20°C R_O%	98.6	94.8
REFLECTIVITY	AT T_F R_F%	96.4	87.6
SPECIFIC HEAT	AT 20°C C_P (CAL/G·°C)	0.921	0.106
THERMAL EXPANSION COEFFICIENT	(1/°C 10^{-6})	16.8	6.5
HEAT OF FUSION	H_F (J/G)	209.34	293.076

Table 2.

It was therefore necessary during the experiment to cover the samples with absorbent coatings to improve the energy transfer efficiently.

An experimental plan was set up to enable us to check the technological possibility of alloying Cr to Cu with a CO2 Laser and to find the effect each treatment parameter has on the result of the process. The parameters and working conditions analysed were:

- Type of Laser wave: in order to obtain different beam powers per unit of surface area and different heating and cooling rates of the workpiece, the Laser was both used with continuous and pulsed wave with different peak powers and pulse time (Table 3).

- Thickness of the electro-deposited chromium: the influence of the thickness of the electro-deposited chromium layer was evalated to find the optimal Cr-Cu coupling conditions which would further the alloying process and to analyse the changes the process underwent.

- Interaction time: the beam-workpiece interaction time determines the quantity of energy that can be transferred during the process. It is given by the ratio of the spot size to the speed rate of the workpiece. The latter was made to vary to evaluate this effect too (60 - 120 mm/min).

CHARACTERISTICS OF LASER BEAMS

\multicolumn{2}{l}{LASER TYPE: CO2 VALFIVRE WITH SLOW AXIAL FLOW FOCUSING LENS: KCL FOCAL LENGTH: 3.5" FOCUS DIMENSION: 450-500 µM BRASS NOZZLE DIAMETER: 12 MM PRESSURE OF COV. GAS : 0.6 ATM FLOW OF COV. GAS: 110 NL/MIN}			
WAVE	CONTINUOUS	PULSE TYPE A	PULSE TYPE B
MEAN POWER (W)	500	70	180
PEAK POWER (KW)	0.5	5.8	4.4
PULSE TIME (µS)	/	60	200
FREQUENCY (HZ)	/	200	200
SPECIFIC POWER (MW/CM2)	0.255	295	223 8

Table 3.

- <u>Absorbent coating</u>: to increase the sample absorption of the Laser beam and reduce phenomena of thermal breakdown due to overheating of the focusing lens by the reflected rays, two types of absorbent coating were used: 1) the first, titanium dioxide, was painted onto the sample to be treated in a quickly-drying liquid form: it is very easy to use, gives an excellent absorption of the Laser beam and it anchors well to the substrate; 2) the second absorbent coating was copper oxide electrolytically deposited onto the chromium so as to create a wafer structure. This second coating was tested because, theoretically, it presents fewer risks of contaminating the melt during alloying process.

- <u>Protection gas</u>: finally, the effect of three different protection gases on the treatment was evaluated. These gases were: air, nitrogen and oxygen.

A diagram of the experimental set-up is given in Table 4, that shows in detail the process parameters used for each sort of experimental sample. Cross-sections of the treated samples were then examined under an optical and scanning electron microscope; the concentration of different elements was measured in specific points using the electron microprobe (EMPA) and along line profile analyses.

DISCUSSION OF THE RESULTS

An analysis of the experimental tests led to several points for discussion and suggested further in-depth research into Cr-Cu alloying. In giving and discussing the results we shall follow the experimental layout previously described, analysing in detail the relative effect of each process parameter considered.

- <u>Type of Laser beam</u>: the reaction of the samples to each type of Laser beam (continuous - samples 21,22,24 - or pulsed - samples

EXPERIMENTAL PLAN AND PROCESS PARAMETERS USED

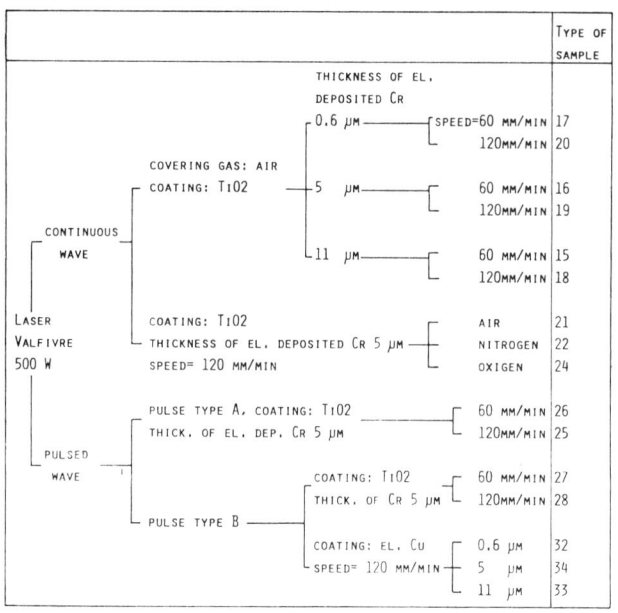

Table 4.

26,27,28,34 -) was very different. In particular, the continuous wave, though giving larger treated zones, was less efficient than the pulsed one in obtaining Cr-Cu surface alloying of a high metallurgical quality.

The pulsed beam has a specific peak power about 10 times greater than what can be achieved with the continuous beam; it gives faster heating and cooling rates, enabling metastable structures to be obtained more easily without altering the grain of the base metal copper.

- <u>Continuous wave</u>: visual examination of the samples treated with a continuous beam shows a thin greyish-green groove about 10 µm deep which is due to the melting and oxidation of the surface chromium. One can also see slight formation of copper oxides produced by the heating of the samples. The Figures 2, 3, 5, 6, 8, 9 show micrographs of the cross-section and the Figures 4, 7 show the result of the analyses done with electron microprobe at different depths.

The microscopic exam shows the complete melting of the electrolytic chromium deposit, which can be found in a finely dispersed form with pale particles a few microns in size up to depths of about 25-30 µm (samples 21,22 - Fig. 2, 3, 5, 6).

The best results from a metallurgical point of view seem to be obtained using compressed air for the covering gas (sample 21 -

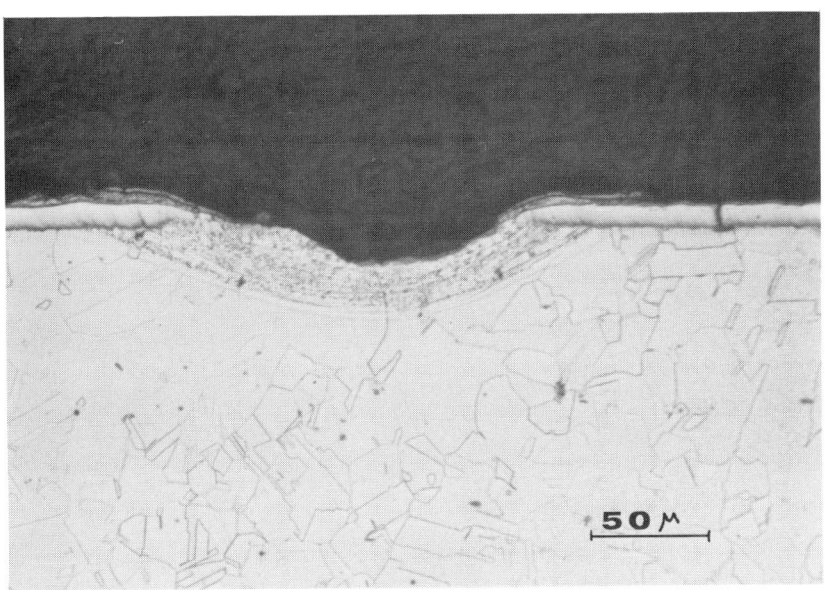

Fig. 2. OM micrograph of sample 21.

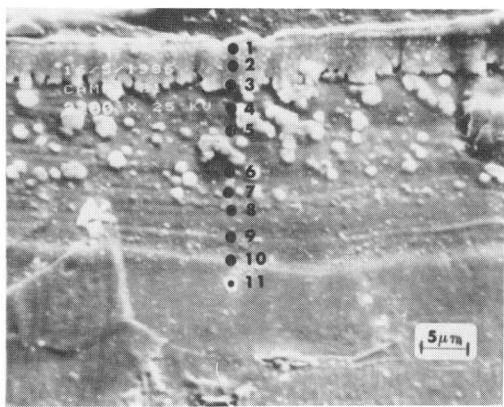

Fig. 3. SEM micrograph of sample 21.

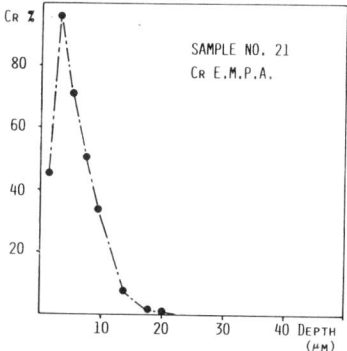

Fig. 4. EMPA micro-probe analysis of sample 21.

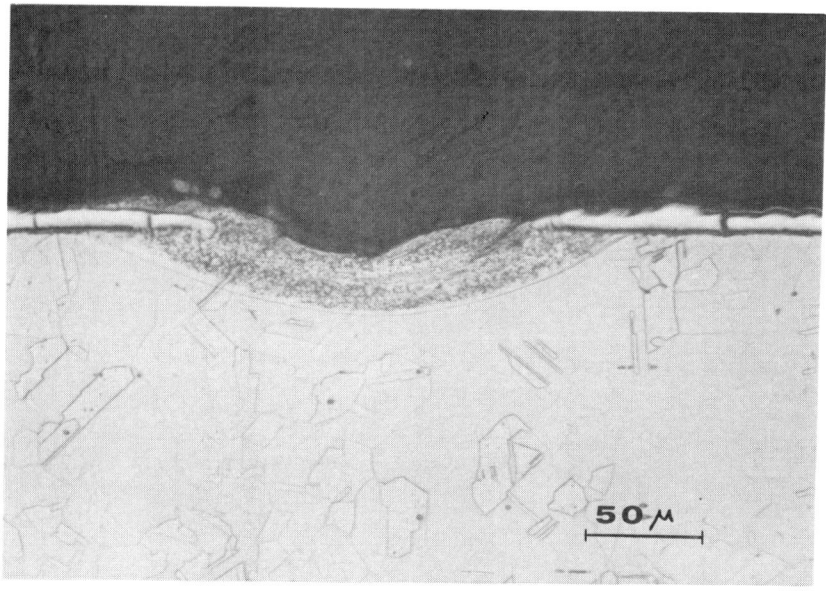

Fig. 5. OM micrograph of sample 22.

Fig. 2, 3) in that, with the nitrogen, one has a more irregular dispersion of the chromium and smaller particles (probably nitrides) included in the melted zone (sample 22 - Fig. 5, 6).

The experiments using an oxiding atmosphere to try to create greater absorption of the Laser radiation, showed that this approach is unadvisable (sample 24 - Fig. 8, 9) in that it causes vaporization of the chromium and considerable oxidation of the copper without giving any alloying.

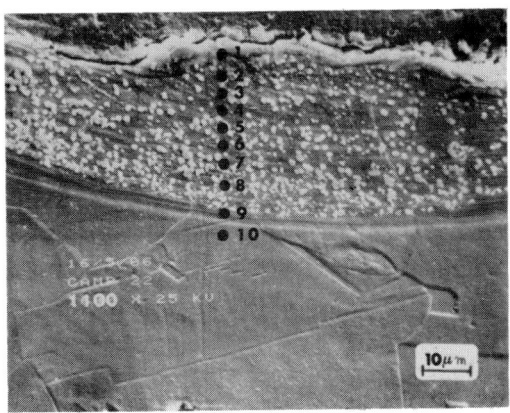

Fig. 6. SEM micrograph of sample 22.

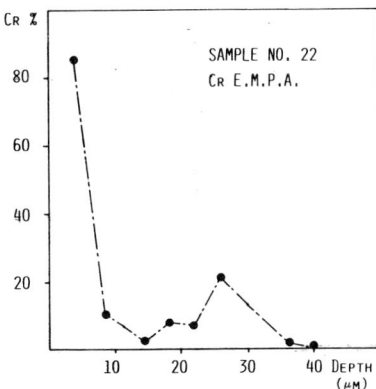

Fig. 7. EMPA micro-probe analysis of sample 22.

- <u>Using a slower treatment speed</u> (greater interaction time) did not give good results in that the greater heat energy given out gives rise to high irregular melted zones.

As far as concerns the samples treated with a continuous wave, appreciable results were only obtained with 11 μm thick electro-deposited chromium layers; the 0.6 and 5 μm thicknesses proved to be unsuitable.

- <u>Pulsed wave</u>: the tests with a pulsed wave laser beam, done with pulses of various durations and peak energies, gave very interesting results. In the treated zones a layer of surface alloying was obtained with structures containing between 40 % and 10 % chromium.

The visual inspection did not show up any thermal or colour alterations in the copper. The Laser path on the chromium was regular and narrower than in the case with continuous wave

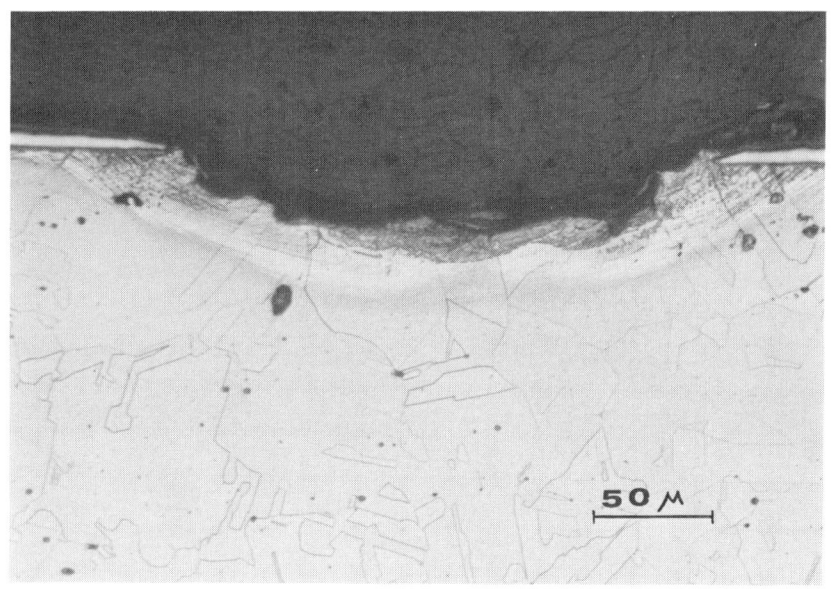

Fig. 8. OM micrograph of sample 24.

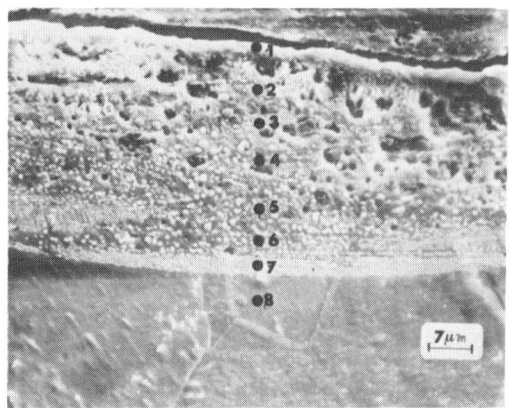

Fig. 9. SEM micrograph of sample 24.

because, with this latter, a greater quality of total energy is transferred.

The optical and electron microscopes showed up smaller melted zones than those obtained with the continuous beam.

- <u>Pulses with higher peak power and shorter pulse time</u> (type A): this form of Laser treatment does not change the structure of the parent metal (copper) because of its small heat contribution (70 W).

The grains did not appear to have changed size and the material did not show signs of annealing. This is obviously a benefit for the technological and mechanical characteristics of the sample. The photographs show a melt which is neither very deep (though greater than the thickness of the chromium layer) nor very extensive.

The treatment speed does not seem to have much effect on the maximum penetration depth of the chromium; however, sample 26 appears to be more evenly alloyed across the width of the radiated zone (Fig. 10, 11).

In the most powerful enlargement a clear-cut separation between the melted zone on the surface (which looks fairly homogeneous) and the copper substrate is visible.

From the microprobe exam and the SEM photo (Fig. 11) one can see a layer about 17 µm thick (as opposed to the initial 5 µm) of penetration of the Cr.

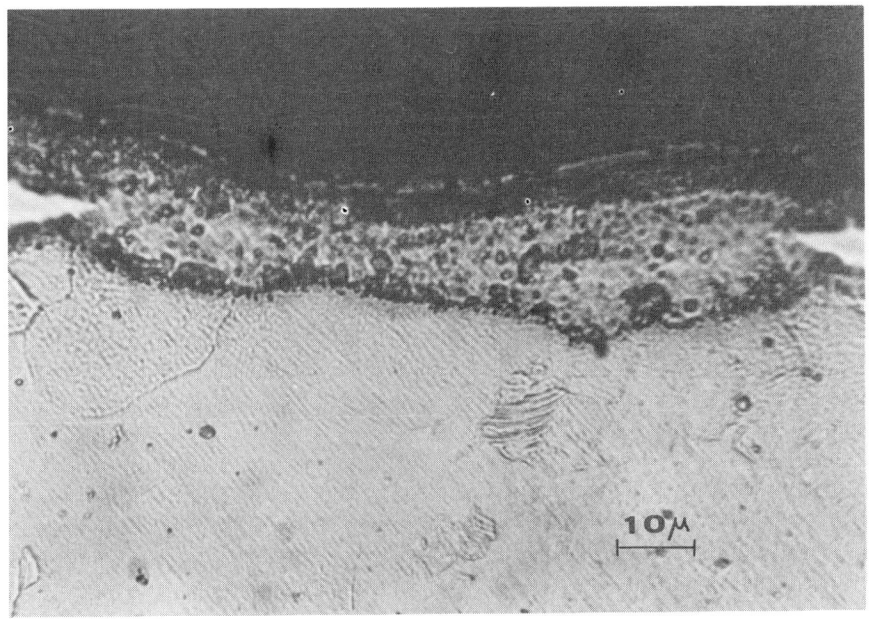

Fig. 10. OM micrograph of sample 26.

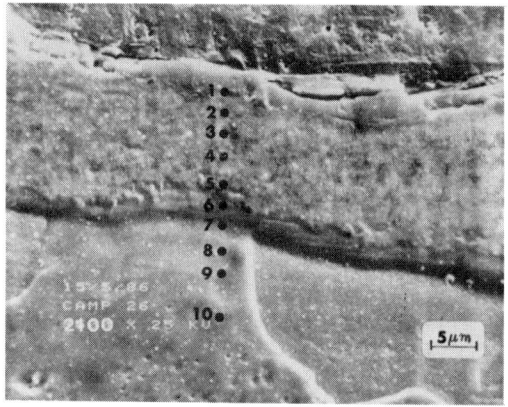

Fig. 11. SEM micrograph of sample 26.

It is worth emphasizing that, in this layer, it is not possible to resolve any metallographic structure.

The high percentages of chromium found therefore indicate the presence of a metastable Cr-Cu alloyed structure.

Sample 26 (treatment speed: 60 mm/min) shows percentages of chromium in the order of 22 % down through half of the melted layer (12.4 µm) and, down to a depth of 17 µm, chromium percentages of over 1 % in weight are recorded (Fig. 12, 13).

The fast cooling and heating speeds (long, narrow pulse), allow a good general structure to be obtained and, if the pale particles noticed in the sample treated with a continuous beam are oxides, the lack of these particles could depend on the reduced influence of the oxygen which probably does not have this negative effect as there is not enough time for its diffusion.

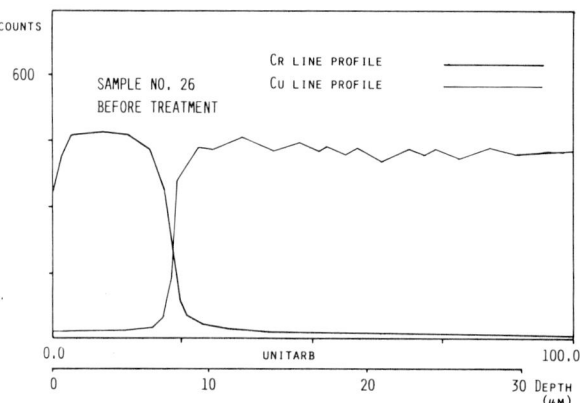

Fig. 12. EMPA micro-probe and line profile analysis of sample 26 before treatment.

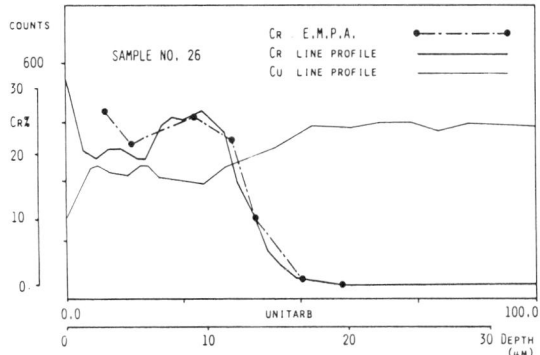

Fig. 13. EMPA micro-probe and line profile analysis of sample 26 after treatment.

- <u>Pulses with lower peak power and greater pulse time</u> (type B): in the samples treated with a greater average power, visual inspection showed a wider and deeper beam pathway on the surface.

The microscopic examination showed that, in some cases, the greater quantity of energy transferred caused turbolence in the melt when working at low speeds (27). In this sample one can, however, see alloying containing Cr percentages of at least 7 % down to about 22 µm from the surface. These samples too do not seem to have undergone any alterations in the Cu base structure (Fig. 14).

At faster treatment speeds (sample 28 - Fig. 15, 16), one can see that the surface zone looks homogeneous, with very slight surface undulation and well-anchored to the substrate. The photos in Figures 15 and 16 show the layer relating to the above-mentioned area where concentration measurements were made in several points and along a line: down to 27 µm the chromium weight values keep above 20 % (point 5). The front penetration is, at most, 35 µm (Fig. 17).

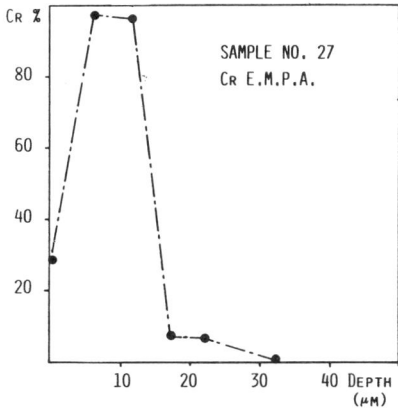

Fig. 14. EMPA micro-probe analysis of sample 27.

70

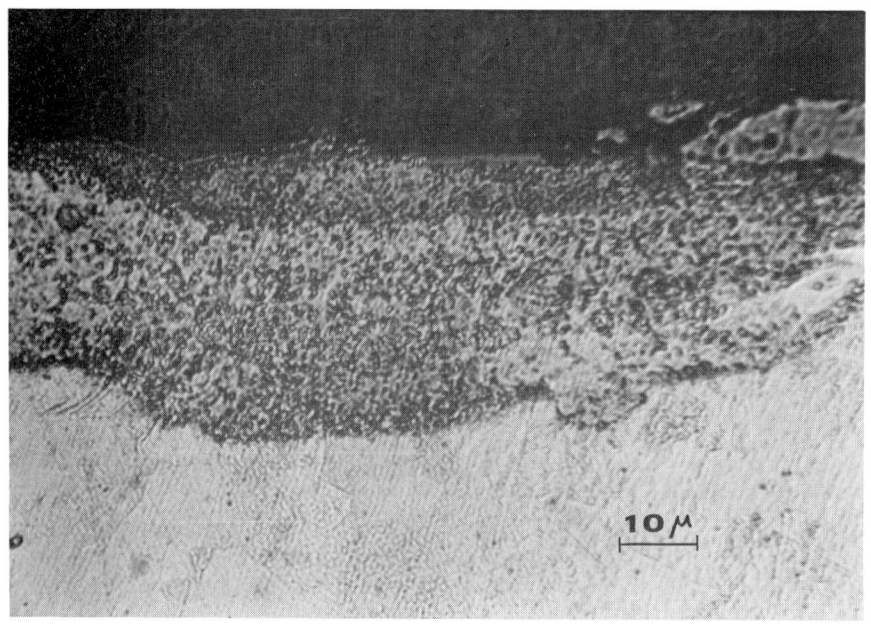

Fig. 15. OM micrograph of sample 28.

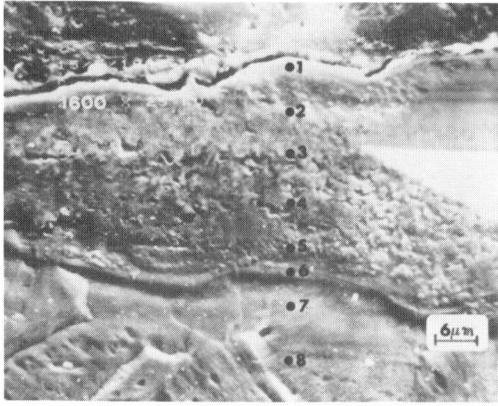

Fig. 16. OM micrograph of sample 28.

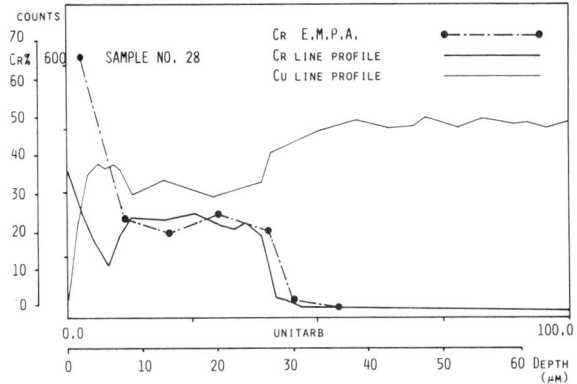

Fig. 17. EMPA micro-probe and line profile analysis of sample 28.

- <u>Influence of the type of absorbent coating</u>: the effect of the different types of absorbent coating is clearly visible from a comparison of sample 28 (TiO2-Fig. 15) and 34 (elect. Cu-Fig. 18) obtained under the same working conditions.

The absorbent coating obtained by electrolytically depositing Cu on Cr, gave its best results using the pulsed Laser with lower peak power and greater duration. By comparing the two samples one can see that the electrolytically deposited Cu seems to encourage the in-depth absorption of Laser radiation; indeed, Cr percentages of about 20 % were found down to 34 μm, as opposed to the 27 μm reached in the sample coated with titanium oxide.

Other positive aspects given by electrolytically deposited Cu are the fact that an initial layer of nearly pure Cr about 5-7 μm thick is kept and there is also less of a risk of contaminating the melt with heterogeneous substances such as TiO2.

- <u>Electro-deposited Cu coatings</u>: during the Laser treatment of the samples bluish-white plumes were formed which indicates the vaporization of the Cu (about 2500 °C). We feel that this is due to the chromium keeping the heat on the surface longer than the Cu as it has a poorer diffusivity and thus enables quick heating of the surface temperature of the sample.

The best thicknesses for the chromium initially deposited was 5 μm (sample 34 - Fig. 18, 19).

The Cu substrate does not appear to be affected by the heat treatment. The melted zone is similar to, though deeper than, that observed in sample 28 (pulsed with TiO2); the maximum depth is 40 μm.

The surface shows a homogeneous layer with a high percentage of chromium (about 95 %), anchored to a thicker layer which has different metallographic characteristics to the previous one (with a structure which cannot be resolved at 1400 x), with Cr

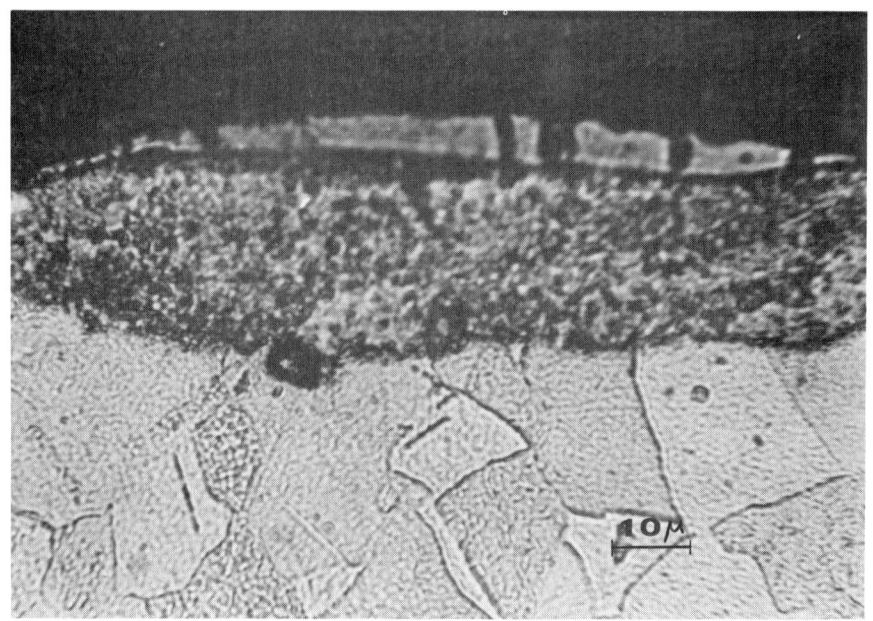

Fig. 18. OM micrograph of sample 34.

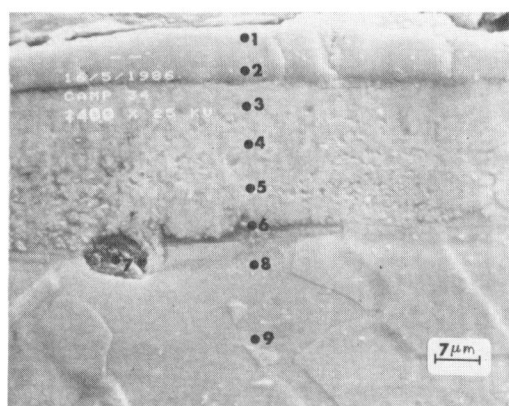

Fig. 19. OM micrograph of sample 34.

percentages varying between 40 % and 22 %. This latter layer seems to be well-anchored to the underlying copper. There are no pale partlicles in the layer affected by the melting.

The chromium in the samples with the thickest electro-deposited layers at the outset (33) seems to have penetrated less homogeneously than in the samples with a 5 µm deposit (34).

The EMPA results, in fact, show a nearly pure Cr layer on the surface, followed by a deeper area with percentages of Cr weight from 31 % down at 1 % at 39 µm beneath the surface (Fig. 20).

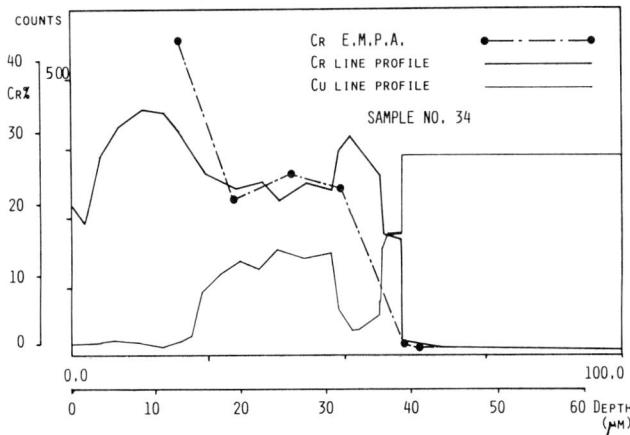

Fig. 20. EMPA micro-probe and line profile analysis of sample 34.

CONCLUSIONS

An analysis of the results shows that it is now possible to surface alloy chromium and copper using low power CO2 Laser sources and process technologies which can easily be adopted in industry.

Previous experiments [16] had shown that this was possible with a Q-switched Nd-Yag Laser working on exceedingly thin wafer multilayer samples of alternating Cr and Cu (a few nanometres thick each); however, this procedure is very difficult to apply on an industrial scale.

The samples treated here show the significant alloying depths obtained (tens of microns), with chromium percentages of between 40 % and 20 %, when using a Laser with pulse waves and electlytically deposited copper absorbent coatings.

The authors are currently finishing another series of experiments using a medium-power CO2 Laser with beams of different characteristics. These experiments are specially geared to studying the sort of absorbent coating which gives the best metallurgical qualities and further improves the treatment technology.

ACKNOWLEDGMENTS

The authors would like to thank the following: Prof. Attilio Alto, full professor of Metals Technology in the College of Engineering of Bari University, for his help and useful advice and the fruitful discussions had with him throughout the research; Mr. Giuseppe Daurelio, researcher at the 'Centro Laser' in Bari, for his vital and enthusiastic contribution during the experiments with the Laser; and Eng. Roberto Paolillo for his constant and competent collaboration throughout the research and, especially, during the microscopic analyses.

REFERENCES

(1) W. M. Steen, C. G. M. Courtney: "Hardfacing of Nimonic 75 using 2 Kw continuous wave CO2 LASER" - Metals Technology, June 1980.
(2) J. F. Ready: "Applicazioni industriali del Laser" Ed. Tecniche Nuove, Milan 1983.
(3) C. W. Draper: "Laser Surface Alloying: the state of the art", Journal of Metals, June 1982.
(4) C. W. Draper, J. M. Poate: "Laser Surface Alloying", International Metal Reviews, 1985, 30 n.2.
(5) D. S. Gnamamuthu, E. V. Locke: "Laser Surface Alloying" - Research Report AVCO Everett Research Laboratory Inc. - Everett Massachussetts.
(6) S. Das, I. Dumler, J. Mazumder: "Laser surface alloyed Fe-Cr-C". Proceedings of the 2nd International Conference on Lasers in Manufacturing, 26-28 March 1985.
(7) T. Takeda, W. M. Steen, D. F. R. West: "In situ clad alloy formation by Laser cladding" - Proceedings of 2nd International Conference on Lasers in Manufacturing, 26-28 March 1985.
(8) J. Mazumder: "Laser heat treatment: the state of the art", Journal of Metals, May 1983.
(9) S. Kou, S. Q. Ksu, R. Mehrabian: "Rapid melting and solidification of a surface due to a moving heat flux" - Metallurgical Transactions B 12B, March 1984.
(10) G. Ruta, M. De Giglio, G. Daurelio: "Alligazione Superficiale al Laser - stato dell'arte" Rel. Prog. Spec. Ric. Scient. Appl. FORMEZ - CENTRO LASER - Bari, July 1986.
(11) R. Paolillo: "Alligazione superficiale mediante laser a CO2": D. Thesis in Mechanical Engineering - Tutors: A. Alto, L. M. Galantucci, University of Bari, July 1986.
(12). N. Ammannati, L. Cento, G. Daurelio, S. Magnanelli, G. Ruta: "Preliminary experiments concerning Laser Surface Alloying of Cu-Cr" - Copper Tomorrow Cu '86 - Technology - Products - Research, 10/12 Sept 1986 Barga (LU) Italy.
(13) CDA Pubb. 1956 51 54.
(14) W. Hielsch, Hamheim: Metall 13th year - Nov. 1959.
(15) S. Sato, K. Nagata (Sumitorno Light Metal Inds.): "Quenching sensitivity of Cu Cr Alloys" - Journal Japan Copper and Brass Res. Assn., 1970 9 (1) 90-97.
(16) C. W. Draper: "The effects of Laser Surface Melting on Copper Alloying" Proceedings of Lasers in Metallurgy in Symposium, 110 th AIME Annual Meeting, Chicago, Illinois, 22-26 Feb. 1981.

LASER SURFACE MODIFICATION OF PLASMA SPRAYED OXIDE CERAMIC COAT LAYER FOR ANTI-CORROSION PERFORMANCE

Mikio Takemoto

College of Science and Engineering, Aoyama Gakuin University
6–16–1, Chitosedai, Setagaya-ku, Tokyo, 157 Japan

ABSTRACT

In order to coat a high quality ceramics on the metallic substrate, laser surface modification of plasma sprayed oxide ceramics is attempted. The surface thin layer up to 100 μm of sprayed coat was remelted by CO_2 laser, utilizing the porous sprayed coat as a crack arrester and/or stress relaxative. Optimum laser irradiation conditions have been examined to yield high quality ceramics free from voids and cracks as a function of laser power scanning intensity: W/V (Kw sec/m) where W is laser power and V is scanning velocity (m/min). Microscopic examination and corrosion test of remelted ceramic coat showed that alumina coat layer was successfully modified, however, titania ceramic suffered severe cracks during laser irradiation. Some potential countermeasures for unsuccessful cases are discussed.

INTRODUCTION

The method to isolate the metallic substrate from severe environment by coating high quality ceramics will be an important engineering surface modification technique. Plasma or gas flame spraying is a most easy method to coat ceramic layer on metallic surface, however, yield a relatively low grade ceramic coat because of their voids and non-melted ceramic particles. Attempts to coat ceramics directly on the metal surface by using laser, however, have been unsuccessful, since cracks and exfoliation of ceramic coat were experienced during laser processing, probably due to the mismatch of thermal expansion coefficient of ceramics and metal.

In view of the extremely low fracture toughness of ceramics, the technical problems to be solved, in applying a high energy density power such as laser, is to yield a high quality ceramic coat free from voids and cracks. Proposed method is to yield a high performance ceramic coat by laser remelting the thin surface of plasma sprayed ceramic coat without spoiling good characteristics of sprayed ceramics such as relatively high adhesivity to substrate and stress relaxative or absorber.

The aim of this study is to search an optimum condition for laser irradiation to sprayed ceramics and to develop a new method for ceramic coating. The characteristics of ceramic coat processed by plasma and laser remelting have been evaluated. Though the study has not been completed, the metallographic characteristics, evaluation of anti-corrosion performance in acidified solution and potential countermeasure to unsuccessful cases will be introduced on three oxide-ceramics, i.e., alumina, titania and calsia stabilized zirconia (PSZ) coated on AISI 304 (SUS304).

CHARACTERISTICS OF PLASMA SPRAYED OXIDE CERAMICS

Before describing the laser remelting technique, the characteristics of plasma sprayed ceramic coat will be introduced in this chapter. In plasma spraying, fine powder with chemical composition listed in Table 1 was sprayed in air directly onto the blasted austenitic stainless steel AISI304. Argon or nitrogen gas was used as primary gas to investigate the effect of plasma gas on the characteristics of ceramics.

Table 1 Chemical composition of ceramic power used for plasma spraying

Titania ceramic

TiO_2	Other oxide
99.0	Balance

Alumina ceramic

Al_2O_3	SiO_2	Other oxide
98.5	1.0	Balance

PSZ ceramic

ZrO_2	CaO	Al_2O_3	SiO_2	Other oxide
93.0	5.0	0.5	0.4	Balance

The condition for plasma spraying is shown in Table 2. As-sprayed coat layer has been shown to contain a number of defects such as non-melted particles and voids (Fig. 1). An underlying path connecting these voids will be a deleterious shortcoming for the purpose of isolating the substrate from environment. For example, the porosity area percentage based on the cross sectional area reach 13 % in sprayed titania, 18 % in alumina and 22 % in partially stabilized zirconia. The porosity percentage of sprayed ceramics seems to increase with the melting point of ceramics, and to be extremely high compared to that of "fine ceramics" processed by HIP or sintering. The adhesive strength of ceramic coat layer to blasted SUS 304 was measured according ASTM C633.

Table 2. Plasma spraying condition

Spraying material	Al_2O_3	Al_2O_3	ZrO_2	ZrO_2	TiO_2	TiO_2
Primary Plasma Gas	N_2	Ar	N_2	Ar	N_2	Ar
Pressure (psi)	50	100	50	100	50	100
Flow (lb/in²)	75	80	75	80	75	80
Secondary Plasma Gas	H_2	H_2	H_2	H_2	H_2	H_2
Pressure (psi)	50	50	50	50	50	50
Flow (lb/in²)	15	15	15	15	15	15
Carrier Gas Flow	37	37	37	37	37	37
Spray Distance (mm)	100	100	80	80	80	80
Spray Gun Type	3MB	3MB	3MB	3MB	3MB	3MB
Arc Voltage (V)	80	70	80	75	80	70
Arc Amps (A)	500	500	500	500	500	500
Nozzle	GH	GH	GH	GGH	G	GH

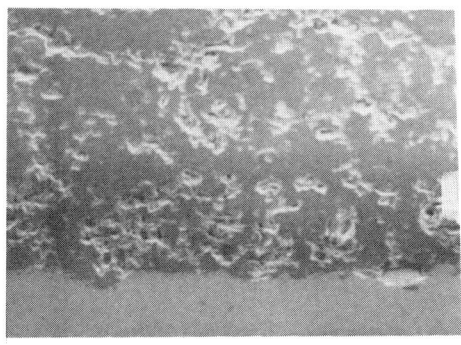

Fig. 1. Transverse section of as plasma sprayed **alumina/** AISI 304.

Results are shown in Fig. 2. There seems to be no discernible effect of plasma gas on the adhesive strength. Though the adhesion of ceramics to substrate metal is thought to be achieved

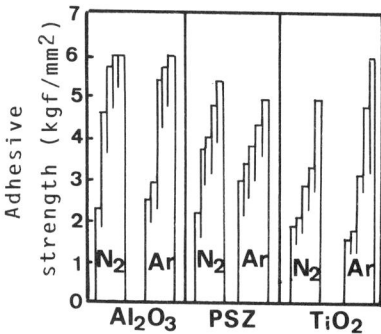

Fig. 2. Adhesive strength of plasma sprayed ceramics to blasted AISI 304.

by pure mechanical anchor effect, some of them have the adhesive strength higher than 6 kgf/mm².

The anti-corrosion performance of specimens with as-sprayed ceramics coat has been evaluated in 4N H_2SO_4-0.4 N NaCl solution (1) in which SUS 304 suffers severe corrosion and also stress corrosion cracking under tensile stresses.

The polarization curve of SUS 304 in 40 °C H_2SO_4 -NaCl solution (Fig. 3) shows that substrate metal suffers extremely high active corrosion (100 mA/cm²) at potential of -230 mV vs. S. C. E.

The current density of specimen which is galvanized with platinum electrode was measured by zero resistance ammeter and used as a parameter to evaluate the anti-corrosion performance of ceramic coat. The corrosion potential of SUS 304 coupled with platinum electrode was about -230 mV vs. S. C. E. which corresponds to the stress corrosion cracking susceptible potential range of SUS 304 under applied tensile stresses.

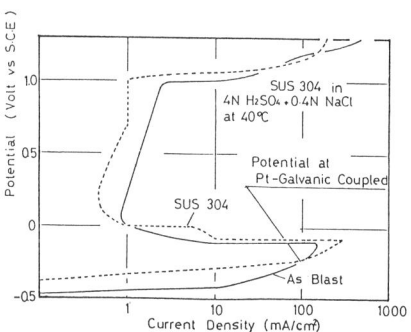

Fig. 3. Polarization curve of **AISI** 304 in 4N H_2SO_4 + 0.4 N NaCl solution.

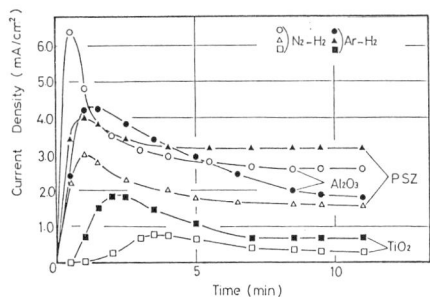

Fig. 4. Change of galvanic current density of as plasma sprayed ceramics/AISI 304.

Figure 4 shows the change of galvanic current density of ceramic coated specimen with the immersion time. The rapid increase of current density at early time, however, suggests that the solution quickly impregnates into the ceramic coat layer and reach the substrate metal. It could be said that the as-sprayed ceramic coat has a relatively poor capability of isolating the substrate metal from environment.

The higher current density of alumina and PSZ which are sprayed using argon-hydrogen plasma gas are thought to be due to the existence of non-stichiometric metal rich oxide. Figure 5 shows the transverse section of coat layer after the galvanic corrosion test. It can be seen that the alumina coat layer is heavily attacked, and contains a number of large voids. Partial exfoliation of alumina coat from substrate metal is also observed. On the other hand, titania and PSZ ceramics keep their original coat layer where a number of voids, however, grow up. Thus, it is obvious that the as-sprayed ceramic coat has a relatively low performance for isolating the substrate. It is indispensable to modify the sprayed ceramic coat to yield a high quality ceramic free from voids and cracks.

LASER REMELTING OF PLASMA SPRAYED CERAMIC COAT LAYER

A CO_2 gas laser with maximum output power of 2.5 Kw was used to remelt the surface thin layer of plasma sprayed ceramic coat. Band melting was done by beam scanning method. Multimode laser beam with power of W KW was oscillated to 8 mm width by beam scanner with 130 Hz frequency. The test specimen sealed with argon gas was moved under the oscillating beam with velocity of V m/sec. In order to search optimum condition for laser irradiation, characteristics of remelted ceramics, such as microstructure, remelted thickness, hardness etc., were examined for specimens with various laser power scanning intensity defined by W/V (Kw min/m). Anti-corrosion performance of some specimens is also evaluated by the galvanic current density as is described in the previous chapter. Studies have been mainly done on the alumina and titania coat specimen.

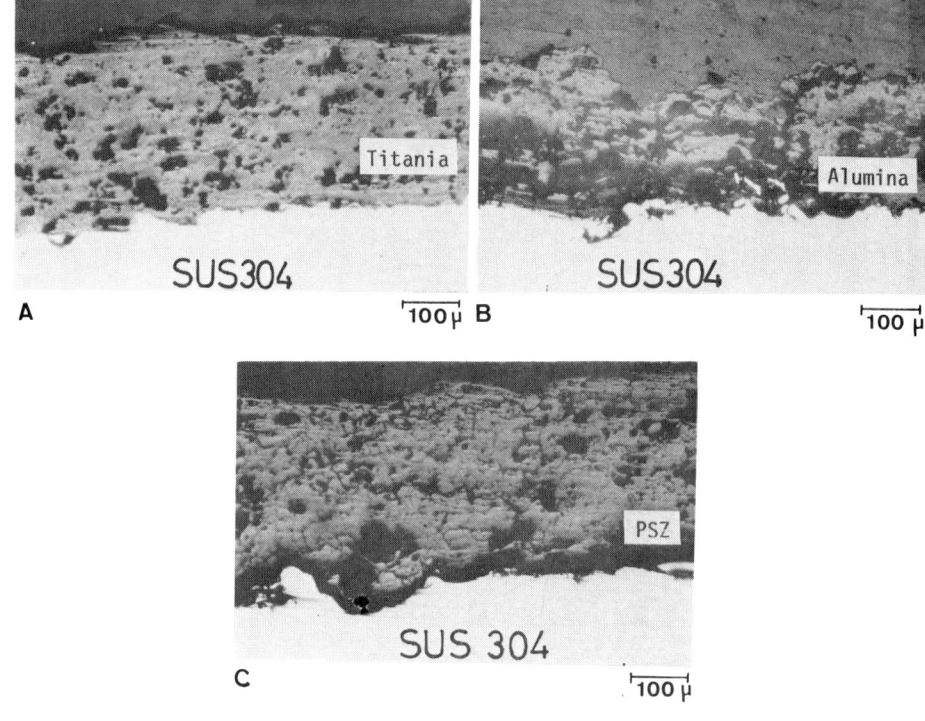

Fig. 5. Optical photographs of transverse section of sprayed titania (A), alumina (B) and PSZ (C) after corrosion test.

LASER SURFACE MODIFICATION OF TITANIA COAT LAYER

Figures 6 and 7 show laser irradiation condition in W-V diagram and the appearance of remelted titania coat. Though zig-zag remelted pattern is seen at higher scanning velocities, this will be solved by increasing the oscillating frequency of beam scanner. In order to avoid the exfoliation of ceramic coat from substrate metal, laser power scanning intensity was controlled as to melt surface thin layer of maximum 100 micrometer.

Figures 8, 9 and 10 show the microstructure of laser processed titania coat. Figure 8 suggests that the laser power scanning intensity: W/V of 0.2 Kw **min/m** is not enough to remelt the titania coat.

Surface layer of titania coat remelted at W/V: 0.3 Kw **min/m** (Figs. 9 and 10) is seen to be continuously remelted and free from voids. All titania coat laser remelted, however, suffered severe cracking during laser processing in spite of W/V values. It is obvious (Fig. 9) that these cracks start from the remelted zone and propagate into coat layer. Cracks are thought to occur during rapid cooling from melted condition, as suggested by the characteristic surface structure with fine needle-shaped crystal.

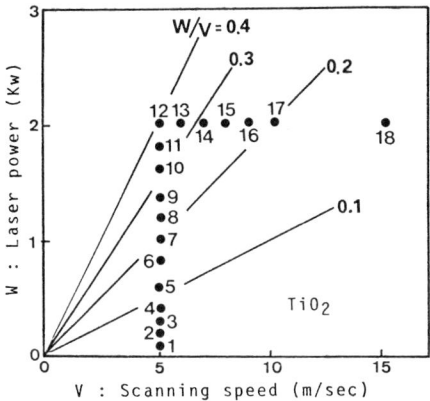

Fig. 6. Laser irradiation points of titania coat.

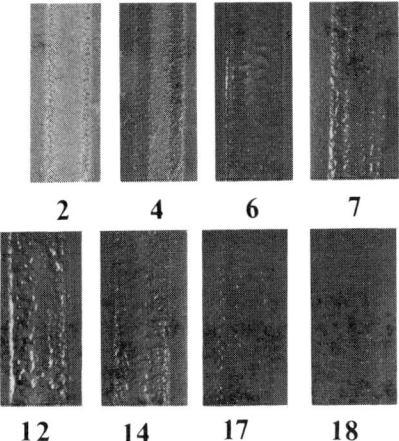

Fig. 7. Surface appearance of laser remelted titania.

The fact that the exfoliation of ceramic coat layer is not observed and cracks stop their propagation before reaching the substrate metal, however, suggests that the as-sprayed coat layer successfully acts as a crack arrester or stress relaxative.

A high cracking susceptibility of titania ceramic during laser remelting will be correlated to the stresses induced by thermal shock and the volumetric change due to the crystallographic transformation. Acoustic emission analysis is now attempted to clarify the mechanism of cracking during laser remelting and to search possible techniques to avoid cracks.

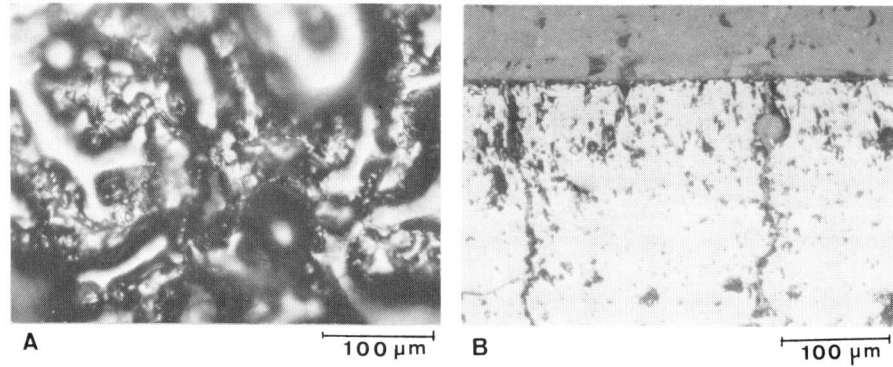

Fig. 8. Optical photograph of surface (A) and transverse (B) of titania laser remelted at W/V = 0.2, No. 18.

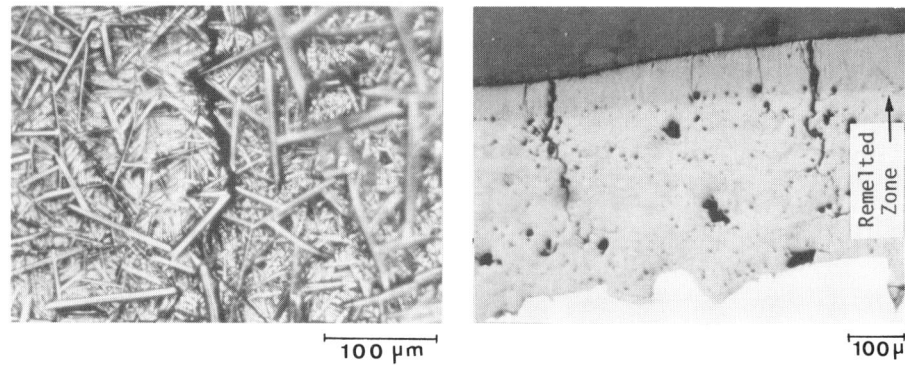

Fig. 9. Optical photograph of surface (A) and transverse (B) of titania laser remelted at W/V = 0.3, No. 9.

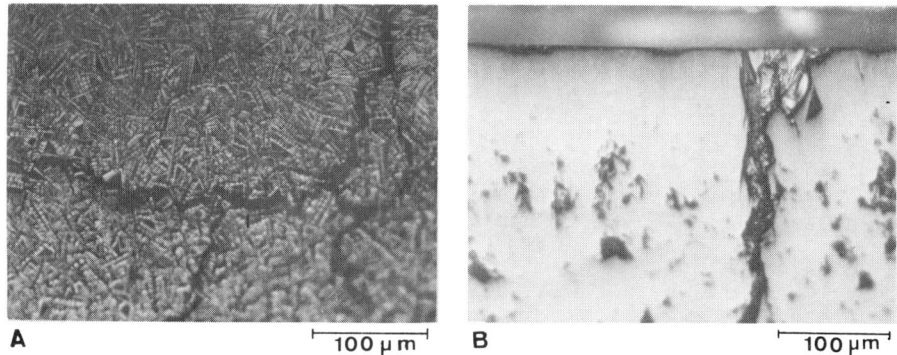

Fig. 10. Optical photograph of surface (A) and transverse (B) of titania laser remelted at W/V = 0.4, No. 12.

Figure 11 shows the change of galvanic current density of specimens with laser remelted titania coat. Specimens 12 and 17 with continuously remelted but cracked ceramic coat have short incubation period, however, finally show high galvanic current density. The specimens 18, 4 and 7 with low W/V, therefore partially remelted ceramic coat showed rapid increase of current just after being galvanized with platinum electrode. In the explanation of current density transient, the existence of metal rich titania should also be taken into account, in addition to the defects in remelted ceramic coat. The titania, sprayed with argon primary plasma gas, has extremely low electric resistance compared to fine ceramics.

Though the steady current density (1 to 10 microampere/cm^2) is extremely small, i.e., two or three order smaller than the passivation current density (1 milliampere/cm^2), and four or five order smaller than that of bare SUS 304 at galvanized potential, however, really accounts of the impregnation of solution to substrate metal. Figure 12 shows the cross-section of ceramic coat layer after the galvanic corrosion test. It can be seen that cracks combine in the coat layer, and propagate into substrate metal.

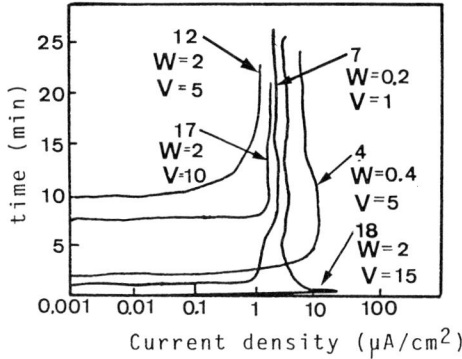

Fig. 11. Change of galvanic current of specimens with laser remelted titania.

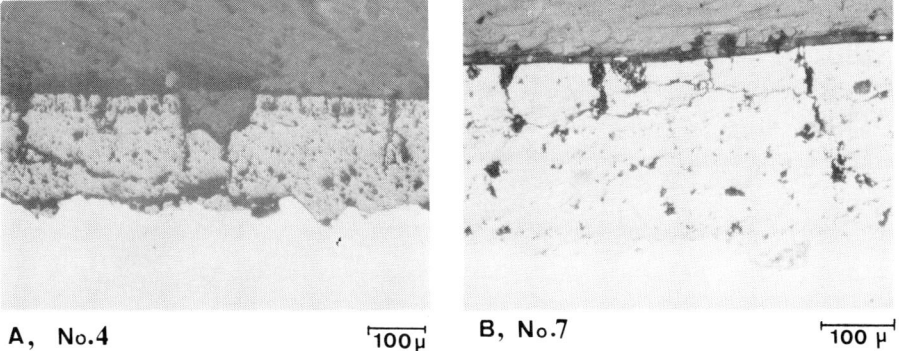

A, No.4   B, No.7

Fig. 12. Transverse section of titania laser remelted at W/V=0.1(A) and W/V=0.2(B) after corrosion test.

The coat layer which is poorly remelted is almost exfoliated. Even one crack in the remelted layer will be fatal to this purpose.

As to the surface modification of titania, alternate method such as laser remelting with pre- and after-heating, or to use compound ceramic of titania aluminate will be necessary. As the melting point of titania is not so high, and titania has electric conductivity, pre- and after-heating by laser or electric heating is possible.

LASER SURFACE MODIFICATION OF ALUMINA COAT LAYER

Figure 13 shows laser irradiation condition of alumina coat layer. Figure 14 compares surface SEM photographs of alumina coat as-sprayed (A) and laser processed (B) at W/V=0.1. A number of circular and angular voids are observed, however, cracks are not observed. This suggests that the laser power scanning intensity of W/V=0.1 is not enough to remelt alumina coat.

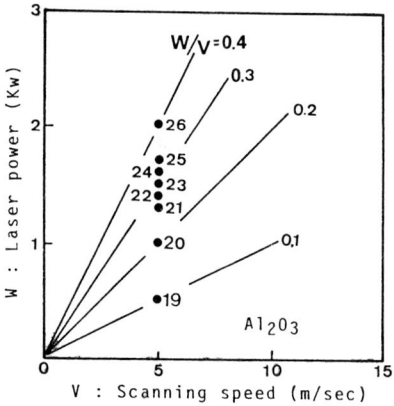

Fig. 13. Laser irradiation points of alumina coat.

Figure 15 shows surface and transverse optical microphotographs of specimen 26 with W/V=0.4. Some large circular voids with radial cracks are observed on surface with fine needle-shaped crystal. This is thought to be the case where ceramic coat is overheated.

On the other hand, any voids and cracks are not observed on specimen 23 which is remelted at W/V=0.3 (Fig. 16). The needle-shaped structure observed on the specimen of titania and alumina with high W/V is not visual on specimen 23. Though the optical microphotograph shows "crack like dendritic" structure in remelted zone (Fig. 16-B), it is confirmed, by the electron microscopic observation (Fig. 17), that these are not cracks but voids induced during cutting and polishing of hardened alumina. The

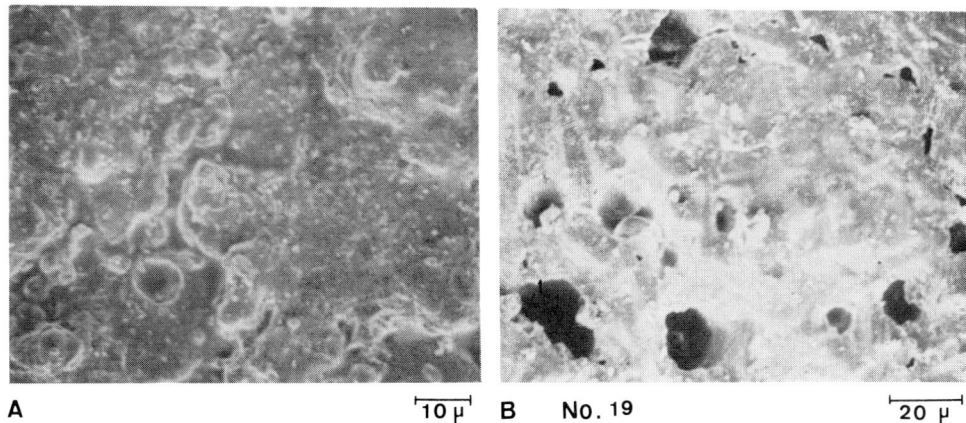

Fig. 14. SEM of as plasma sprayed (A) and laser remelted (B) at W/V=0.1 alumina coat surface.

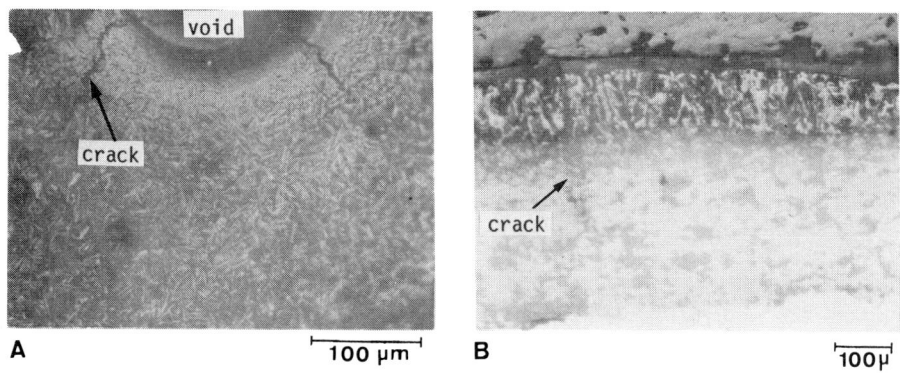

Fig. 15. Optical photograph of surface (A) and transverse (B) of alumina laser remelted at W/V=0.4, No. 26.

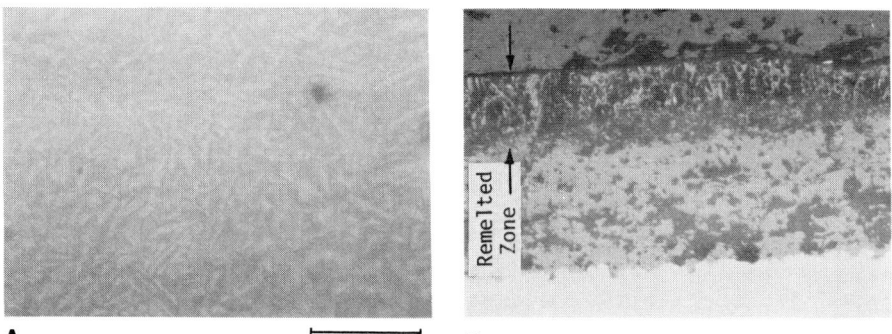

Fig. 16. Optical photograph of surface (A) and transverse (B) of alumina laser remelted at W/V=0.3, No. 23.

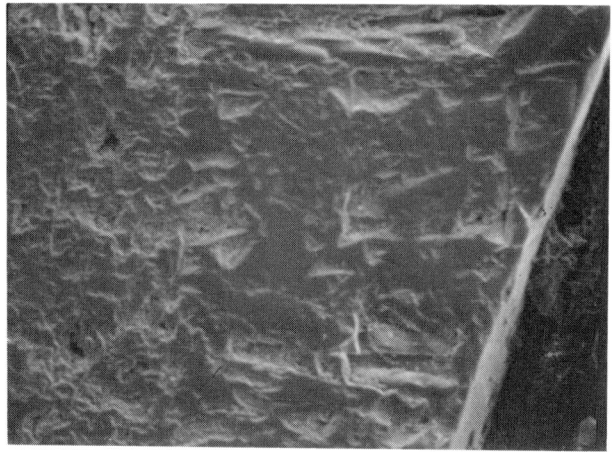

Fig. 17. Transverse SEM of laser remelted alumina. W/V=0.3, TP No. 23.

Vickers hardness of laser remelted zone increases to 1800 from 1200 at as-sprayed condition. Though the melting point of alumina is higher than that of titania, alumina is well melted at smaller W/V, i.e., 0.2 to 0.3, than that for titania. Appropriate laser irradiation conditions will be affected by other characteristics of ceramics, such as reflexibility or absorbability of laser.

It could be said that alumina generally has a low cracking susceptibility during laser remelting, and is successfully qualified under appropriate irradiation conditions.

An evaluation of anti-corrosion performance has been carried out by the same method used for titania ceramic coat, and the results are shown in Fig. 18. The specimens 20 and 23 which are remelted at W/V of 0.2 and 0.3 respectively, showed almost perfect anti-corrosion performance. The current density of 0.001 microampere/cm^2 is six order smaller than the passivation current density. The coat layer after the galvanic corrosion test were free from voids and cracks. Specimens 19 and 26, which are laser processed at unappropriate W/V show high current density due to the penetration of solution into coat layer. Figure 19 shows transverse section of specimen 26 after the galvanic corrosion test. It can be seen that the remelted ceramic coat layer is partially exfoliated along the melted/non-melted boundary. Anti-corrosion performance of this case is also easily spoiled by only one crack or void in remelted layer. As a matter of course, it should also be taken into account that the anti-corrosion performance is affected by the corrosion resistance of ceramics, such as some kinds of transformed phases, impurities and non-stoichiometric ceramics induced by rapid quenching.

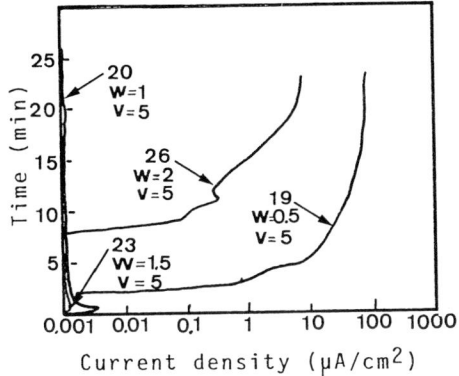

Fig. 18. Change of galvanic current of specimens with laser remelted alumina.

Fig. 19. Transverse section of laser remelted alumina (W/V=0.4, No. 26) after corrosion test.

CONCLUSION

Plasma sprayed oxide ceramics were remelted by CO_2 laser to yield a high quality ceramic layer, to isolate the substrate metal from severe environment. The study has not been completed, however, following results are obtained;
1) As-sprayed ceramic coat contains a number of voids, and showed very poor anti-corrosion performance.
2) As-sprayed porous ceramic coat layer will be usefully used as a crack arrester and/or stress relaxative during laser surface remelting.
3) Titania ceramic has an extremely high cracking susceptibility during laser remelting, therefore, can not be modified by simple laser irradiation. Alternate modification method involving pre- and after-heating will be necessary.

4) Alumina ceramic coat layer was sucessfully qualified under appropriate laser irradiation conditions, and showed excellent corrosion resistance.

ACKNOWLEDGMENT

The author would like to express his gratitude to Tokyo Metalicon Co. Ltd. for the preparation of plasma sprayed ceramic coat specimen. Laser remelting was done under the cooperation of Mitsubishi Electric Co. Ltd. (Nagoya). The author acknowledges their close cooperation.

REFERENCE

(1) F. Mazza and N. D. Greene, Comptes Rendues du 2-eme les Inhibiteurs de Corrosion, p. 401, U. of Ferrara (1965).

EFFECT OF LASER MELTING ON EUTECTIUM OF BORIDING LAYER ON STEEL

Zhang Luting, Shao Huimeng and Zhu Jingpu

Shenyang Polytechnic University, China

ABSTRACT

This paper is a comparative study of both processes of interphase boundary melting and solidification of boriding layer on steels treated by normal eutectium treatment and by laser-melting. The differences of structure morphology, phase composition and property of hardfacing layer using both techniques of normal eutectium and laser-melting are investigated by means of microscopic analysis, X-ray diffraction and microhardness testing from the point of view of metallography. The results of test provide a scientific base for developing the applicable area of boriding technique.

INTRODUCTION

It is past test results that by boriding technique the surface layer of steel possesses high hardness of 1600--1800 HV in FeB layer and 1200--1500 HV in Fe2B layer, and is provided with high resistance to wear abrasion, adhesive wear, corrosion and high red hardness. However the applicable area of boriding technique is limited because of large brittleness. For the reasons given above, the boriding technique of one phase layer with Fe_2B and the eutectium technique of boriding layer have been investigated for decreasing the brittleness, of the boriding layer on steel. With the later technique, the toughness of hardfacing layer is evidently increased, but the hardness number drops to 500--800 HV and the liquid phase flows out of borid shell structure during eutectium process. For above reason, the new technique of surface laser-melting has been adopted and confirmed to be effective on improving the technique on eutectium of boriding layer and the toughness of hardfacing layer. This test stresses the effect of laser-melting on process of interphase boundary melting and solidification, change of structure morphology and phase composition in eutectium process of boriding laser and searches for ways to improve the eutectium technique and to develop its application.

EXPERIMENTAL CONDITIONS

Specimens

The specimens were prepared with 45 steel and their size is 50 x 20 x 12 mm^3, with boriding layer of 100-150 μm consisting of the structure of $Fe_2B + Fe_3(CB)$.

Testing methods

A half of the specimens were heated at 1180 °C for different times in a box-type electric furnace, another of the specimens were laser-melted with a 2 KW-CO_2 laser at the scanning rate of 9-100 mm/sec.

The eutectium process, the structure morphology, the phase composition and the microhardness have been studied using an optical microscope, a SEM, an X-ray diffraction, etc.

EXPERIMENTAL RESULTS

On mechanism of interphase boundary melting in eutectium process

The melting process occurs first at interphase boundary between the teeth-like Fe_2B and the r-Fe substrate with borided specimens of pure iron, that is, $Fe_2B(8.8\%B) + r-Fe(0.0021\%B) \underline{1149°C} L (3.8\%B)$. But with 45 steel specimens, there is a quantity of $\overline{Fe_3(CB)}$ phase at the tooth top or the middle between teeth, as shown in Fig. 1. At first, a little liquid phase appears due to the reaction of $Fe_3(CB) + r-Fe \underline{1143°C} L$, and then, the quantity of liquid phase is gradually increased, Fe_2B and r-Fe dissolve into liquid melt, which leads to liquid melt expanding into Fe_2B layer and r-Fe substrate. The process above is called an interphase boundary melting.

Fig. 1. Structure of boriding layer on 45 steel   X 400

The rate of interphase boundary melting is controlled by diffusion of boron element, as shown in Fig. 2. It is apparent that there is a difference of boron concentration in the liquid phase which causes diffusion of boron atoms from Fe_2B--L interface to L--r-Fe interface in liquid melting, leading to disrupting the concentration equilibrium. After that the concentration equilibrium is recovered depending on the continuous dissolution of Fe_2B and r-Fe substrate. The two processes above are changed alternately in the process and the liquid melt expands into boriding layer and steel substrate.

If controlled suitably, the thin shell of Fe_2B outside eutectium structure can be saved from damage, as shown in Fig. 3. But generally, the liquid phase seeps out along teeth of Fe_2B and the thin shell of Fe_2B is damaged, as shown in Fig. 4.

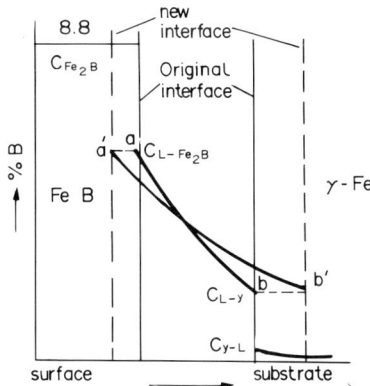

Fig. 2. Schematic diagram of liquid phase expanding controlled by diffusion of boron element.

Fig. 3. With complete thin shell of borid   X 400.

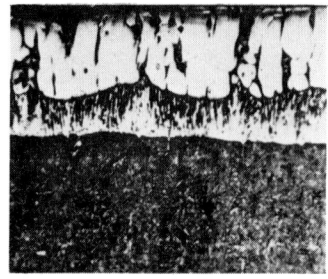

Fig. 4. With damaged shell of borid X 400.

Structure transformation in solidification

It is made clear by means of high temperature metallography observation that the solidification of the liquid melt on substrate is finished in a temperature interval. At first, the crystals of Fe_2B precipitate from liquid melt below liquidus plane in Fe-B-C diagram. And then the liquid phase reacts with the precipitating phase of Fe_2B earlier and two new types of phases are formed, that is $Fe_2B+L \leq 1100°C$ $r-Fe+Fe_3(CB)$. As a results the flower-like structure composed of $Fe_2B+r-Fe+Fe_3(CB)$ is obtained at room temperature, as shown in Fig. 5. When there is an excess of Fe_2B in above reaction, the fish-bone structure composed of $Fe_2B + r-Fe$ appears round them, as shown in Fig. 6. Below 910 °C, the r-Fe transforms to α-Fe and $Fe_{23}(CB)_6$ phase appears, which is explained with the isothermal section diagram shown in Fig. 7. The test of X-ray diffraction verifies that phase composition of the surface structure in room temperature is $Fe_2B + \alpha-Fe + Fe_3(CB) + Fe_{23}(CB)_6$, as shown in Fig. 8. The above result is illustrated with the effect of non-equilibrium solidification and original boriding structure enriched in carbon component in the test condition.

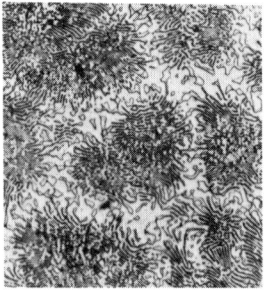

Fig. 5. Surface structure of complete eutectium layer X 400.

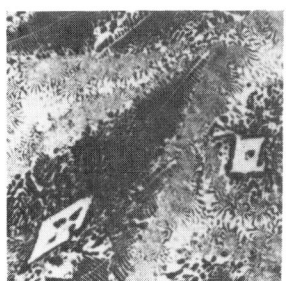
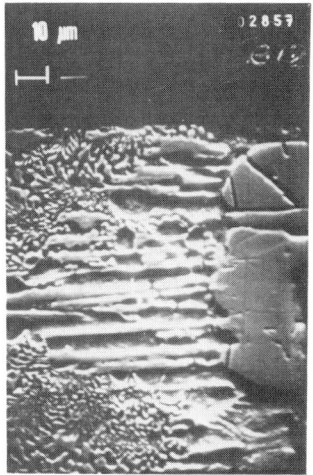

Fig. 6. Structure of non-complete eutectium layer

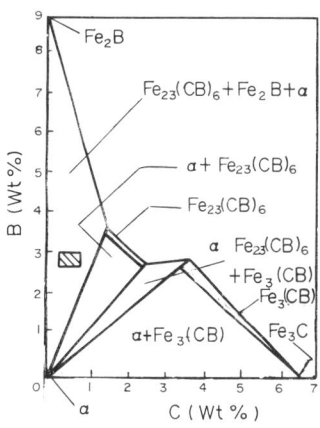
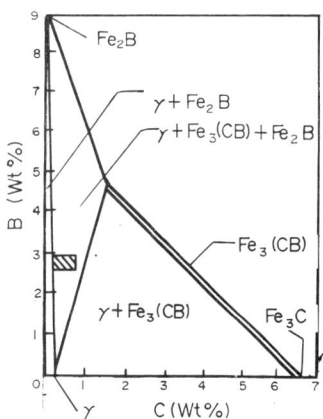

a) At 700 °C   b) At 950 °C

Fig. 7. Isothermal diagram of Fe-B-C alloy.

The melting process of boriding layer by laser-melting

When heated by laser scanning, the interphase boundary melting takes place between toothtip and steel substrate probably, but the rapid heating gives a priority of fracture and decomposition to the boriding layer. The boriding layer is gradually broken into pieces which are subdivided into granular particles based on eutectic-like structure shown in Fig. 9 and space shape of

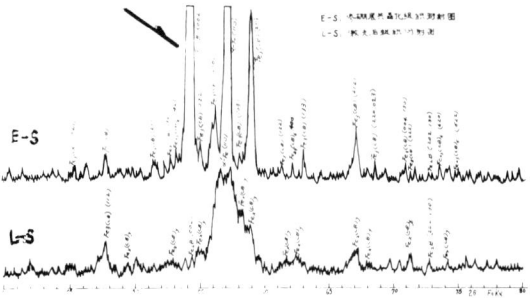

Fig. 8. X-ray diffraction curves of boriding layer by normal eutectium and by laser-melting.

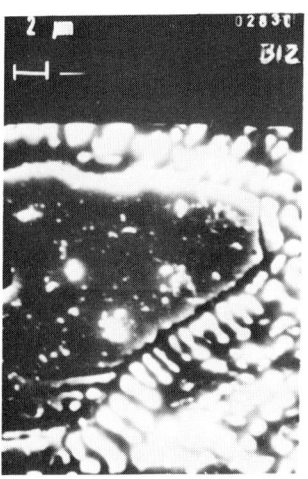
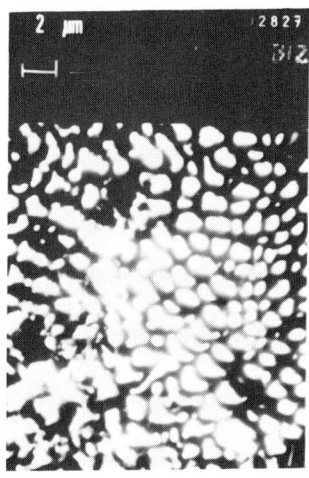

a) Fragmentation of boride   b) Granular structure of surface
Fig. 9. Structure observation of decomposition process of boriding layer.

the particles like columns vertical to substrate, as shown in Fig. 10, a kind of mosaic structure is formed on the steel substrate.

If the boriding layer is completely melted by laser scanning at lower rate or by reciprocal scanning, the Fe_2B particles with regular shape can precipitate from liquid melt firstly, and arrange in lines along heatflow direction, as shown in Fig. 11. This kind of mosaic structure possesses high hardness of 1200-1500 HV, and the good toughness and the high wear resistance from the view of metallography.

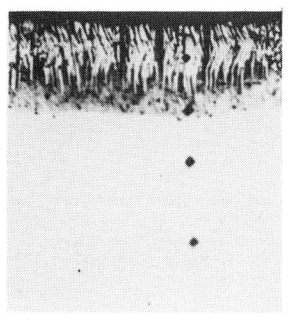 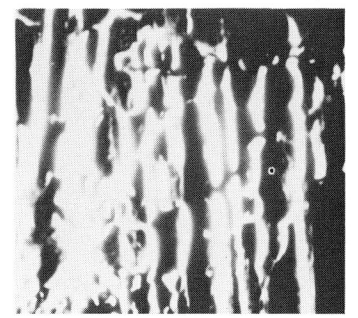

a) By optical microscope (x 400)　　　　　　　　b) By SEM (x 2000)

Fig. 10. Section structure of decomposited borid layer.

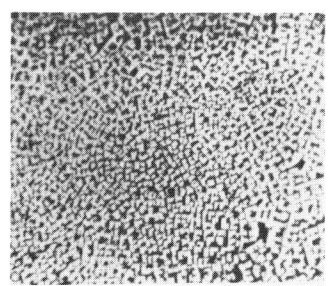 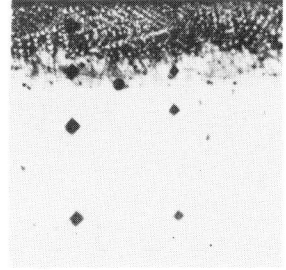

a) Surface morphology　　　　b) Section morphology

Fig. 11. Granular structure of Fe_2B precipitating from liquid melt   X 400.

Comparison of growth of liquid phase into steel substrate

In normal eutectium process of boriding layer, the concentration difference of boron component at L--r-Fe interface makes boron atoms diffuse into steel substrate. It is taken notice in the test that the grain boundaries of substrate provide the rapid permeating passage. When boron atoms seep into the boundaries, the boundaries are enriched in boron component, which leads to lowering melting temperature of the boundaries and gives a priority melting to the boundaries. Nearby L--r-Fe interface, the substrate grains are rounded by liquid phase enriched in boron component and dissolves in liquid melt followed by homogenising process of composition in liquid melt. With above process discontinued at transitional stages, the traces of boron atoms seeping into the boundaries **and** precipitating from that as well as substrate grains embezzled by liquid melt is observed, as shown in Fig. 12.

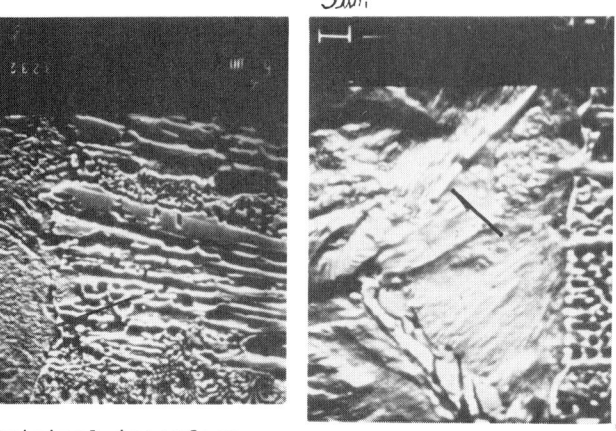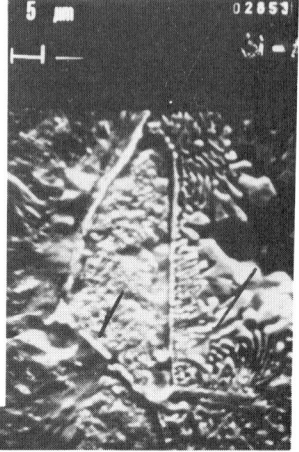

Original interface or L-r phases Trace of boron seeping into the boundaries Trace of substrate grain embezzled by L

Fig. 12. The trace of liquid phase expanding to steel substrate by SEM.

In the laser heating condition, the phenomenon of boron atoms diffusing along the boundaries in advance is not observed in the test. In the hardfacing layer, the crypto-crystalline structure observed with optical microscope is eutectic-like structure with honeycomb morphology observed with the SEM, as shown in Fig.13. Its phase composition is determined as α'-Fe + Fe_2B + $Fe_3(CB)$ + $Fe_X(CB)$ by means of X-ray diffraction, as shown in Fig. 8 formerly. Its hardness is more than 1200 HV, much higher than that of normal eutectium structure.

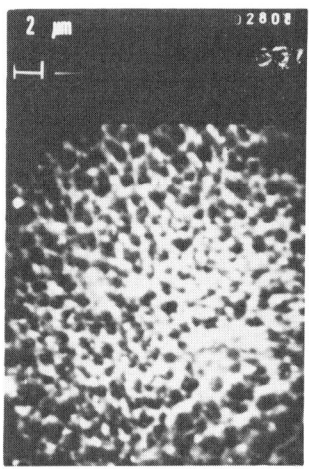

Fig. 13. Eutectic-like structure by laser-melting.

Influence of normal eutectium and laser melting on substrate structure

By normal eutectium, the coarse-grain structure and Widmanstatten structure are often obtained under cooling in air as shown in Fig. 14. By laser-melting, the finer crypto-crystalline martensite appears at the thermal affect area between the hardfacing layer and original structure and possesses hardness of about 800 HV. And the granular structure is observed **in the half melting** area nearby L--r-Fe interface with about 800--900 HV of hardness, as shown in Fig. 15. The layer with finer martensite and granular structure acts as a supporting layer to the hardfacing layer.

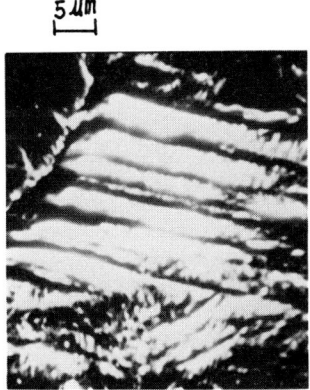

Fig. 14. W-structure of 45 steel substrate.

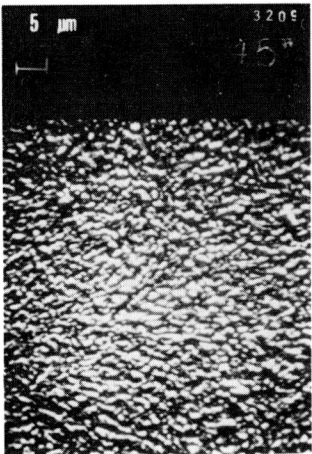

Fig. 15. Structure of half melting area of 45 steel substrate.

CONCLUSION

Normal eutectium treatment of boriding layer contains interphase boundary melting and solidification. In the process of liquid expanding to substrate, the substrate grains are embezzled and melted continually, for the reason of substrate boundaries giving a diffusing passage to boron element.

The effects of rapid heating followed by rapid solidification using laser-melting technique can change the features of eutectium process of boriding layer. It is observed that the boriding layer fragmentates first of all and subdivides into column-like particles vertical to substrate. Even if the borid layer melts completely, the composition homogenisation of liquid phase has not enough time to be put in progress, which gives the priority of precipitation from liquid melt to Fe_2B particles. And the trace of boron element diffusing along substrate boundaries is not observed under the condition of laser-melting.

The difficulty in obtaining complete thin shell of Fe_2B outside eutectium structure and the grain-coarsing of steel substrate restricts the application of normal eutectium technique in industrial production. Depending on the laser-melting, the mosaic structure with granular or column-like particles of Fe_2B based on eutectic-like structure is obtained. And the structure morphology and the hardness distribution of specimen section is favourable to working conditions from the view of metallography.

The work is opening a new area for tough hardening treatment of boriding structure on steel.

REFERENCES

(1) Shao Huimeng: On the Methods of Powder Boriding and the Structure Feature of Boriding Layer, "Heat Treatment of Metal", Vol. 1, 1979.
(2) Shao Huimeng: Eutectium of Boriding Structure on Steel, "Journal of Heat Treatment of Metal", Vol. 2, 1982.
(3) Zhang Luting, Shao Huimeng and Zhu Jingpu: On Structure of Surface Coating of Eutectic Alloys on Steel by Laser-Melting, "Journal of S.P.U.",VVol. 3, 1982.
(4) H. Frenke: Metastable Phase in Rapidly Quenched Fe--B Alloys, R.Q.M.III, 1978.
(5) P. R. Strutt: Structural Analysis of Laser Surface Melted Tool Steels, R.Q.M.III, 1978.

Chapter 3

THE MECHANISM OF SURFACE MODIFICATION BY LASER

THE EFFECT OF LASER IRRADIATION ON STRUCTURE AND MICROHARDNESS OF AISI 1045 STEEL

M. Riabkina-Fishman and J. Zahavi

Israel Institute of Metals, Technion, Haifa 32000, Israel

INTRODUCTION

Laser beam processing has been found to produce surface hardening in carbon and low-alloyed steels as a result of local heating, above or below the melting point, of a thin surface layer followed by rapid resolidification and/or self-quenching [1].

In this investigation we have studied the feasibility of laser surface hardening of medium carbon steel and the effect of laser irradiation on the structure of 1045 steel. The laser-affected region was produced by a single beam pass as well as by successive passes with varying degrees of overlapping.

EXPERIMENTAL PROCEDURE

The study was conducted on 5 mm thick specimens cut in the transverse direction from a 1" rod of commercial AISI 1045 steel in as received condition. Irradiation was carried out by means of a continuous CO_2 laser with maximum power 1 KW; the beam was formed by a lens with focal distance 95 mm, and its diameter was 200 μm at the focal distance.

To protect the lens against contamination and overheating and to minimize specimen oxidation under the beam, the flow of argon or nitrogen at surplus pressure 1 atm directed at the irradiated surface was used. The specimens were fixed on a rotating table, and the transverse velocity of the beam over the specimen surface could be regulated by changing the rotation speed of the table. Provision for step-shifting of the table enabled the specimen to be advanced a certain distance in the radial direction with each revolution of the table, whereby a series of equidistant beam passes could be obtained.

Irradiation parameters were varied by changing laser power (in the range of 300 to 900 W), the distance of the specimen surface from the focal plane (up to 3 mm), and the specimen velocity (in the range of 4 to 90 cm/sec). In addition, antireflection coatings (manganese phosphate, chromium powder - both with an organic binder, - a conducting carbon coating or an oxidation film produced by 20 % orthophosphoric acid) were applied to the specimens is some experiments.

The structures produced by the processing were analyzed by means of optical microscopy, scanning electron microscopy (SEM) and x-ray diffractometry. Prior to preparation of cross-sectional samples for metallographic observation, a layer of Ni coating was electroplated on some specimens. The hardness was measured with Matsuzawa DMN2 microhardness tester under a 50 g load.

RESULTS AND DISCUSSION

We first present and discuss the structure and hardness of the laser hardened region in the case of a single (or isolated) pass and then those of an extended surface area processed by series of closely spaced or partially overlapping passes.

Single Pass Laser Beam Irradiation

A low magnification SEM image presented in Fig. 1a shows the free surface of a specimen with two tracks of laser irradiation. In appearance, the tracks resemble a weld seam and have a well-pronounced ripple structure. The dendrite structure of the ripples, revealed under higher magnification (Fig. 1b), indicates that melting has taken place in the region. Dendrite width is less than 1 μm, the absence of dendrite branches of higher orders indicates that solidification velocity was very high.

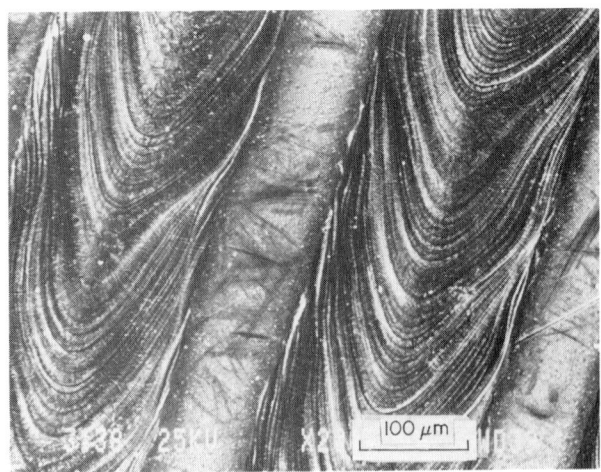

Fig. 1a. Low magnification SEM image showing ripple structure of two laser irradiation tracks on free specimen surface.
Irradiation conditions: 700W, 40cm/sec.

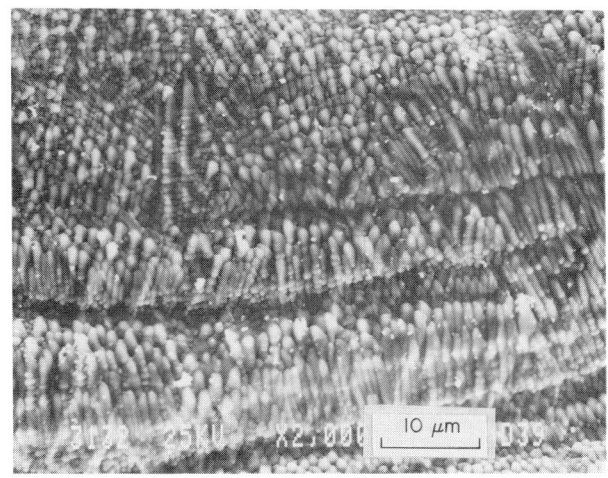

Fig. 1b. SEM image of dendrites on a free surface of irradiated zone.
Irradiation conditions: 700W, 40cm/sec.

Optical microscopy of initial specimen structure (Fig. 2a) reveals dark regions of pearlite and white grains of ferrite. In SEM image (Fig. 2b) pearlite exhibits a lamellar structure.

Laser affected region (Fig. 3) consists of two distinct zones [2], namely, a melt zone surrounded by austenitization zone.

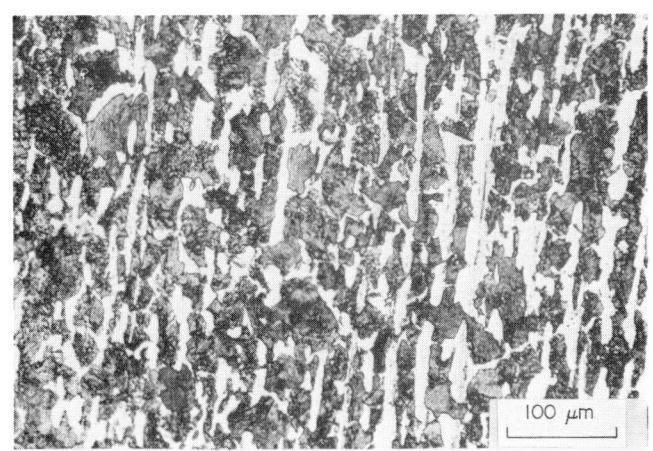

Fig. 2a. Optical microscopy of initial specimen structure.

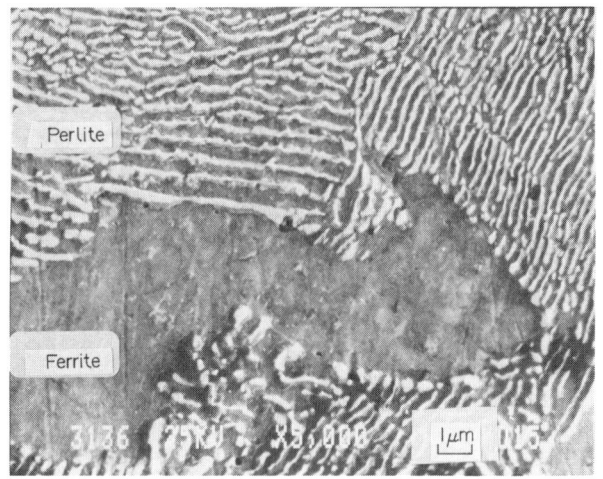

Fig. 2b. SEM image of initial specimen structure.

Fig. 3. Optical microscopy of laser affected region. Irradiation conditions: 300W, 4cm/sec.

(Although the latter is referred to in the literature as "heat affected zone" we prefer a more specific term indicating explicitly that the temperature increase in this zone was sufficient for austenitization to take place). The melt zone has a dendrite structure revealed by both optical metallography and SEM. The austenitization zone has a duplex structure with former pearlite and ferrite region readily recognizable within it. However, the former pearlite regions show much lower etchability than the original pearlite and even lower than the melt zone. In the SEM image (Fig. 4) former pearlite regions have rather smooth appearance but the plate-like structure typical of pearlite is still discernible in some places. The most interesting feature of the former ferrite grains within the austenitization zone observed in Fig. 3 is that their boundaries become fuzzier on approaching the melt zone boundary. As a result, smaller grains situated in the inner part of the austenitization zone appear as contourless light spots.

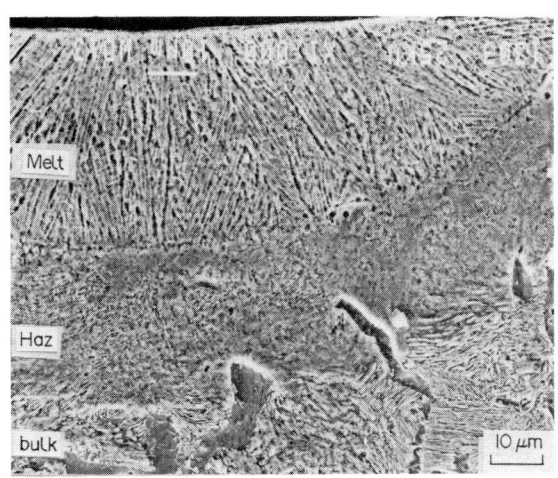

Fig. 4. SEM image of laser affected region.
Irradiation conditions: 700W, 40cm/sec.

Though no morphological features of martensite transformation were revealed by optical or scanning electron microscopy, the presence of martensite was confirmed by x-ray diffraction pattern taken from the surface swept by a series of parallel passes; the presence of some amounts of retained austenite was also found.

There is no systematic variation of microhardness within the melt zone. The values observed are in the range of HV730-760 when argon flow is used and HV850 over the whole depth of the melt zone if nitrogen is used. It should be noted that no effect of nitrogen on hardness is observed in coated specimens but the chromium itself increases the hardness of melt zone up to HV900 (Table 1).

Higher hardness of melt zone in Cr-plated specimen was accompanied with a finer dendrite structure (Fig. 5). A uniform chromium content of about 0.5 % was found within the melt zone (Fig. 6).

Hardness values obtained within the austenitization zone depend primarily on whether the measurements are taken in the regions of former ferrite or pearlite. In former pearlite regions hardness is higher than in the melt zone in the case of argon flow and reaches HV900, at the former ferrite grains hardness drops

Table 1. Maximum hardness values (HV) of AISI 1045 steel specimens treated under different conditions

Hardness of conventional structure components	Ferrite	Pearlite	Martensite (Water Quenched)
	180	300	700
Hardness in Laser affected regions	Austenitization Zone		Melt Zone
	Former Ferrite	Former Pearlite	
Uncoated specimen, Ar	250	900	750
Cr-coated specimen, Ar	250	900	900
Uncoated specimen, N_2	250	900	850

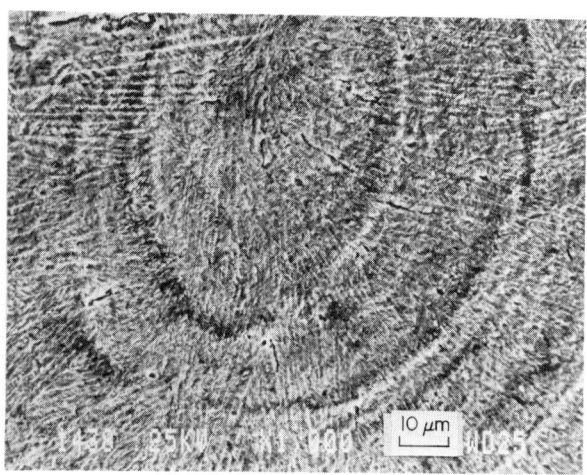

Fig. 5. Microstructure of melt zone in Cr plated specimen (SEM image).
Irradiation conditions: 700W, 40cm/sec.

to HV250. For comparison, outside the laser affected region pearlite and ferrite have hardness values of HV300 and HV180, respectively, whereas the same steel after austenitization and conventional hardening by water quenching has hardness HV700 (Table 1).

The uniform hardness within the melt zone is indicative of a uniform carbon distribution achieved during the laser induced remelting, as a result of a vigorous agitation of molten material within the melt pool [3]. The higher hardness of the martensite formed in the melt zone can be attributed to the very high cooling rate achieved in this case.

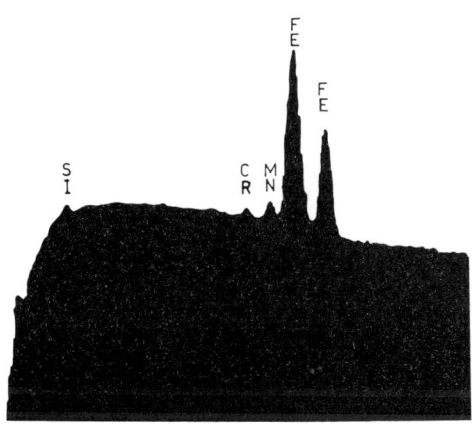

Fig. 6. X-ray microprobe analysis of the composition in melt zone of Cr-plated specimen. Irradiation conditions: 700W, 40cm/sec.

As evidenced by the distinct contrast between former ferrite and pearlite regions within the austenitization zone, the austenite condition within this zone was too shortlived for appreciable long-range carbon diffusion to take place, and only the onset of this process could be traced in the vicinity of the melt zone, due to the higher temperatures (approaching the melting point) which prevailed there. As a result, the austenite formed during heating had a very low carbon content in the former ferrite regions and one close to the eutectoid level (0.8%C) in the former pearlite regions. The higher carbon content in the latter case seems to account for the higher hardness of the martensite formed in these regions compared with the martensite within the melt zone. As for the higher hardness of the former ferrite regions within the austenitization zone compared with that of the substrate ferrite, it has been reported that under high cooling rates produced by laser processing, martensite-type transformation takes place in Armco iron resulting in hardness values of HV380 in the melt zone [4] and of HV250 when no melting occurs [5].

Overlapping Laser Irradiation Passes

In most industrial applications of laser hardening the surface to be processed is too large to be covered with a single pass. One of the feasible ways to produce laser hardening effect on an extended surface is by sweeping it with a series of successive passes with the workpiece advanced after each pass.

We have studied the structure and hardness of surface layers produced by this technique at various degrees of overlapping between successive passes. The results are illustrated in Figs. 7a and b which refer to an uncoated specimen swept by series of

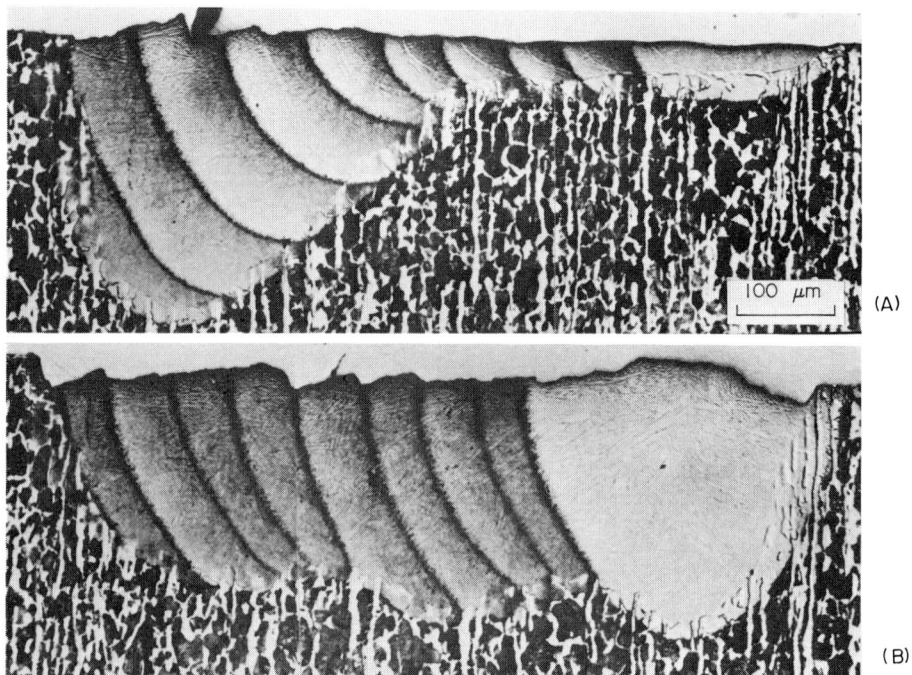

Fig. 7. Cross-section of a specimen covered by series of ten equidistant passes (optical microscopy). Irradiation conditions: 30cm/sec, (A) 900W; (B) 700W.
The specimen shift between passes is 100 μm.

passes displaced by 100 μm with beam power 900 and 700W, respectively, and Figs. 8 a,b,c which refer to **a Mn-phosphate coated specimen** processed by six passes of a 700W laser beam, with the advance of the specimen between passes being 300, 200 and 100 μm, respectively.

In spite of the fact that all irradiation parameters such as power, traverse velocity and specimen distance from the focal point were kept constant during each series of passes, in some cases, as can be seen in Fig. 7a, the size of laser affected region swept by 900W beam was largest at the first pass and then decreased to a greater or lesser extent with each subsequent pass. We have found that this effect appears only when the beam power exceeds a certain threshold level that depends on other experimental conditions. For an uncoated specimen and traverse speed 30cm/sec the effect is very strong when 900W beam is used (Fig. 7a) and vanishes at beam powers of 800W and less (Fig. 7b). The observed attenuation of laser-induced effect during first three passes was so drastic that, rather paradoxically, the subsequent depth of the laser-affected region was much smaller than the fairly steady depth produced under similar experimental conditions by a beam of 700W only (Fig. 7a,b).

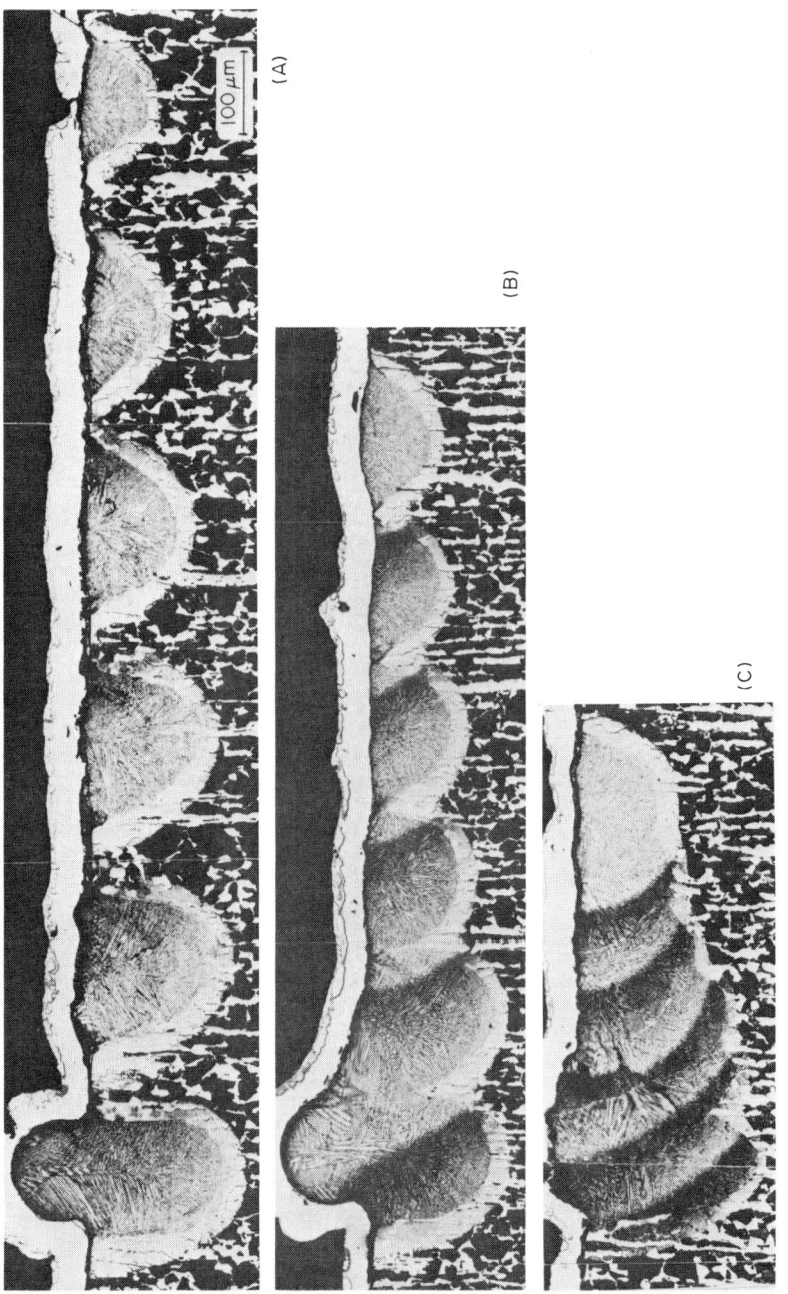

Fig. 8. Cross-section of a specimen covered by a series of six equidistant passes (optical microscopy). Irradiation conditions: 700 W, 40 cm/s; manganese phosphate coatings. The specimen shift between passes is (A) 300 μM; (B) 200 μM; (C) 100 μM.

In our opinion, this phenomenon can be attributed to the beam attenuation caused by absorbing plasma that forms near the irradiated surface and propagates along the direction of the beam [6]. This effect, often referred to as a laser-supported absorption, appears in the case of excessive metal heating and evaporation and its initiation is characterized by a definite threshold value of laser power density.

Another unusual phenomenon observed under conditions of high power densities is the formation of large-scale undulations on the solidified surface with the amplitude comparable to the depth of the melt zone (Fig. 9). These undulations appear only at the initial stage of irradiation while a large depth melt zone is produced; later rather a smooth transition to the regular ripple morphology occurs as can be seen in the second pass in Fig. 9. In the transverse cross-section these undulations exhibit a regular semicircular shape as can be seen in Fig. 8a,b. Antireflection coatings, **by** increasing the fraction of beam power absorbed by the specimen, reduce the threshold level and the effect is quite pronounced already at 700W (Fig. 8).

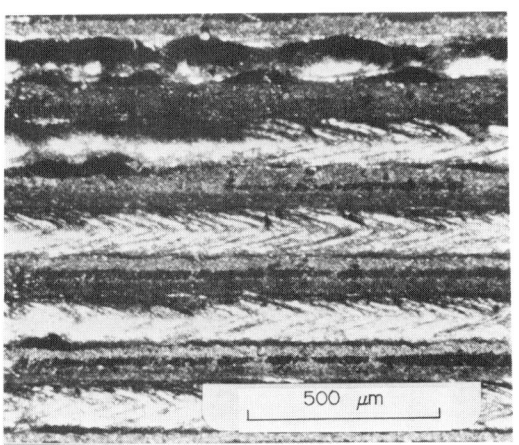

Fig. 9. Optical photograph of free specimen surface showing large-scale undulations in the first laser beam track (at the top) and transition to ripple structure in the second track.
Irradiation conditions: 700 W, 30 cm/sec, manganese phosphate coating

Figure 8 also brings out the structural changes undergone by laser affected region as the overlapping of the passes increases. When the austenitization zones of adjacent passes are in contact (Fig. 8a) the only change is a gradual increase in etchability of both the melt and, especially, the austenitization zones on approaching the boundary of the next pass in the series; this effect becomes more pronounced as overlapping increases (Fig. 8b). In the case of a substantial overlapping (Figs. 7a,b and 8c) high etchability characterizes the whole beam affected region, except the area of the last pass. However, an appreciable contrast still remains at the outer boundaries of all austenitization zones.

Another change concerns the structure of the austenitization zone. The part of the zone that overlaps the melt zone of the preceding pass inherits its existing dendrite structure and can be distinguished from the latter only owing to its lower etchability. The "primary" duplex-structure austenitization zone survives only where there is no overlapping between it and the melt zone of the preceding or subsequent passes. As a result, the surface layer formed by a series of overlapping passes appears, on the first glance, as a continuous melt zone showing a dendrite structure surrounded by a cusped band of the continuous duplex-structure austenitization zone (Fig. 8b,c).

It should be noted that there is a partial correspondence in dendrite configuration on both sides of the boundary between two overlapping melt zones (Fig. 10). It means that remelted material can solidify **epitaxially** on the dendrite structure at the interface with the previously formed melt zone.

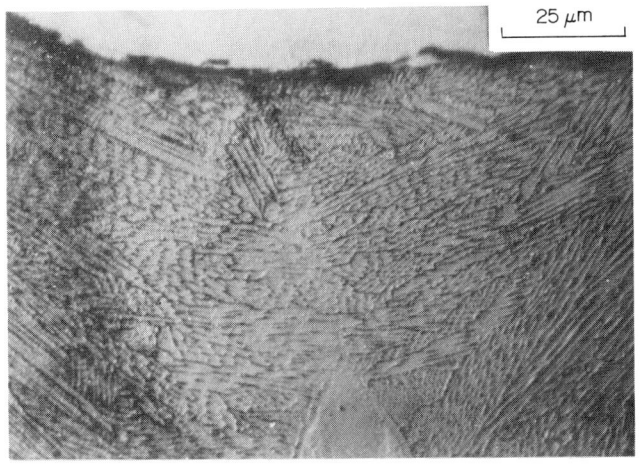

Fig. 10. Enlarged view of first two passes **presented** in Fig. 8a (optical microscopy). Note partial correspondence of dendrite configuration on the both sides of the boundary between two overlapping melt zones.

As regards to the influence of overlapping on the hardness distribution, we first consider the case of two overlapping passes illustrated by Fig. 11. In the scheme we use numbers 1 and 2 to denote, respectively, the processes of melting and austenitization as well as to designate the "primary" melt and austenitization zones. By using the same symbols, the area of overlapping in Fig. 11 can be considered as composed of four zones designated as 11 - double melt zone, 12 - zone of melting followed by austenitization, 21 - remelted austenitization zone, and 22 - double-austenitization zone.

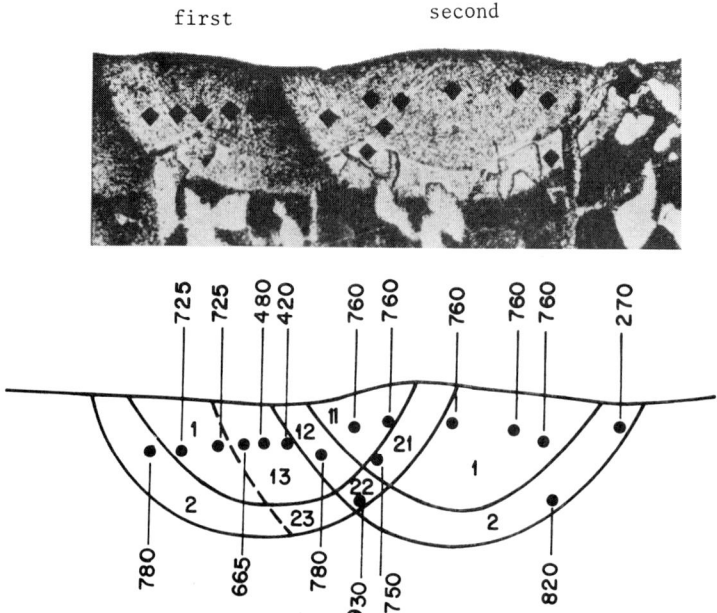

Fig. 11. The structure and hardness distribution observed in the case of two partially overlapped passes: 1 - primary melt zone; 2 - primary austenitization zone; 11 - double melted zone; 12 - austenitized melted zone; 21 - remelted austenitization zone; 22 - double austenitization zone; 13 - tempered melt zone; 23 - tempered austenitization zone.

As hardness measurements in this and other specimens have shown, double melted zone 11 as well as remelted austenitization zone 21 have the same hardness as primary melt zone 1. Zone 12 shows, in this specimen as well as in others, a slight but steady increase in hardness over the melt zone. And, finally, the highest hardness is observed in double austenitization zone 22.

However, the most important change in hardness occurs within zones 1 and 2 of the first pass: hardness decreases on approaching the boundary of the second pass and in some cases values as low as HV400 have been measured in the vicinity of the boundary. This sharp decline in hardness (accompanied by an increase in etchability) is due to a tempering effect produced by each beam pass on the martensite of the earlier formed melt and austenitization zones. Quite obviously, the effect was the strongest near the austenitization zone boundary of the superimposed pass where the temperature, albeit for a short time, approached the austenitization level. Assigning the number 3 to the tempering effect produced by the laser beam beyond the boundary of the austenitization zone, we have designated in Fig. 11 as 13 and 23 the regions of tempered martensite within the melt and austenitization zones, respectively. As hardness measurements have shown, the tempered zone with hardness below HV700 was 65 µm wide when melt zone was 165 µm wide and reached 100 µm when the latter was 245 µm, i.e. the width of the tempered zone is about 40 % of that of the melt zone. At such a ratio, even if melt zones of adjacent passes are merely in contact, about one third of the hardened surface would be tempered to a greater or lesser extent. Due to the tempering effect, high hardness values typical of laser hardening cannot be achieved over the whole surface covered by a series of overlapping passes. Instead, there is a periodic hardness distribution whose features depend on the degree of overlapping. Figure 12 presents schematically the hardness distribution pattern in the cases when the melt zones are overlapped by 30 %, 50 % and 75 % of their width. It can be seen that with a high degree of overlapping, corresponding approximately to 100 µm specimen shift in our case, the whole hardened layer turns into tempered martensite with a periodic hardness pattern reflecting the distribution of tempering temperatures.

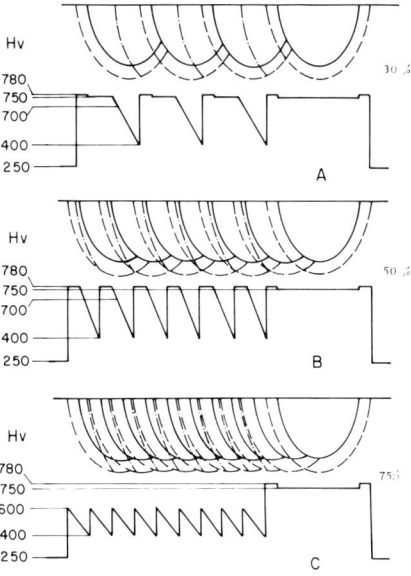

Fig. 12. Examples of hardness distribution across the surface of a specimen covered by series of equidistant passes overlapping by (a) 1/3; (b) 1/2 and (c) 3/4 of their melt zone width. ——— melt zone boundary; - - - - austenitization zone boundary; .-.-.- tempering zone boundary.

An attempt to minimize tempering effect by avoiding overlapping even between austenitization zones leads, as can be seen in Fig. 8a, to the depth of hardened layer being very shallow where austenitization zones meet. In addition, low hardness regions of former ferrite within austenitization zones can emerge at the surface in this case. It seems that the degree of overlapping yielding the optimum result is to be determined depending on particular application requirements. It is obvious, however, that the full potential of laser hardening cannot be realized over an extended surface when the technique of overlapping passes is used.

SUMMARY

The effect of continuous CO_2 laser processing on the structure and hardness of AISI 1045 steel has been investigated in the cases of isolated and partially overlapping passes. The laser affected region produced by an isolated pass consists of two distinct zones: (i) a melt zone surrounded by (ii) an austenitization zone having duplex structure representing former pearlite and ferrite regions. Hardness of the melt zone (about HV750) exceeds that of conventionally hardened 1045 steel; even higher hardness (up to HV900) is observed in former pearlite regions of austenitization zone, that of former ferrite regions being about HV250. Hardness of melt zone can be increased by performing laser treatment in nitrogen atmosphere or by applying chromium antireflection coating.

In the case of an extended surface processed by a series of equidistant passes, no essential changes are observed within the partially overlapping melt and austenitization zones. However, due to a heating effect produced by each pass in a previously hardened layer, a relatively wide region of tempered martensite is formed with minimum measured hardness values down to HV400. As a result of this interaction the full potential of laser hardening cannot be realized when the technique of successive passes is used and the surface layer thus formed shows a periodic hardness distribution pattern whose features depend on the degree of overlapping.

It has been also found that the use of a high power beam can be counter-productive because a drastic decrease in the hardening depth occurs when the beam power density exceeds some threshold level.

ACKNOWLEDGMENTS

The authors are grateful to Dr. M. Bamberger for his help at the initial stage of the investigation and to Dr. L. Zevin for x-ray diffraction measurements.

This research was supported by a grant provided by the National Council for Research and Development, Israel, and the Directorate-General for Science Research and Development of the Commission of the European Community.

REFERENCES

(1) Gnanamuthu, D. S., in Proceedings of the Conference on

Applications of Lasers in Materials Processing, Washington D.C., April 1979, edited by E. A. Meltzbower (ASM, Metals Park, Ohio, 1979) p. 177.
(2) Rawers, J. C., J. Mat. Sci., 20 (1985) 1929.
(3) Moore, P. G., and Weinman, L. S., "Laser Application in Materials Processing", edited by Ready, J. F., Proc. Soc. Photo-Opt. Instr. Eng., 198 (1979) 120.
(4) Borodina, G. G., Kraposhin, V. S., Romanov, Yu. A., and Kosyrev, F. K., Metallovedenie i termischeskaya obrabotka metallov (MiTOM) (1983) n4, p. 14 (in Russian).
(5) Benedek, J., Shachrai, A., and Levin, L., Optics and Laser Technology, Oct. 1980, p. 247.
(6) Ready, J. F., Proc. IEEE, 70, (1982) 533.

MICROSTRUCTURE AND MICROCRACKING OF LASER GLAZED Fe-B-C SURFACES

Sabine Staniek[*] and Erhard Hornbogen

*Institut für Werkstoffe,
Ruhr-Universität Bochum, D-4630 Bochum, FRG*

INTRODUCTION

The glass forming ability (gfa) of alloys depends in a very wide range on its chemical composition. A practical way to establish the gfa is to measure the maximum thickness of a ribbon which can be obtained as a glass by a defined melt-spinning procedure. The gfa is related to the structure and thermodynamic stability of the amorphous and crystalline phases, but depends predominantly on the mechanisms and kinetics of crystallization. Research on conventional rapid cooling techniques (e.g. melt-spinning, splat-cooling) has indicated that there are three principle crystallization reactions, namely, primary, massive and eutectic crystallization [1]. The first two produce one new phase while the third produces two crystalline phases simultaneously. The primary and the eutectic reaction are relatively slow because they imply long range diffusion. Massive crystallization can be very rapid as the propagation of its crystallization front requires only on atomic hop. The resulting microstructures consisting of amorphous or microcrystalline metastable phases are known to have interesting physical or mechanical properties. Due to the limited thickness of such ribbons the technical application is still limited.

A somewhat more recent technique of rapid solidification is the melting and resolidification of thin surface layers of bulk material by means of high power lasers [2, 3]. Purpose of this investigation is to compare the data known from melt-spinning with laser surface melting and subsequent self-quenching. In principle similar reactions are anticipated. However, there should be differences due to different nucleation conditions as well as different temperature gradients and cooling rates.

As an additional feature the formation of internal stresses is anticipated depending on the geometry of the laser heated zones.

---

*Present address: DFVLR, D-5000 Köln-Porz, FRG

Such internal stresses are well known from different types of surface treatments and from conventional welding. These stresses will have to be considered in context with the microstructures and phases which originate from rapid solidification.

There is a two-fold purpose of this work: first to acquire a principle understanding of the processes which occur during laser surface melting of well selected alloys, and, secondly, to use this knowledge for an accessment of the possibility to produce glassy surfaces with high hardness and fracture toughness and, consequently, high wear resistance.

MATERIALS AND EXPERIMENTAL PROCEDURE

The influence of the alloy composition on the microstructure of laser treated surfaces of eutectic binary $Fe_{83}B_{17}$ and $Fe_{83}C_{17}$ as well as their ternary mixtures $Fe_{83}(B_xC_y)_{17}$ (x:y = 25:75; 50:50; 75:25) have been investigated. The good (Fe-B) and negligible (Fe-C) gfa in this system are well established [4]. Thus the chosen alloys (the chemical compositions are given in table 1) were expected to cover a wide range of behaviour.

These alloys were subject to different laser treatments using a 0.5 kW and 5 kW-CO_2-laser. Pulse melted spots and continuously melted single laser traces as well as their overlappings were produced by moving the as-cast samples on a rotating disc within the velocity v_z through the focus of the laser beam (Fig. 1). Laser parameters, i.e. power density p and interaction time t, were varied. Typical values were: $p = 4 \times 10^6$ Wcm^{-2} and $t = 10^{-4}$ s (referring to specimen velocities v_z of about 2 ms^{-1}).

Table 1: Investigated alloys and their analyzed chemical composition

Alloy	at % B	at % C
$Fe_{83}B_{17}$	17.06	-
$Fe_{83}B_{13}C_{4,5}$	12.85	4.52
$Fe_{85}B_{9,5}C_5$	9.71	4.97
$Fe_{83}B_8C_{8,5}$	8.35	8.65
$Fe_{85}B_6C_9$	6.04	9.23
$Fe_{82}B_4C_{14}$	4.36	14.20
$Fe_{83}B_3C_{14}$	3.05	13.79
$Fe_{83}C_{17}$	-	17.20

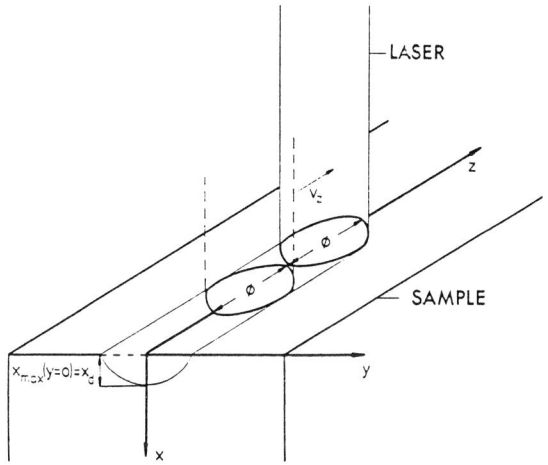

Fig. 1. Schematic experimental set up for laser treatment.
$x_d = x_{max}$: maximum melt depth
v_z = relative laser beam velocity
Different metallographic sections given in the
following figures refer to indicated (xyz)
direction.

Resulting microstructures were analyzed by a combination of different microscopic methods. A qualitative information on the mechanical properties of the phases and microstructures was obtained by microhardness indentations. They could also be used to explore mechanisms of crack initiation and propagation.

EXPERIMENTAL RESULTS

The shape and geometry of the laser affected zone predominantly depends on the laser parameters, to a much lesser degree on the alloy composition and the primary microstructure. The maximum melt depth x_d shows a dependence on the alloy properties (Fig. 2). In general, x_d increases with increasing substitution of boron by carbon.

A plot of melt depth versus laser power p for different duration of interaction t is shown for the binary $Fe_{83}B_{17}$ alloy in Fig. 3. The dotted curve has been plotted for constant interaction times $t = 2.2 \times 10^{-4}$ s.

The known difference in gfa between Fe-B and Fe-C alloys is highly reflected in the microstructures obtained after laser treatment. The $Fe_{83}C_{17}$-alloy is fully crystallized as well as the alloys containing less than 75 % B (~ 13 at % B). It is evident that larger borides/carbides stay partly undissolved in the liquid solid interface (Fig. 4a). The 75 % B boron-alloy only occasionally showed smaller areas of metallic glass.

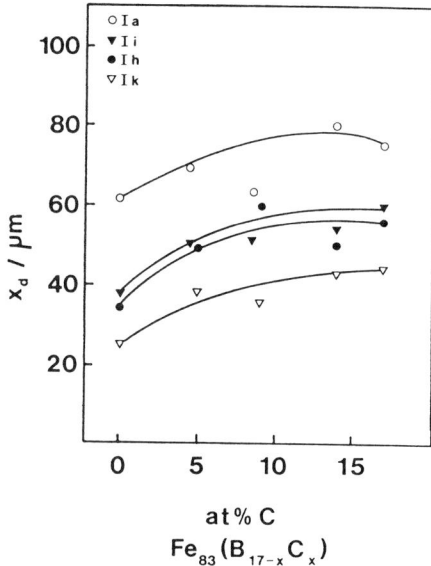

Fig. 2. Maximum melt depth x_d for different alloy composition.
Laser parameters:
Ia: $p = 4.07 \cdot 10^6$ Wcm^{-2}, $t = 2.2 \cdot 10^{-4}$ s
Ii: $p = 2.25 \cdot 10^6$ Wcm^{-2}, $t = 1.7 \cdot 10^{-4}$ s
Ih: $p = 5.53 \cdot 10^6$ Wcm^{-2}, $t = 7.5 \cdot 10^{-4}$ s
Ik: $p = 5.53 \cdot 10^6$ Wcm^{-2}, $t = 5.0 \cdot 10^{-5}$ s

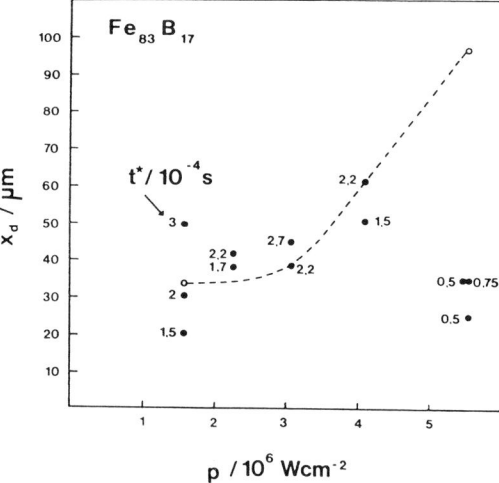

Fig. 3. Maximum melt depth x_d of $Fe_{83}B_{17}$ traces versus power densities p for different interaction

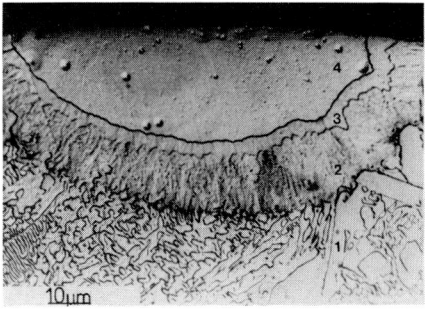

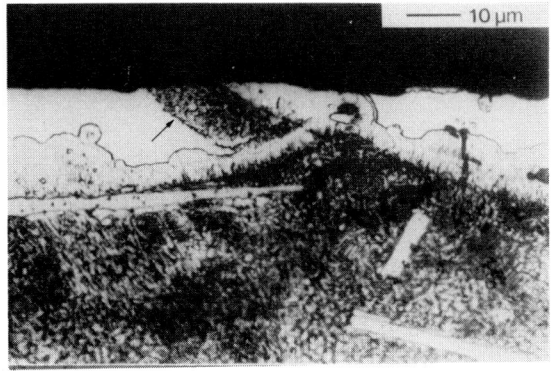

Fig. 4. x-y-section of laser traces: a), b) single traces, c) overlapping pulsed spots.
  a) $Fe_{83}B_{12.5}C_{4.5}/p = 5.53 \cdot 10^6$ $Wcm^{-2}/t = 7.5 \cdot 10^{-4}$ s
     Fully crystallized in radial direction. Close to melt/solid interface undissolved $Fe_{23}(B,C)_6$ phase (light) is visible. There is no evidence for epitaxial growth.
  b) $Fe_{83}B_{17}/p = 2.25 \cdot 10^6$ $Wcm^{-2}/t = 1.7 \cdot 10^{-4}$ s
     1: primary unmelted microstructure,
     2: metastable eutectic,
     3: metastable polymorphous zone,
     4: amorphous zone.
  c) $Fe_{83}B_{17}/p = 3.4 \cdot 10^4$ $Wcm^{-2}/t = 5 \cdot 10^{-4}$ s
     Arrow indicates crystallized zone due to overlapping. Cracks only appear in crystalline phase.

Typical for the binary $Fe_{83}B_{17}$-alloy is a four-zone-microstructure (Fig. 4b):
1. the unmelted primary microstructure including a heat affected zone which shows evidence for martensitic transformation;

2. a eutectic zone (α-Fe + orthorombic o-Fe$_3$B);
3. a zone containing a homogeneous crystalline phase (tetragonal t-Fe$_3$B);
4. metallic glass (amorphous, extending to the surface up to 20 μm thickness).

In overlapping traces no remarkable features were found in the fully crystalline carbon-containing alloys. In the four zone-structured binary Fe-B-alloy an additional heat affected zone is caused by the subsequent trace. It includes a new crystallization front which extends into the originally glassy zone, and a zone of massive crystallization (Fig. 4c).

Light microscopy was not sufficient to clarify all microstructural features. As an example for TEM investigations the identification of the submicron metastable eutectic (zone 2) is shown in Fig. 5.

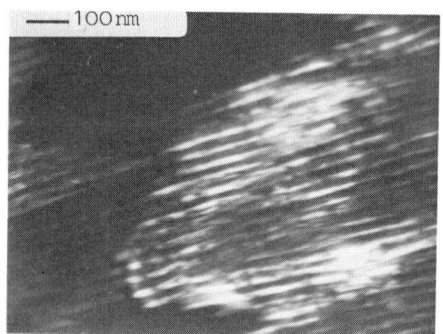 

Fig. 5. TEM-investigation of eutectic zone 2 (s. Fig. 4). Left: dark field image, right: SAD pattern with typical spikes originated by submicron lamellas.

During the resolidification cracks form in the laser treated surfaces. Fully crystallized single laser spots basically show three typical crack patterns:
o circumferential, i.e. parallel to the melt interface (Fig. 6a);
o radial, mostly crosswise through the spots (Fig. 6b);
o induced by the primary microstructure (Fig. 6c).

Principally the same types of cracks are observed in single laser traces. Here the crosswise radial cracks of spots correspond to longitudinally and transverse cracks (Fig. 7).

In the four-zone-structured Fe-B-spots with amorphous surface layer exclusively radial cracks appear in the crystalline zones. At the interface to the amorphous zone cracks are stopped and this zone remains crack-free (Fig 8). Accordingly, corresponding crack formation occurs in overlapping spots (Fig. 9).

For a qualitative analysis of the individual fracture mechanical properties of the different phases and interfaces cracks have

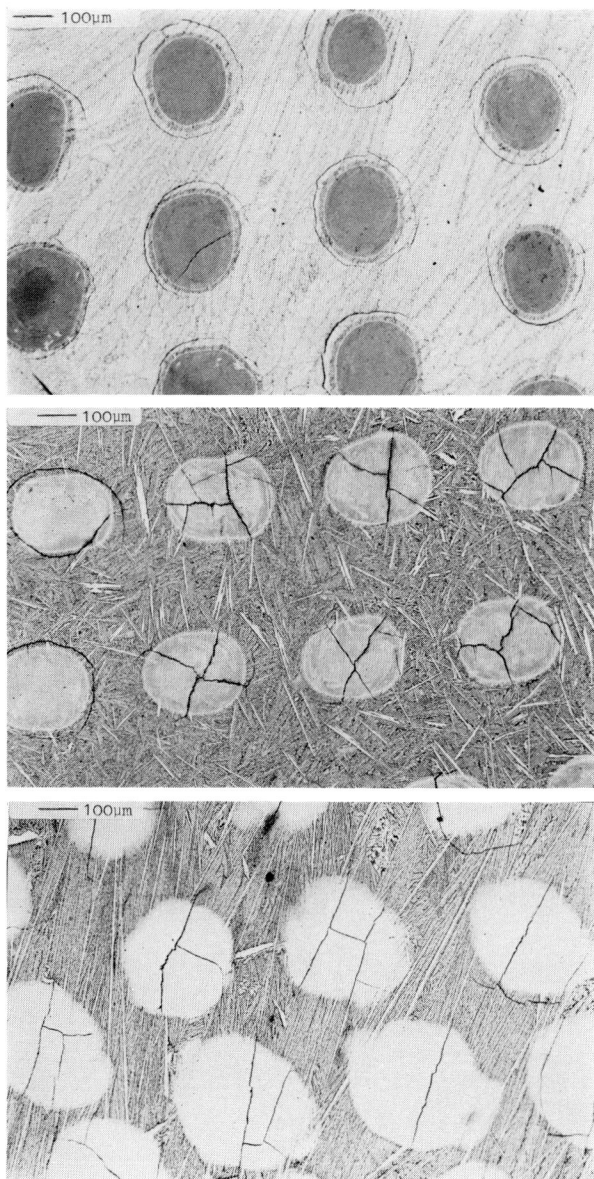

Fig. 6. a) $Fe_{85}B_{10}C_5$, b) $Fe_{85}B_6C_9$, c) $Fe_{83}C_{17}$
y-z-section, $p = 3.4 \cdot 10^4$ Wcm^{-2}, $t = 5 \cdot 10^{-4}$ s
Typical crack pattern of laser spots:
a) circumferential
b) radial
c) induced
by $Fe_{83}C_{17}$-primary microstructure in α-Fe/Fe_3C-interfaces.

Fig. 7. x-y-section/$Fe_{83}B_{8.5}C_{8.5}$/$p = 5.5 \cdot 10^6$ Wcm^{-2}/$t = 7.5 \cdot 10^{-4}$ s
Longitudinally and transverse cracks in fully crystalline single laser trace.

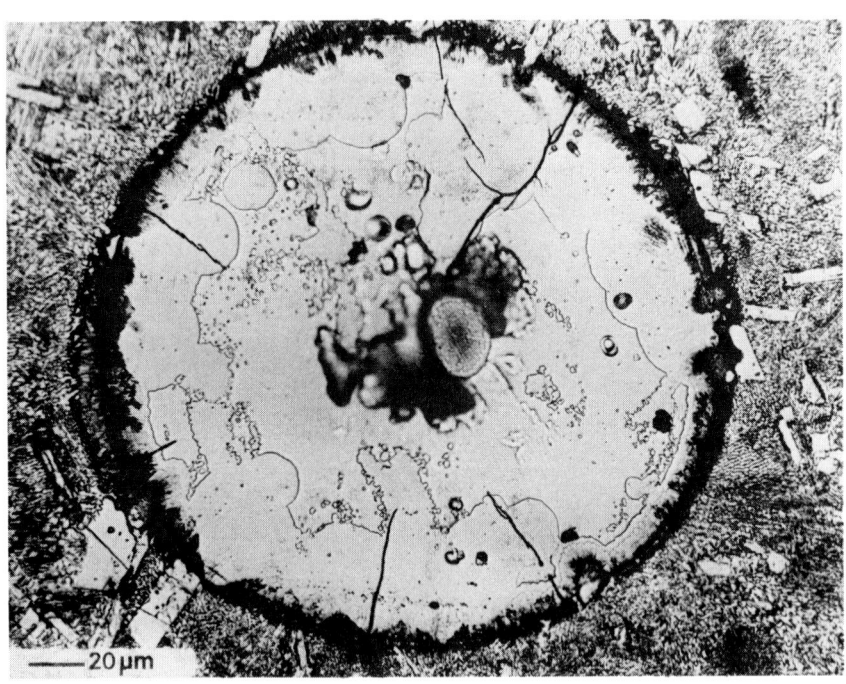

Fig. 8. $Fe_{83}B_{17}$/$p = 3.4 \cdot 10^4$ Wcm^{-2}/$t = 5 \cdot 10^{-4}$ s/y-z-section
Pulsed laser melted spot with amorphous crack-free zone. Radial cracks appear only in the crystalline zones.

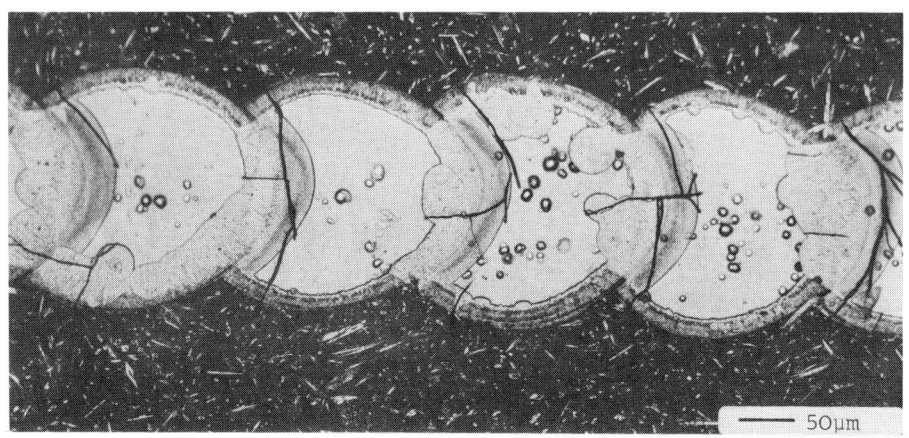

Fig. 9. $Fe_{83}B_{17}/p = 3.4 \cdot 10^4$ $Wcm^{-2}/t = 5 \cdot 10^{-4}$ s/y-z-section
Cracks only appear in overlapping crystalline zones.

been initiated by Vickers-microhardness indentations. With loads of 400p cracks could be induced in the crystalline microstructure of the laser spots of the boron-rich alloys (Fig. 10).

Microhardness indentations in amorphous zones cause plastic deformation but no crack formation has been observed (Fig. 11). The zone which is most sensitive to crack formation is the **amorphous** zone as-crystallized after annealing. Proof loads of only 25p already lead to crack initiation.

DISCUSSION AND CONCLUSIONS

For an interpretation of the microstructure in addition to the three crystallization mechanisms the course of temperature with time at the sites x,y,z is required. A schematic plot of the temperature profile in x-direction at the end of the melting process is given in Fig. 12. At this time the solid/liquid-interface has reached the maximum melt depth x_d. The maximum temperature T_{max} at the surface (x = 0) is limited by the vaporization temperature. The temperature T_i at the solid/liquid interface may well exceed the eutectic temperature T_e (which corresponds to the melting temperature T_m). During heating the liquid/crystalline interface propagates into a heterogeneous solid. The steepness of the temperature gradients and consequently the high temperature velocity allows considerable overheating of the melt as well as of the crystalline phases. The difference between Fe-C- and Fe-B-alloys is caused by the absolute insolubility of boron in iron and the different thermodynamic stability of Fe_3C and Fe_2B. This explains the results in Fig. 2: the stable borides retard the melting process. Undissolved borides are frequently found floating in the melt. As a result the cooling rate for Fe-B will be higher than for Fe-C after exposure **to equivalent** laser conditions. A short time after the laser was turned off the liquid/solid

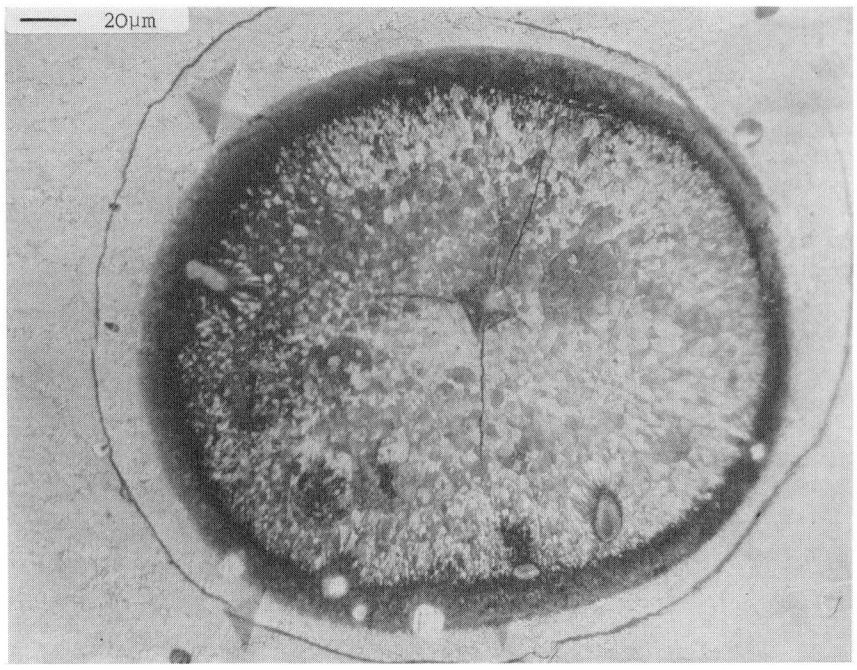

Fig. 10. $Fe_{83}B_6C_9$/y-z-section/$p = 3.4 \cdot 10^4$ Wcm^{-2}/$t = 5 \cdot 10^{-4}$ s
Cracks induced by microhardness indentations (400p) in a crystalline laser spot. In the primary microstructure indentations did not result in crack formation.

interface starts to recede after sufficient undercooling is reached. In $Fe_{83}B_{17}$ all basic solidification reactions occur (Fig. 13): In the beginning of solidification the eutectic reaction $L \rightarrow \alpha'' + \beta'$ (α-Fe + metastable orthorhombic Fe_3B) will take place which originates at the original microstructure. If the liquid is bounded by the glassy phase (overlapping spots/traces) exclusively the massive reaction $1 \rightarrow \alpha'$ (metastable tetragonal Fe_3B) will take place. In addition, the crystallization reaction $a \rightarrow a' + \alpha'$ occurs in the annealed part of the (solid) amorphous zone. This provides evidence for the fact that heterogeneous nucleation conditions affect the nature of the crystallization reactions. The eutectic reaction is predominantly a crystal growth process. According to the known course of eutectic growth velocity with temperature this reaction stops at a certain undercooling. The lamellar spacing of the eutectic amounts to 10-20 nm which provides evidence of considerable undercooling $T_E - T$ up to approximately 400 K. When these conditions are reached, individual nucleation of tetragonal Fe_3B takes place in the immediate vicinity of the reaction front followed by the growth of the massive reaction. This reaction continues for a short path into the liquid and forms the second zone. As it is frozen-in the prerequisites are reached for amorphous solidification.

Fig. 11. $Fe_{85}B_{10}C_5$/y-z-section/$p = 3.4 \cdot 10^4$ Wcm^{-2}/$t = 5 \cdot 10^{-4}$ s
Crack initiation in the crystalline phase while cracks could be stopped in the amorphous (light) zone.

The analysis of the crack patterns has shown that they depend on the shape of the laser affected zone and on the particular primary microstructure. The shape of the laser traces and consequently the distribution of the residual thermal stresses may be compared with conventional welds. The maximum self-contained tensile stress is located in the center of the traces or spots. As the coefficients of thermal expansion of the phases in the eutectic of the primary microstructure are quite different, additional stresses will be caused, which lead to the observed cracks induced in proximity of the primary microstructure. In this context the formation of the metallic glass is of special interest as amorphous Fe-B has a negative thermal expansion coefficient [5]. This reverses the sign of the stress as compared to the crystalline traces.

The effect of the microstructure on cracking is characterized by large local differences between fracture mechanical properties of the phases and interfaces. The metallic glass can show a high fracture toughness, i.e. relatively high values of K_{IC}. All crystalline zones are brittle. Its brittleness is only surpassed by that of the crystallized zone in the heat affected glass.

The local tendency for crack formation is inverse proportional to the critical crack length a_c.

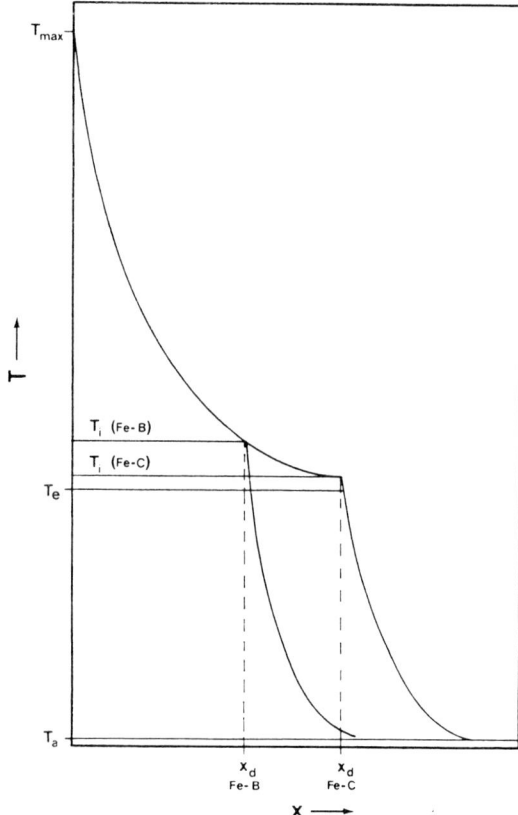

Fig. 12. Schematic temperature profile at the end of the melting process.

$$a_c = \frac{K_{IC}^2}{\Pi\sigma^2}.$$

If it is small the probability for cracking is high. Valus of a_c which surpass the dimensions of the microstructural zones imply that no cracks can form.

The local stresses σ, the fracture toughness (K_{IC}), and the related critical crack length (a_c) for a crystalline and amorphous laser zone is schematically shown in Fig. 14. This approach allows an interpretation and a prediction of the crack pattern. The minimum of the critical crack length a_c, in the center of the crystalline zone, defines the **sites of crack initiation.** **The** high fracture toughness of the glassy zones combined with small tensile stresses σ and consequently high critical crack lengths a_c prevent cracking. The glassy zones even arrest cracks in crystalline zones.

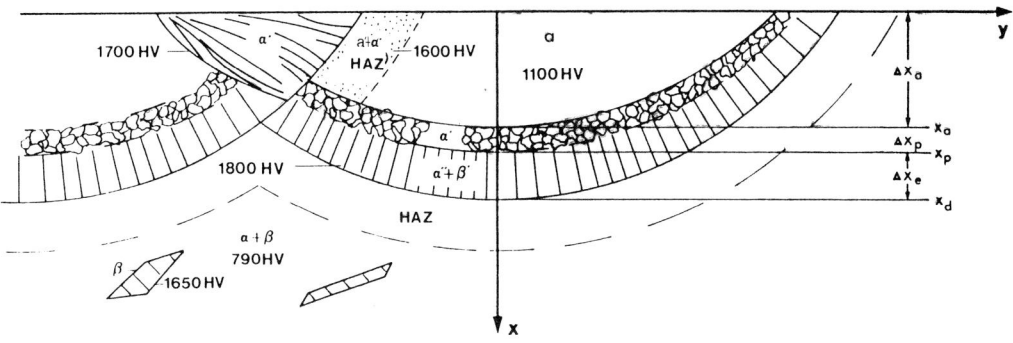

Fig. 13. Microstructures and reactions in four-zone-structured overlapping $Fe_{83}B_{17}$ laser traces (schematic).
HAZ: heat-affected zone, modified original microstructure $\alpha + \beta$
Δx_e: eutectic zone, originated by the eutectic reaction $1 \to \alpha'' + \beta'$
Δx_p: polymorphous zone, originated by the massive reaction $1 \to \alpha'$
Δx_a: amorphous zone, originated by the amorphous solidification $1 \to a$
HAZ': heat-affected (amorphous) zone, originated by the reaction $a \to a' + \alpha'$
In addition, the microhardness (HV) of the phases is indicated.

In conclusion it can be stated that it is possible to produce surface layers of metallic glasses of considerable thickness with the desired combination of high hardness and toughness. The formation of such layers predominantly depends on the properties of the particular alloy and to a much lesser degree on the laser parameters. In case of good gfa the only problem left is the mutual effect of parallel traces which causes crystallization resulting in embrittlement of this area. A modification of the laser parameters is one way by which this situation may be improved. It is more likely that the search for alloys with optium primary microstructure, an improved gfa, and higher thermal stability of the amorphous phase will provide the most promising perspective for further improvement of the quality of laser glazed surfaces.

ACKNOWLEDGMENT

The support of this work by the DFG (Ho 325/19) is gratefully acknowledged. Laser treatments were carried out in cooperation with the Bremer Institut für Angewandte Strahlforschung, BIAS (Director: Dr. G. Sepold) by R. Becker. Preparation of centrifugally cast alloys was carried out by Dr. K. Fritscher and H. Mangers, DFVLR Köln, Institut für Werkstoff-Forschung (Director: Prof. Dr. W. Bunk).

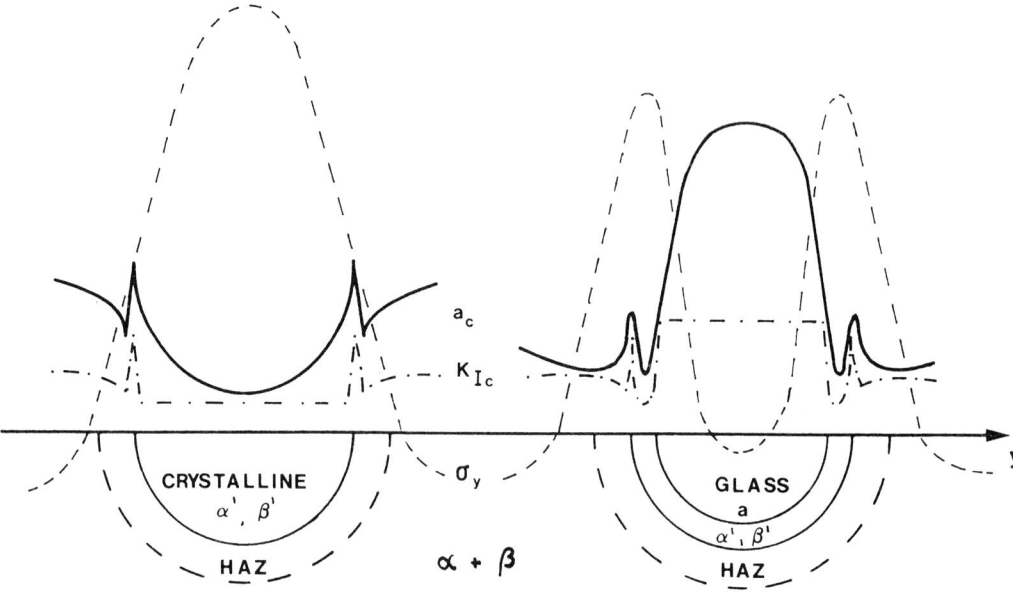

Fig. 14. Qualitative course of local stresses (σ_y), fracture toughness (K_{Ic}) and the related critical crack length $a_c = \dfrac{K_{Ic}^2}{\pi \sigma^2}$ fr a crystalline (left) and amorphous (right) laser zone.

(1) E. Hornbogen: The Origin of Microstructures of Rapidly Solidified Alloys; Z. Metallkde. 77 (1986) 306.
(2) B. H. Kear, E. M. Breinan, L. E. Greenwald: Laser Glazing - A New Process for Production and Control of Rapidly Quenched Metallurgical Microstructures; Metals Technology (1979) 121.
(3) K. Mukherjee, J. Mazumder (eds.): Lasers in Metallurgy, AIME Conf. Proc., The Metallurgical Society of AIME, Warrendale, Pa. (1981).
(4) E. Hornbogen, I. Schmidt: Forming Ability and Thermal Stability of Metallic Glasses, in: Liquid and Amorphous Metals, E. Lüscher, H. Cofal (eds.), Sijthoff & Noordhoff (1980) 353.
(5) G. Hunger, H. W. Bergmann, B. L. Mordike: Thermal Expansion of Amorphous Iron-Chromium-Boron Alloys, in: Rapidly Solidified Amorphous and Crystalline Alloys, G. H. Kear, B. C. Giessen, M. Cohen (eds.), Elsevier Science Publ. Co., (1982) 231.

EFFECT OF LASER HEATING ON THE SUBSTRUCTURE OF 0.4% CARBON STEEL

M. C. Seegers*, S. Mandziej and J. Godijk****

Twente University of Technology, Department of Mechanical Engineering,
*Precision Mechanics Section
**Materials Science Section
P.O. Box 217, 7500 AE Enschede, Netherlands

SYNOPSIS

Laser surface treatment by CO_2-laser is expected to be useful for industrial applications where improvement of wear resistance is required on selected areas of already accurately machined parts. Available precision of hardening of such surfaces by laser is an additional advantage. Moreover it is possible to obtain these desirable properties out of low-cost construction steels.

The aim of this work is to determine the microstructural changes within the laser heated zones of commercial grade medium carbon C45 steel. For this reason transmission electron microscope studies have been performed on thin foils taken from laser treated zones. The substructure components are described and discussed as regards the influence of rapid laser heating on austenite formation and its transformation into martensite.

INTRODUCTION

The power density of the laser beam at the workpiece, the time during which the beam interacts with the material and the thermal properties of the treated surface determine if the surface will be heated, melted or vaporized. These three possibilities form the three main areas of laser surface treatments, within these areas one can make a further subdivision (see Figs. 1 and 2). (1, 2).

From all the possible laser surface treatments, Fig. 1, transformation hardening is one of them which has found its way to industry, e.g. the automotive industry. Localized areas submitted to wear as insides of ring grooves or parts of crankshafts are hardened in this way. (2,3).

To harden the surface of a material, the material must be able to form a hard structure after rapid quenching, as e.g. martensite in steels. To apply a CO_2 laser, it is necessary that the material properly absorbs the laser energy. To meet the last

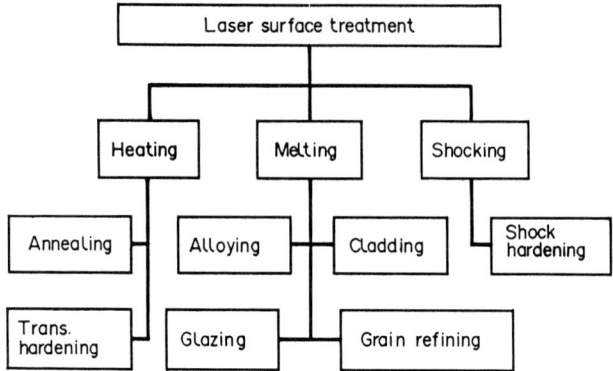

Fig. 1. Schematic representation of the various laser surface treatments.

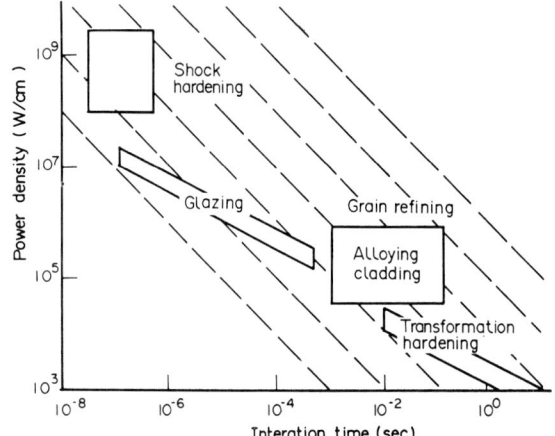

Fig. 2. Power density and interaction time for laser surface treatments

requirement the hardenable steel surface should be coated by a absorptive layer, like manganese-phosphate or graphite. Other wise the steel surface will absorb only about 10 % of the ener of the laser beam (for 10, 6 µm wavelength). (4).

The hardening process consists of two parts, the heat-up and cool-down. During the heat-up a homogeneous austenite must be formed. This is difficult to obtain because of the very short interaction time (appr. 10^{-1} s); To increase the change of forming a homogeneous austenite phase, it is advisable to star from an optimized initial structure of the steel. Sorbite see to be a good candidate, because of its homogeneous distributic of fine carbides and unstable structure. (5,6).

It should be noted, that high heating rates of about 1000 - 10.000 °C/s raise the temperature of transformation. For heat

rates between 100 and 7.000 °C/s, reached with resistance and
induction heating, the transformations in steels have been studied
and described in detail in literature (7). According to these
data the overheating of the A_1-temperature is typically of about
50 to 60 °C for heating rates from 1.000 to 3.000 °C/s if the
initial structure of the 0.4 %C steel is sorbite. Other data
show that the A_1-temperature can be shifted up to 900 °C for
heating rates of about 1.000 °C/s for a 100 Cr6 steel, see Fig.
3 (8).

During the subsequent cool-down the austenite in the hot areas
transforms diffusionless into martensite or related structures.
Around this area a heat influenced zone, in which no transformations has occurred, is created.

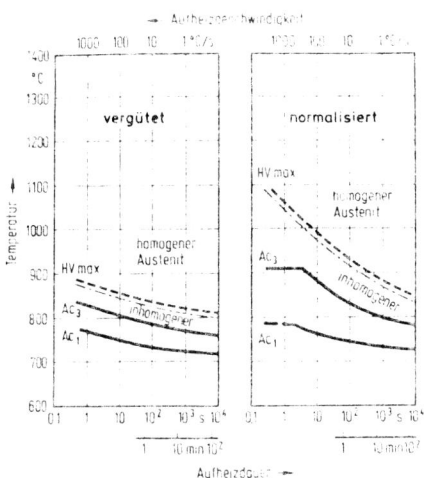

Fig. 3. Time-temperature-austenitization diagram for
100 Cr6 steel.

EXPERIMENTAL

The material which was used in this investigation was commercial
grade medium-carbon C45 steel of 8 mm thickness, its chemical
composition is given in table 1.

In order to obtain a sorbite structure the specimens were
austenitized at 840 °C for 30 min., then water quenched, subsequently tempered at 600 °C for 30 min. and finally air cooled.
After the heat-treatment the surfaces were ground, lapped and
coated with a 8 μm thick graphite layer.

The laser treatment was carried out with a CO_2-laser working in
a gaussian energy distribution, TEM_{00}-mode, and focussed on the
surface using Schwarzschild optics. A pattern of hardened lines

was generated on the surface. The experimental conditions of the transformation hardening are given in table 2.

Table 1. Chemical composition of the C45 steel in weight percent

C	Si	Mn	P_{max}	S_{max}
0.42	0.17	0.5	0.04	0.04

Table 2. Experimental conditions

Spot radius	323 μm
Distance between the lines	0.9 mm
Scan speed	4 mm/s
Power	80 W
Shielding gas*	N_2
Absorption	75 %
Power density	$2.2 \cdot 10^8$ W/m²
Focussing optics	Schwarzschild optics

* N_2 was used to protect the mirrors in the Schwarzschild optics.

After the laser treatment the coating was removed with ethanol and the samples were then ultrasonically cleaned. Sections were cut perpendicular to the traces of the laser beam for optical microscopy. These were ground, polished and etched in 2 % nital solution.

Thin slices were cut parallel to the surface for transmission electron microscopy. The slices were ground to a thickness of approximately 200 μm. Thinning was continued chemically in a mixture of 60 vol % hydrogen peroxide and 40 vol % orthophosphoric acid at approximately 30 °C. Discs of 3 mm diameter were then punched out and electropolished in a 10 vol % perchloric acid solution in methanol at minus 30 °C. The resulting foils, which represent material from approximately 30 μm beneath the laser treated surface, were examined with a JEOL 200CX (200 kV) TEM.

RESULTS AND DISCUSSION

RESULTS AND DISCUSSION

By means of light microscopy the laser heated zone was identified and measured to be about 500 μm in diameter, i.e. slightly smaller than the diameter of the laser beam (Table 2), and about 80 μm in depth (Fig. 4).

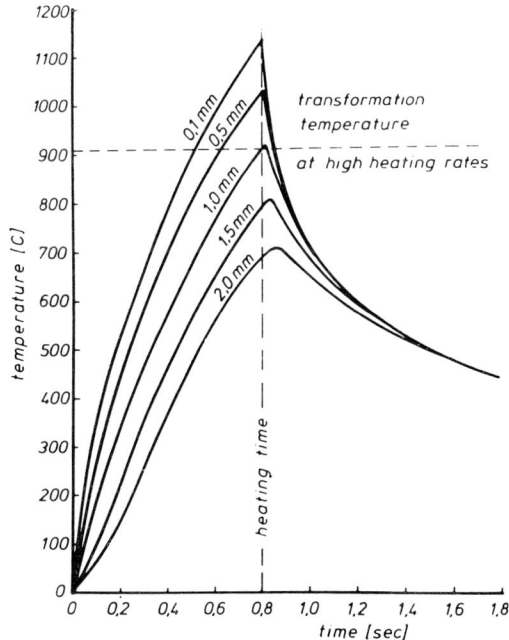

Fig. 4. Optical micrograph.
Cross-section of two laser heated zones.

At larger magnifications the structure of the laser heated zone was studied and it seemed to consist of a well pronounced martensite at the surface with a shift to some finer structure towards the middle of the laser heated zone. At the bottom of the affected zone, a mixture of fine martensite with initial sorbitic structure is visible. This structure is, however, too fine to determine precisely its different components (Fig. 5). A shallow surface layer of ca. 2 μm thickness with a large amount of retained austenite is obviously produced by diffusion of carbon from the graphite coating into the bulk martensite during the laser heat treatment. The carbon content in this layer was shown by microanalysis to be twice as high as in the base material. Our transmission electron microscope observations were performed on a series of thin foils, taken over a range from the base material to the structures well within the laser heated zone. The TEM observations allow to distinguish areas of different phase content in the laser heated zone as is schematically shown in Fig. 6. The initial sorbitic structure contains spheroidal carbides (cementite) of different sizes, distributed rather uniformly in a ferrite matrix. The substructure of the ferrite is formed of

Fig. 5. Optical micrograph.
Cross-section of a laser heated zone showing the retained austenite, fine martensite, transient zone and sorbitic structure.

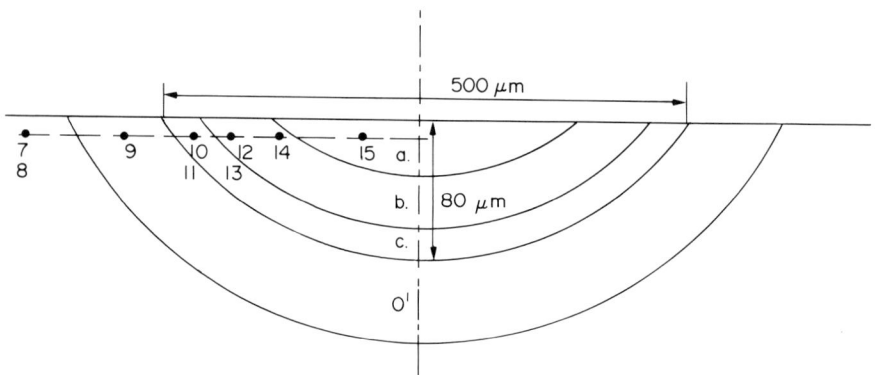

Fig. 6. Schematic representation of different areas in the laser heated zone:
    a. homogeneous bulk martensite,
    b. unhomogeneous martensite,
    c. martensite and ferrite/sorbite mixture,
    d. laser heat influenced sorbite.

subgrains which frequently have polygonized dislocation boundaries. The arrays of subgrains retain the shape of the former martensite needles. Along the boundaries of these former martensite needles the cementite particles are larger and form characteristic chains while the cementite precipitates inside the subgrain are smaller (Figs. 7, 8). Towards the interface between the base material and the laser heated zone no structural changes are observed. The only visible effects in the interface zone are the formation of recrystallization boundaries and consequently larger ferrite grains as well as further coagulation of carbide particles. Such structure seems to be relevant to temperatures about A_1 (Fig. 9).

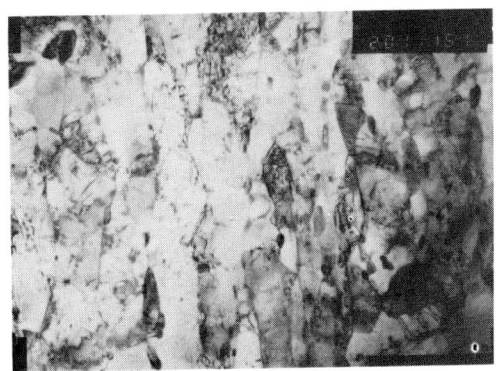

Fig. 7. TEM micrograph.
The sorbitic structure.

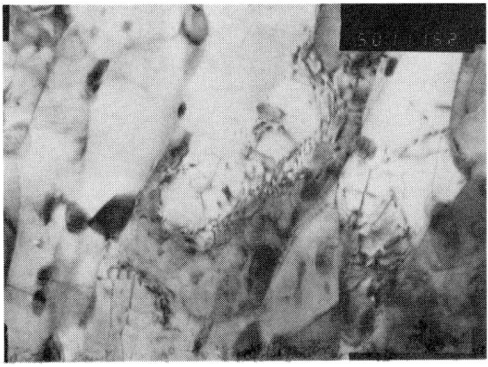

Fig. 8. TEM micrograph.
The sorbitic structure shows a polygonized dislocation boundary.

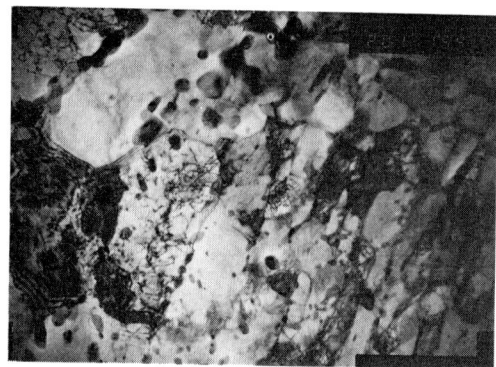

Fig. 9. TEM micrograph.
The interface shows coagulated carbides and formation of recrystallization boundary.

Towards inside this zone the first traces of martensite are observed. This martensite forms at larger carbides and their conglomerates (Fig. 10), and has the form of islands in the ferrite grains. Some of the martensite between the larger carbides shows a twinned substructure (Fig. 11), characteristic for its high-carbon content.

Towards the centre of the laser heated zone the amount of martensite increases and the subgrain structure of the ferrite gradually disappears. The larger martensite islands in this area are surrounded by ferrite grains with high dislocation densities (Fig. 12). The martensitic areas are mixtures of different components

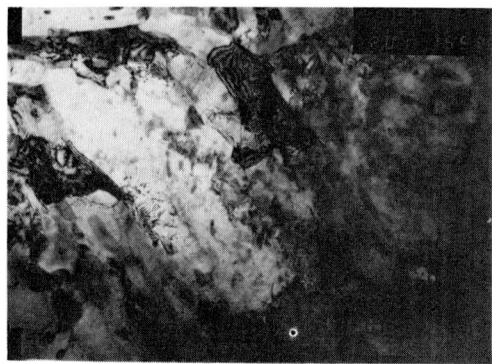

Fig. 10. TEM micrograph.
Outer part of the laser heated zone containing traces of martensite at larger carbides.

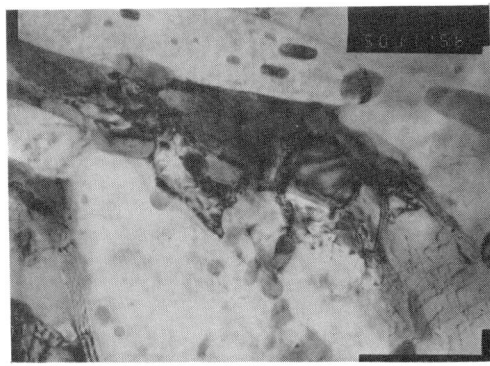

Fig. 11. TEM micrograph.
Small islands of martensite with a twinned substructure.

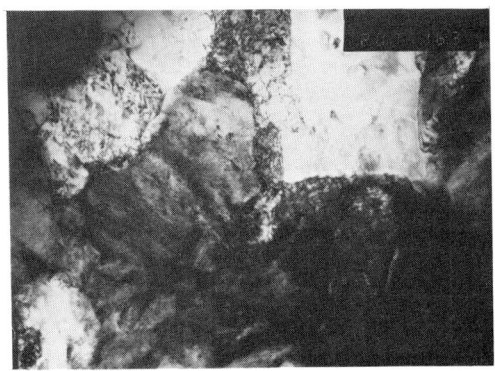

Fig. 12. TEM micrograph.
Towards inside the laser heated zone the amount of martensite and the dislocation density in the ferrite increase.

i.e. they comprise fields of lath-like (low-carbon), bulk (medium-carbon) and twinned (high-carbon) martensite (Fig. 13), in which some irregularly shaped ferrite grains are embedded (Fig. 14).

At the centre of the laser heated zone, i.e. at a depth of about 30 μm from the surface, the martensite is rather homogeneous and shows a characteristic bulk substructure, typical for a medium-carbon steel (Fig. 15). At that depth, no retained austenite was found in the centre of the laser heated area, nor anywhere further up towards the surface.

In order to relate the temperature distribution in the heated spot to the occurring transformations attempts have been made to

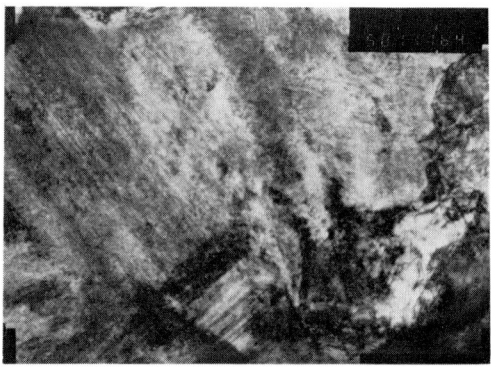

Fig. 13. TEM micrograph.
Martensite of different morphology due to a
difference in carbon content.

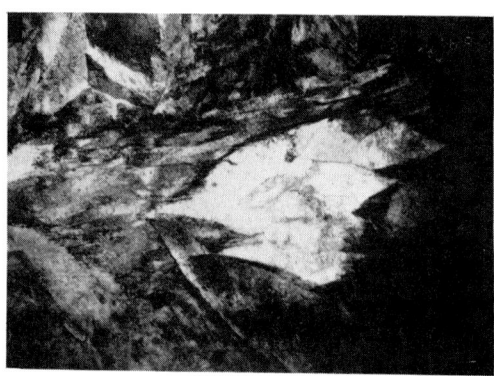

Fig. 14. TEM micrograph.
Irregularly shaped ferrite grain embedded in
the martensite.

calculate this distribution. Most models however do not cover
properly our case of a small and thin heat affected zone. To
know the maximum temperature reached at the surface an approximation, which incorporates the heat losses due to conduction,
was used. This approximation gives a maximum temperature at the
surface of about 1200 °C. With a heating time of 0,16s, this
results in a heating rate of about 7.500 °C/s. (9).

The effects we have observed in TEM allow to describe the
following characteristic stages of the transformation caused by
rapid laser induced heating:

1. The recrystallization phenomena of the initial sorbitic

Fig. 15. TEM micrograph.
Characteristic bulk, medium-carbon martensite.

structure are retarded up to the A_1-temperature and the first mobile recrystallization boundaries start to migrate at the same temperature at which nucleation of austenite on carbide/ferrite interfaces occur.

2. Then, at higher temperatures, austenite grains start to grow while ferrite recrystallizes. Up to a 1:1 ratio between austenite and ferrite, there still remain sorbite subgrains within the ferrite, indicating that the recrystallization is not complete.

3. Rapid laser induced heat input causes nucleation of austenite in areas where the available carbon supply is high enough because of local concentrations of carbides of larger than average sizes. This high-carbon austenite results in the formation of twinned martensite. In this range the dissolution of the carbides is faster than the austenite grain growth.

4. When the percentage of austenite and ferrite have become equal, the austenite is inhomogeneous as to its carbon content, which gives rise to different morphologies of martensite. The growth of the austenitic grains does not seem to depend anymore on the local carbon concentration. In this region, the formation of martensite generates a high dislocation density in the remaining ferrite.

5. The last retained parts of the ferrite have irregular, even acicular shapes which suggest their formation from a very low carbon austenite. They are often embedded in the inhomogeneous martensite.

6. The homogeneous bulk martensite formed in the top central part of the laser heated zone (down to a depth of about 30 μm) consists of 'blocks' of 0.5 to 3 μm in diameter and originates from 10 to 20 μm austenitic grains. The latter is an evidence of retardation of the grain growth of austenite as the estimated temperature at the surface has reached about 1500 °C.

The result of this laser surface treatment is a line hardened surface consisting of a relatively soft sorbitic matrix with hardened zones, which are slightly above the surface due to the volume increase by the martensitic transformation. These hardened zones are created by the used laser processing conditions, producing firstly a fine-grained homogeneous austenite which transforms into fine and homogeneous bulk martensite at the surface.

Between the fully martensitic and sorbitic structure, a transient zone is revealed by the TEM. This zone, resulting from concurrent ferrite recrystallization, austenite nucleation and austenite homogenization phenomena, always contains softer and harder components. These components are: different martensites and acicular ferrite at the top, and sorbite, recrystallized ferrite and small islands of high-carbon martensite at the bottom of the treated zone.

REFERENCES

1. D. S. Gnanamuthu: Opt. Eng., (1980), 19, (5), 783 - 792.
2. P. J. Oakley: The Welding Institute Research Bulletin, Jan. (1981), 4 - 11.
3. A. S. Bransden, S. T. Gazzard, B. C. Inwood, J. H. P. C. Megaw: 'The Laser Hardening of Ring Grooves in Medium Speed Diesel Engine Pistons', Report CLM - P745, Culham Laboratory, Abingdon, Oxfordshire, (1985).
4. V. G. Gregson: in 'Laser Materials Processing', (ed. M. Bass), 201 - 233; (1983), Amsterdam, North-Holland Publishing Company.
5. G. Stähli: H.T.M., (1974), 34, (2), 55 - 63.
6. R. Chatterjee-Fischer, R. Rothe, R. Becker: H.T.M., (1984), 39, (3), 91 - 98.
7. I. N. Kidin: 'Fizyczeskie osnowy elektrotermiczeskoj obrabotki melallow i spawow', 58 - 153; (1969), Moscow, Metallurgia. (in Russian).
8. W. Englisch and R. Baumert: in 'Laser in Industrie und Technik', 45 - 122; (1985), Sindelfingen, Expert Verlag.
9. O. M. Roessler: in 'The Industrial Laser Annual Handbook, 1986 Edition', 16 - 30; (1986), Oklahoma, Penwell Books.

LASER SURFACE REMELTING AND ALLOYING OF ALUMINIUM ALLOYS
P. L. Antona, S. Appiano and R. Moschini
Centro Richerche Fiat, Italy

1. INTRODUCTION

Following are some of the most outstanding results of this work, which aims at local modification of the properties of an aluminium alloy for castings by laser surface remelting.

This objective can be attained through local modification of the microstructure with remelting followed by fairly fast solidification.

If however, the surface of the aluminium alloy is to possess mechanical and metallurgical properties not inherent in the original alloy, modification of the microstructure must be accompanied by modification of its chemical composition, also through remelting but in this case with addition of alloying material to the molten bath before solidification.

This paper covers certain aspects of the process which tends on the whole to be fairly complicated, as the laser treatment of aluminium alloys gives rise to a large number of metallurgical problems, connected with the peculiar physical properties of the material; these tend to make the actual process critical (high thermal conductivity, low density, low melting temperature compared with the added materials).

2. MATERIALS AND TREATMENTS

The tests were made using G-Al Si 7 Cu 3 (DIN) aluminium alloy as substrate in the form of plane specimens or components.

The experiments were made using the 15-KW AVCO HPL 6 laser as source and consisted in a set of penetration tracks, with variation of the parametric conditions each time.

3. SURFACE REMELTING-RESULTS AND DISCUSSION

The surface remelting tests were assessed on the basis of metallographic inspections to point out the maximum depth, the hardness, soundness and microstructure of the remelted layer.

Figure 1 shows the melted depths obtained with a single-pass, detected at the centre of sections drawn at 20 and 40 mm from the starting point of the treatment.

As expected, the high thermal conductivity of the material led to a noticeable increase in the depth of penetration (and generally in the melted volume) as the treatment proceded; this increase became more and more significant as the time of interaction increased.

From a metallurgical point of view, the most remarkable effect of fast solidification induced by laser is the refinement of the dendritic structure, as revealed by the metallographic examinations (Fig. 2).

Generally it is possible to distinguish three different areas with different microstructure in the treated part : an interface between the remelted area and the basic material, with a somewhat coarse structure, with a tendency towards concentrations of microporosities and interdendritic shrinkages; an area close to the surface with a very fine structure with equiaxial growth of the dendrites and an intermediate area between the two outlined above which reveals a basically columnar behaviour of the dendritism.

There, both metallographic and ultrasonic (C-Scan) inspections showed almost complete removal of the pre-existent porosities.

Tests with partially overlapping tracks showed structural and sclerometric faults at the edge of the overlapped area (Fig. 3) where there is also a tendency towards the formation of fairly large porosities. However, hardness values proved to be, on average, 25 % higher than in the basic material.

4. SURFACE ALLOYING-RESULTS AND DISCUSSION

This process differs from the simple remelting process in that it modifies the chemical composition of the irradiated material, through application of pre-established quantities of alloying powders melted together with the basic material on its surface.

In our experiment, the iron-based powders were sprayed on; laser parameters were set as follows : incident power 5 KW with speeds between 60 and 120 cm/min.

The single-pass were obtained by reducing the scanning speed every 40 mm during the course of the treatment. This condition must always be taken into account when interpreting the results given below in that each step treated is affected by the preheating induced by the previous steps (a similar effect to that already found in remelting).

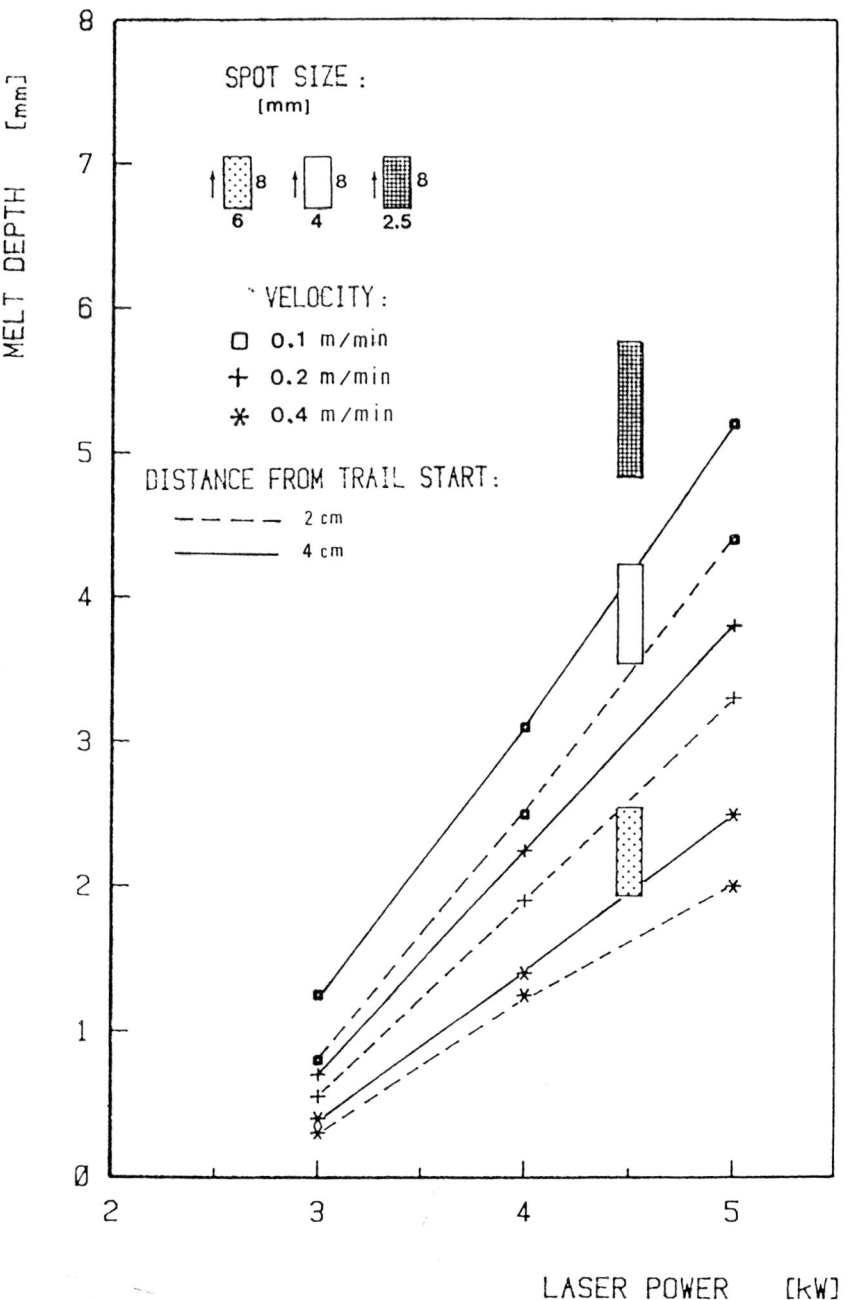

Fig. 1. Laser surface remelting. Melt depth with regard to process parameters.

146

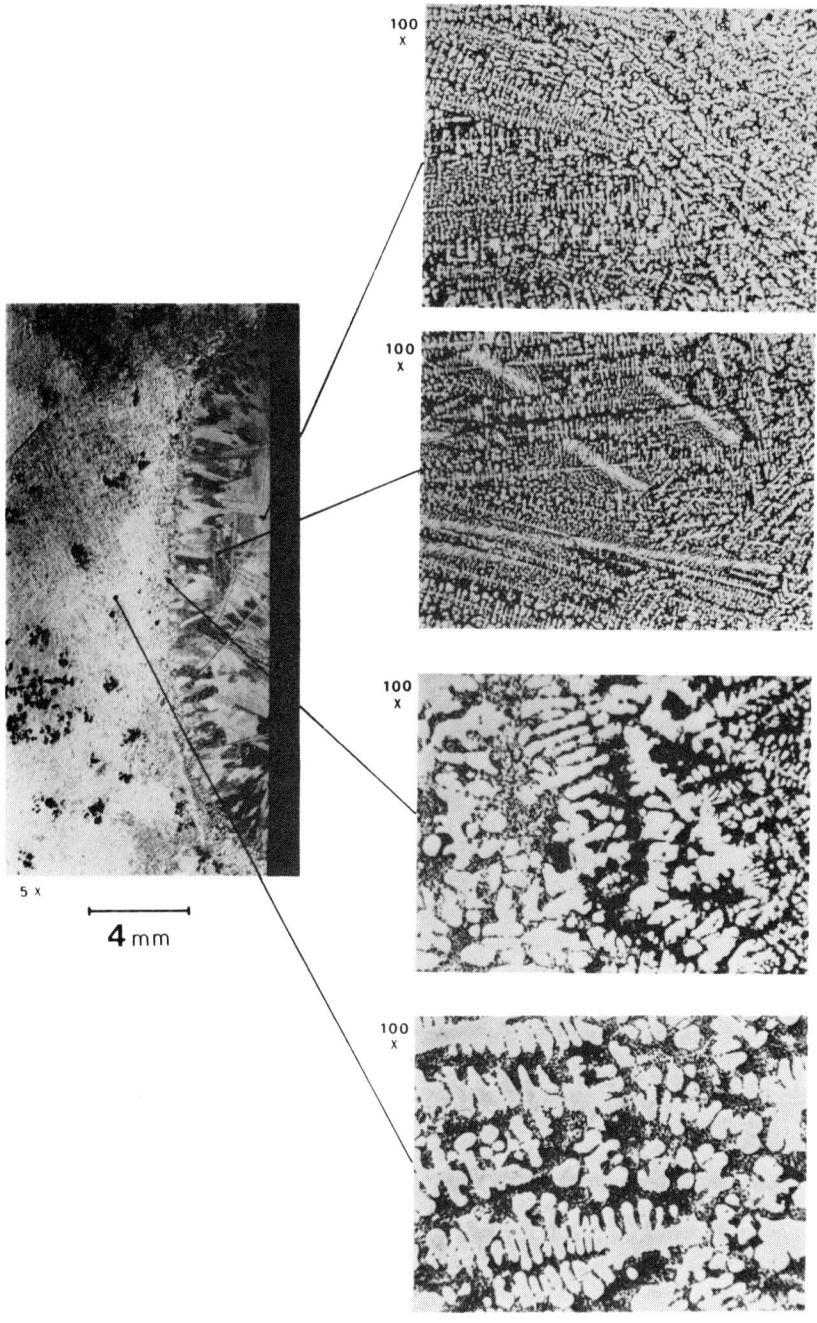

Fig. 2. Laser surface remelting. Optical micrographs.

b)

INSPECTION DEPTH : 1.5 – 2.5 mm

SCANNING : Ø.1 mm

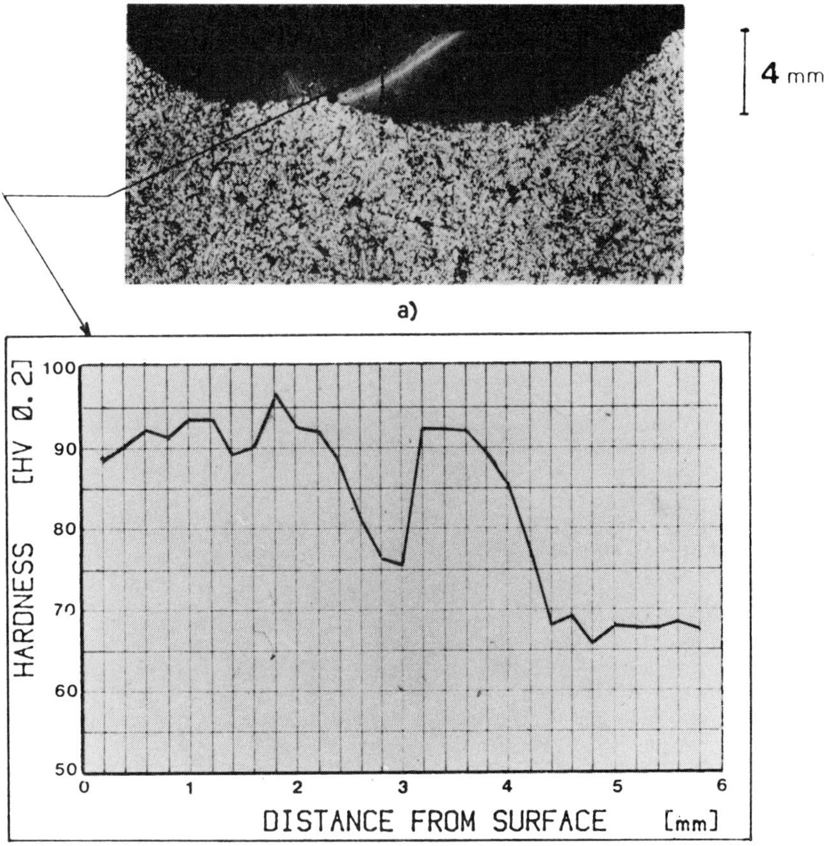

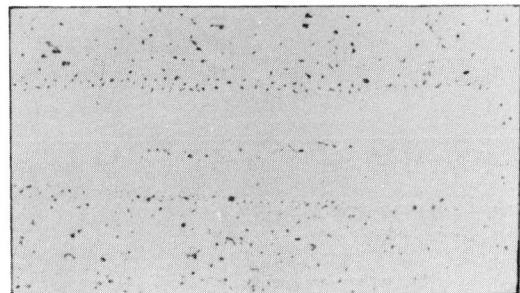

Fig. 3. Laser surface remelting: a) overlap of 2 trails: b) C-Scan inspection.

Figure 4 shows the optical and electronic micrographs of a typical alloyed layer.

4.1. HARDENING MECHANISM

Tests made with an electronic microprobe on the various test specimens treated always showed a rather high percentage of aluminium (over 60 %) in the alloyed layer, even though high hardness values were achieved (up to 600 HV).

This is possible because the rapid solidification induced by laser makes it possible to obtain supersaturated solutions of the alloying elements in aluminium characterized by a wide range of hardness values. In the case under consideration, taking into account the melted masses and volumes, the cooling rate was estimated to be around 10^4 K/s.

Figure 5 shows as an example the hardness and total of alloyed element trends according to the distance from the surface in a typical laser alloyed layer.

Hardening by supersaturated solutions must be considered, at present, one of the possible hypothesis but not the only one able to justify the hardening mechanism. Another hypothesis, strengthened by new observations, may be based on the formation of needlelike phases that are always observed in all the alloyed layers. Exams by electronic microprobe pointed out that such phases are characterized by nearly constant chemical composition indifferently to be localized in alloyed layers very differently hardened. Observe on the subject the micrographs and the table of Fig. 6. So we can't exclude that such phases are, as a matter of fact, intermetallic compounds that confer hardness according to their distribution density and not, by definition, according to their level of saturation.

In any case it should be noted how the cooling rate at the solid-liquid interface ($\dot{\varepsilon}$) and the sum of the concentrations of all the alloying elements (Σc_i) represent the independent variables responsible for the hardness:

$$H = H(\Sigma c_i, \dot{\varepsilon}) \qquad [1]$$

with

$$\dot{\varepsilon} = \dot{\varepsilon}(\vec{r}, V_f) \qquad [2]$$

where $\vec{r}$ is the vector that indicates the single positions inside the treated area and V_f is the melted volume.

The high density laser power used for the alloying treatment leads to a negligible functional dependence between $\dot{\varepsilon}$ and $\vec{r}$; just as negligible is the dependence of $\dot{\varepsilon}$ on the amount of the melted volume, the variations of which, at least in the cases under consideration, were very limited. Therefore, in our case, there is a mere functional dependence between the hardness values and the sum of the concentrations of the alloyed elements, that is:

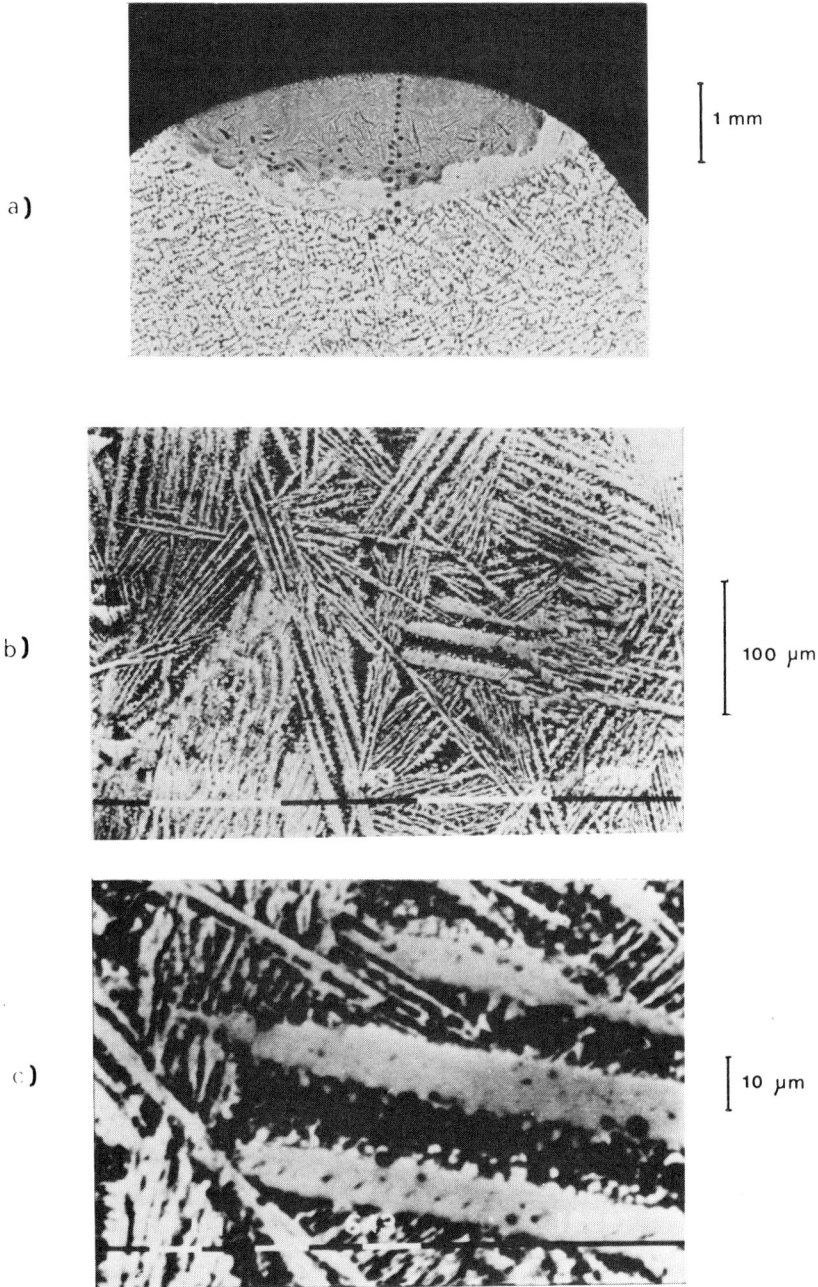

Fig. 4. Cross-section of an alloyed trail: a) optical micrograph: b) c) SEM micrograph.

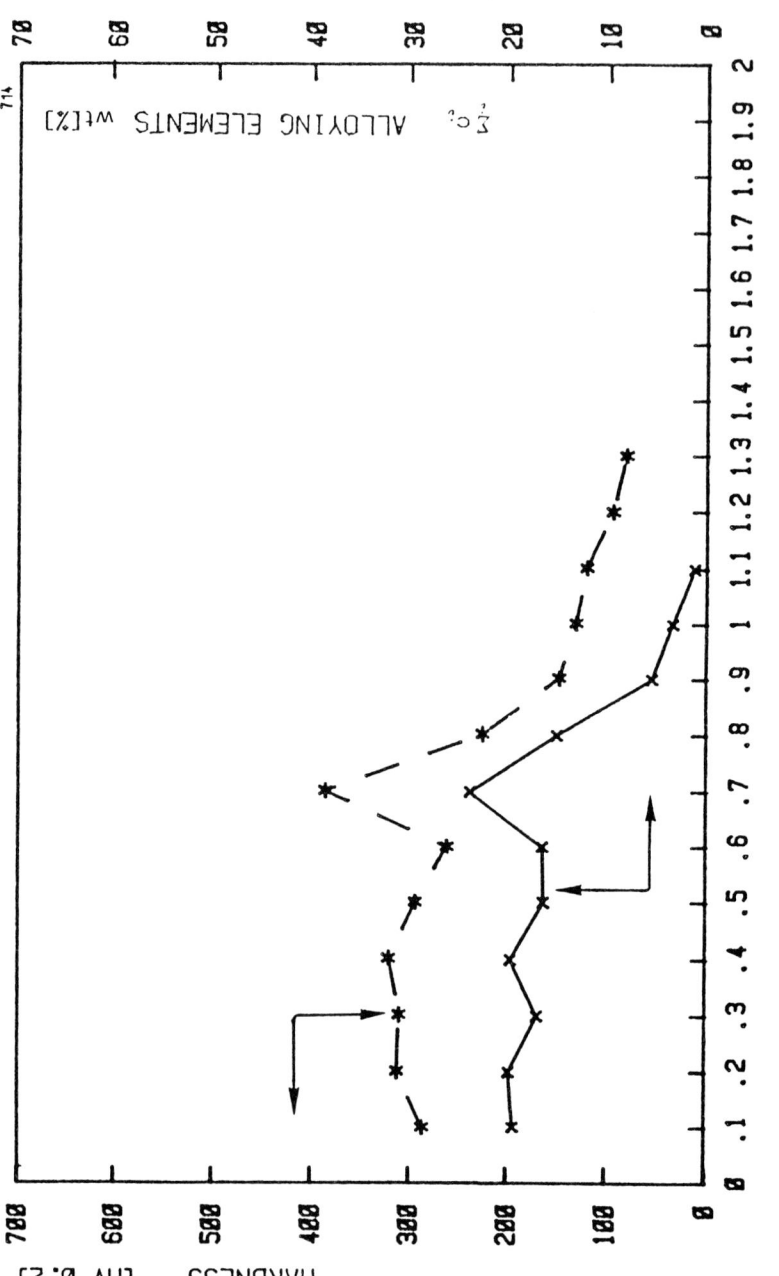

Fig. 5. Comparison between microhardness and the alloying element concentrations in the same points.

	SUBSTRATE			ALL. ELEMENTS
%	Al	Si	Cu	Σc_i
A	60.69	2.83	1.38	35.1
B	60.57	3.11	1.53	34.79

710

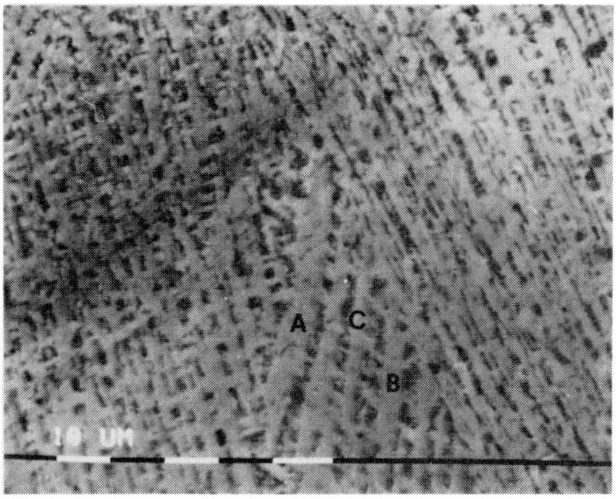

⟨630⟩ $HV_{0.2}$

713

⟨232⟩ $HV_{0.2}$

	SUBSTRATE			ALL. ELEMENTS
%	Al	Si	Cu	Σc_i
A	62.17	2.55	1.38	33.90
B	67.14	3.75	1.72	27.39

Fig. 6. Analysis by electronic microprobe on needlelike phases.

$$H = H(\Sigma c_i) \qquad [3]$$

where Σc_i clearly depends on the amount of powder used and on the melted volume which in turn depends on all the independent variables of the process.

Figure 5 confirms the one-to-one correspondence between hardness and composition of the layer as defined by [3]. Note that the composition values given in the graph were obtained through microanalyses made inside the hardness impressions.

Figure 7 shows the hardness values taken on transverse sections of a **single alloyed track, drawn at a distance of 10 mm from each** other beginning from the start of treatment area.

The fairly constant behaviour of the hardness values in the single sections is particularly worthy of note, confirming structural homogeneity (see the micrographs of Fig. 4) as a result of the strong convective flows that are generated in the liquid pool.

A comparison of the three curves shows that the mean hardness values tend to drop off gradually as the treatment proceeds. This phenomenon can be ascribed not to the effect of postheating which leads to redistribution of the elements in the various phases or a partial resolubilization, but rather to a greater **dilution** of the alloying elements due to the gradual increase in the melted volume during the course of the treatment. This phenomenon is obviously emphasized in the case of specimens with limited mass (as in Fig. 6) while, with larger specimens, there is no great difference between the mean hardness of the individual sections.

4.2 CORRELATION OF HARDNESS/SCAN RATE

Figure 8 gives the mean values and the standard deviations of the hardness values measured on the transverse sections of alloyed layers, obtained by varying the traversing speed of the specimens during treatment. The two curves refer to two different amounts Q of sprayed powder.

Note that for a given amount of powder the hardness of the alloyed layer increases, as the traversing speed is reduced. This is to be ascribed to the combination of two contrasting effects : if we reduce the scan rate, while keeping the amount of powder sprayed constant, the amount of powder involved in the process increases and this leads to enhancing the hardness; however if we reduce the scan rate the melted volume increases, the concentration of the alloyed elements reduces and by consequence the hardness decreases. **The predominance** of one or other of the two factors depends on the mass of the specimen involved : if the mass is fairly small, the second factor tends to dominate and the hardness is progressively reduced; otherwise, if the mass is large, the first factor prevails and there is a notable increase in hardness.

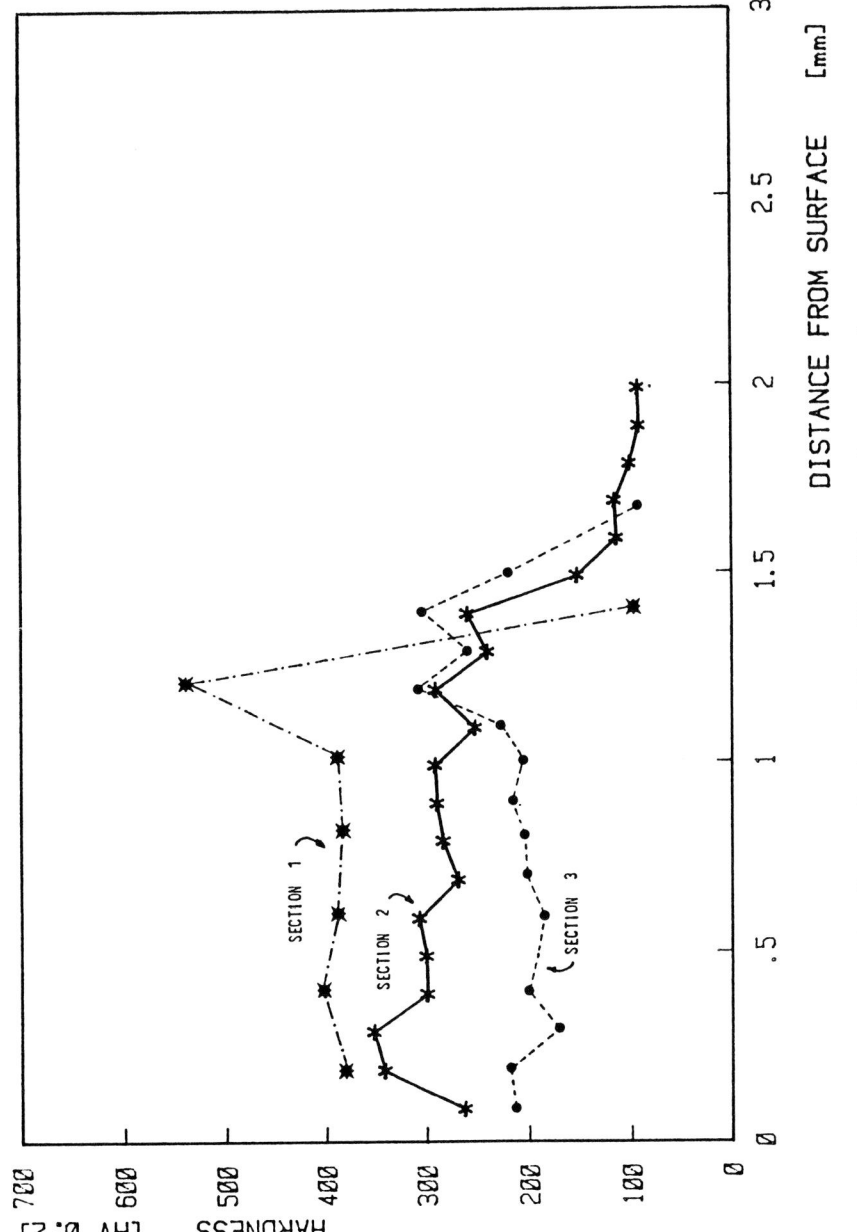

Fig. 7. Hardening of the alloyed layer at progressive distances [1 → 2 → 3] from start.

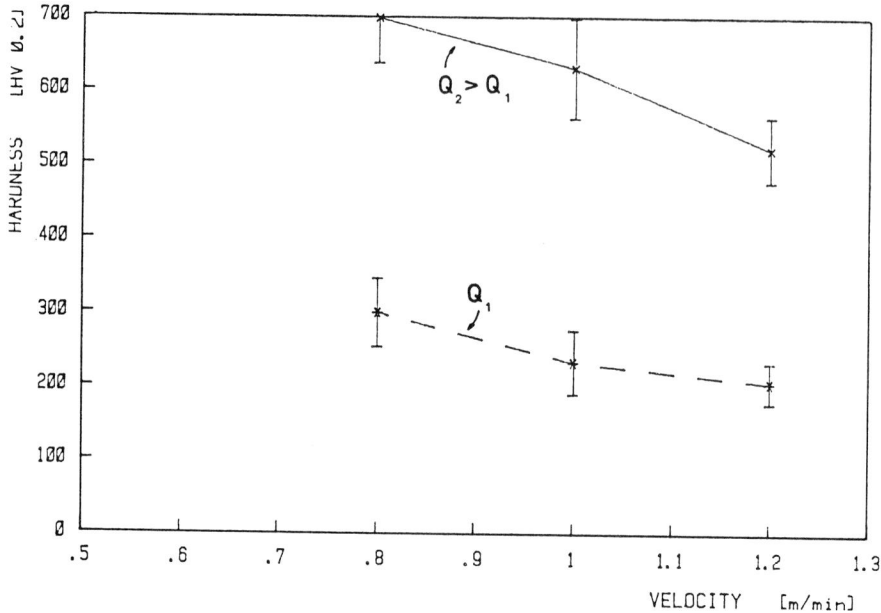

Fig. 8. Hardness versus velocity at different powder mass flows.

The case shown in Fig. 8 refers to a large-size specimen in which there is a very limited increase in melted volume as the treatment speed is reduced.

Figure 9 gives the sums of the concentrations of the alloyed elements according to the treatment speeds. Note that a 25 % increase in the amount of powder sprayed ($Q_2 = 1.25\ Q_1$) does not lead to a proportional increase in concentration. This phenomenon can be explained considering that a higher amount of powder tends to enhance the screening effect on the radiated surface and this leads to a lower melted volume. Similar conclusions to those set forth above make it possible to expect the effect of variation in the size of the laser spot. It is clear from what has been outlined above, that if the diameter of the spot is reduced, the treatment speed and the amount of powder sprayed remaining constant, there is a reduction in the melted volume with consequent increase in the concentration of alloyed elements and hence in hardness (Fig. 10, 11).

Figure 12 shows hardness trends according to concentration of alloyed elements. The curve plotted is the best fit of the experimental points referred to the various process parameters considered (variation of treatment speed, amount of powder and diameter of the laser spot).

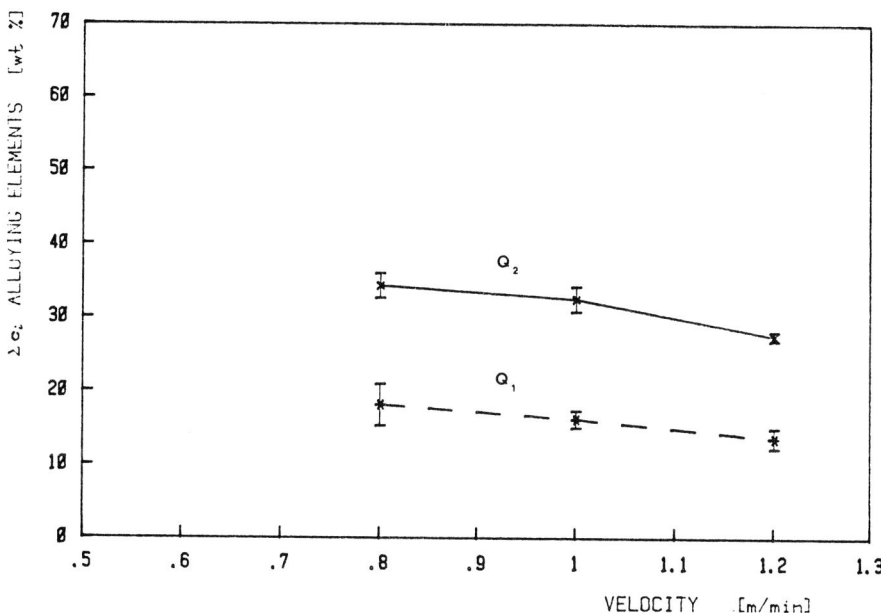

Fig. 9. Alloying element concentrations versus velocity at different powder mass flows.

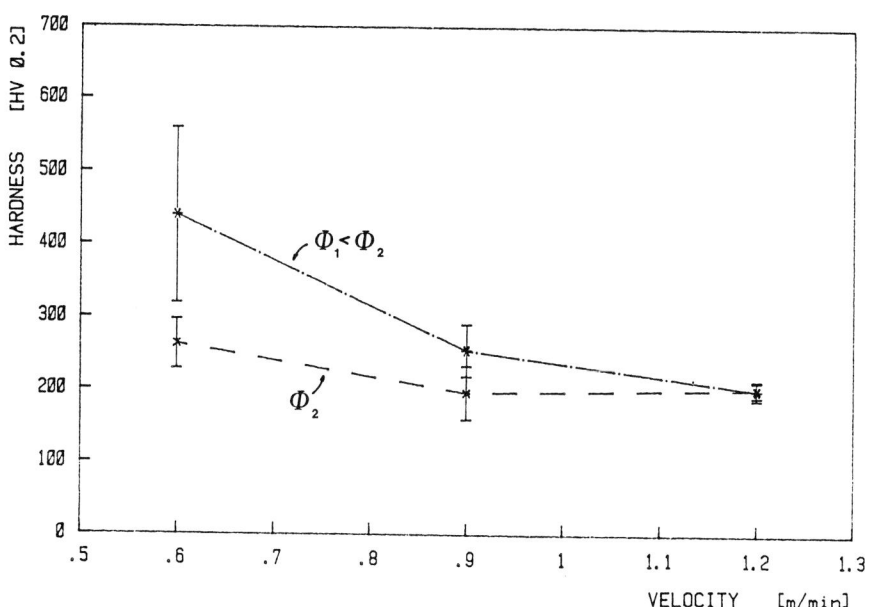

Fig. 10. Hardness versus velocity at different spot laser diameters.

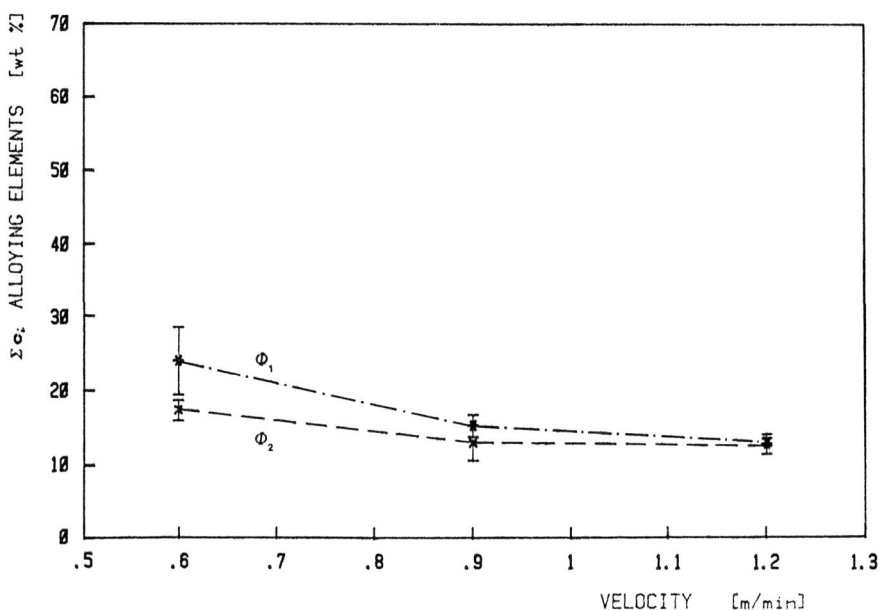

Fig. 11. Alloying element concentrations versus velocity at different spot laser diameters.

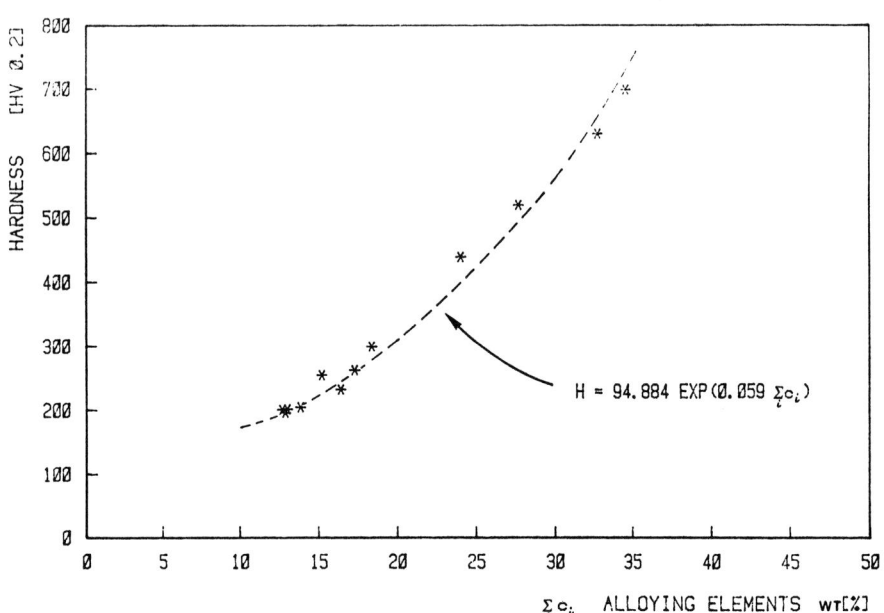

Fig. 12. Hardness versus alloying element concentrations. Powder: Fe Base. Substrate: G-Al Si7 Cu3 (DIN).

The analytical equation which expresses the functional dependence between hardness and concentration of alloyed elements, valid for the type of powder used here, is as follows:

$$H = 94.884 \exp (0.059 \Sigma c_i)$$

5. CONCLUSIONS

An examination of some of the metallurgical aspects of the laser surface alloying and remelting processes revealed a number of correlations between the characteristics of the alloyed layer and the treatment conditions.

It is clear that, at least as far as alloying is concerned, the properties of the treated layer can be directly ascribed to the degree of **dilution** of the alloying elements; it follows therefore that if the correlation between this and the process parameters is known, it is possible to "design" the **alloyed** layer according to the way in which the component is to be used.

COATING DESIGN FOR LASER TRANSFORMATION HARDENING

E. W. Kreutz[*,**] and K. Wissenbach[*,***]

*Institut für Angewandte Physik, Technische Hochschule Darmstadt, Schloßgartenstraße 7, D 6100 Darmstadt, FRG
**Present address: Institut für Lasertechnik, Rheinisch-Westfälische Technische Hochschule Aachen, Drosselweg 87, D 5100 Aachen, FRG
***Present address: Fraunhofer-Institut für Lasertechnik, Drosselweg 87 D 5100 Aachen, FRG

ABSTRACT

The physical background of laser transformation hardening is discussed with particular attenuation being focused on the function of energy-absorbing coatings. The absorptivity, the stability, and the energy coupling of various coatings are systematically investigated by static and dynamic measurements of incident, diffusely and specularly reflected laser power, by metallurgical examination, and by surface replication technique as a function of processing variables. The performance data, the case depths achieved, and the metallurgical structures observed allow for technical applications by optimizing coating selection the matching of the thermal response of the alloys to the rate of delivered beam energy. Different analytic and numerical models on the base of the three-dimensional heat conduction equation are reported for the mathematical analysis of the multiparameter problem laser taansformation hardening providing a testing comparison with the physical picture and an easy estimation of process parameters in technical applications.

INTRODUCTION

Laser transformation hardening of iron base alloys is gaining increasing acceptance with a number of established applications in production [1] to improve strength, wear resistance, and fatigue properties. In addition to the ability to apply the hardening only at those areas, where it is needed, even on a complex shape of workpiece, the laser offers a versatile tool with further advantages over conventional hardening methods as the minimization of bulk heating, thereby increasing efficiency in terms of processing energy, reducing distortions of the surrounding material, and eliminating possibly the need for an external quenchant [2].

For intensities $I<I_M$ (I_M intensity of melting) with temperatures below melting point metal surfaces mainly reflect (90 - 98 %) the

the incident radiation [3-6]. Although near infrared energy
(1.06 μm Nd:YAG) is absorbed more efficiently than far-infrared
energy (10.6 μm CO_2) in this temperature range most of it is
still wasted. For intensities $I \gtrless I_M$ and $I \gtrless I_V$ (I_V intensity of
evaporation) the absorptivity of metal increases from less than
10 % at room temperature to 30 - 40 % at the melting point and
to as high as 90 - 95 % at the vaporization point [3-6]. Laser
transformation hardening is performed without the onset of melting together with the lack of surface damage, hence, the applicable metal reflectivities are high and the Drude Free-Electron
Theory holds. As such, in all practical cases a thin layer of a
highly absorbing material is applied as a surface coating to enhance the absorptivity and to increase the use of the available
power [4,7].

Laser transformation hardening involves complex physical processes and interactions that would require considerable experimental and theoretical effort to understand [8,9]. Few systematic studies have been reported [7,10,11] comparing coatings.
The present paper reports on measurements of the properties on
coated materials under surface conditions supplied by commercial
vender or of various surface finish as a function of various
process parameters in order to get information on the properties
of coatings for an efficient and reproducible energy transfer to
the substrate. The simultaneous quantitative analysis and optimization of transformation hardening, which would be of great
benefit, offers a powerful addition to an otherwise purely experimental approach, in order to get more information on the fundamental advantages and/or limitations of the process. To obtain
a satisfactory physical model, information is required on the
absorption coefficient of the laser beam with the workpiece, the
temperature dependence of both the absorption coefficient and
thermophysical properties of the workpiece. As another objective
of this investigation a numerical analysis of the three-dimensional heat conduction equation is accomplished, in order to
calculate temperature distribution and elevation, heating and
cooling rates, and hardened case depth for a given beam profile,
laser power, and processing speed.

BASIC PRINCIPLES OF LASER TRANSFORMATION HARDENING

Metallurgical considerations

Despite the shorter processing times the transformation process
during laser hardening is not fundamentally different from the
conventional hardening processes because the same metallurgical
reactions occur, which have been studied in great details (for
further information, see the technical literature [12]). The
solid-state transformations in iron and steel simply involve
heating a surface, or a localized region on a surface, above a
transformation temperature, and then allowing the transformed
region to cool very rapidly through self-quenching by the mass
of the surrounding bulk. The basic reaction is brought about by
heating to temperatures at which the steel is austenitized (Fig.
1), i.e. where carbon is present as a solid solution of iron
carbide in fcc(γ)-Fe. Subsequent rapid cooling results in a

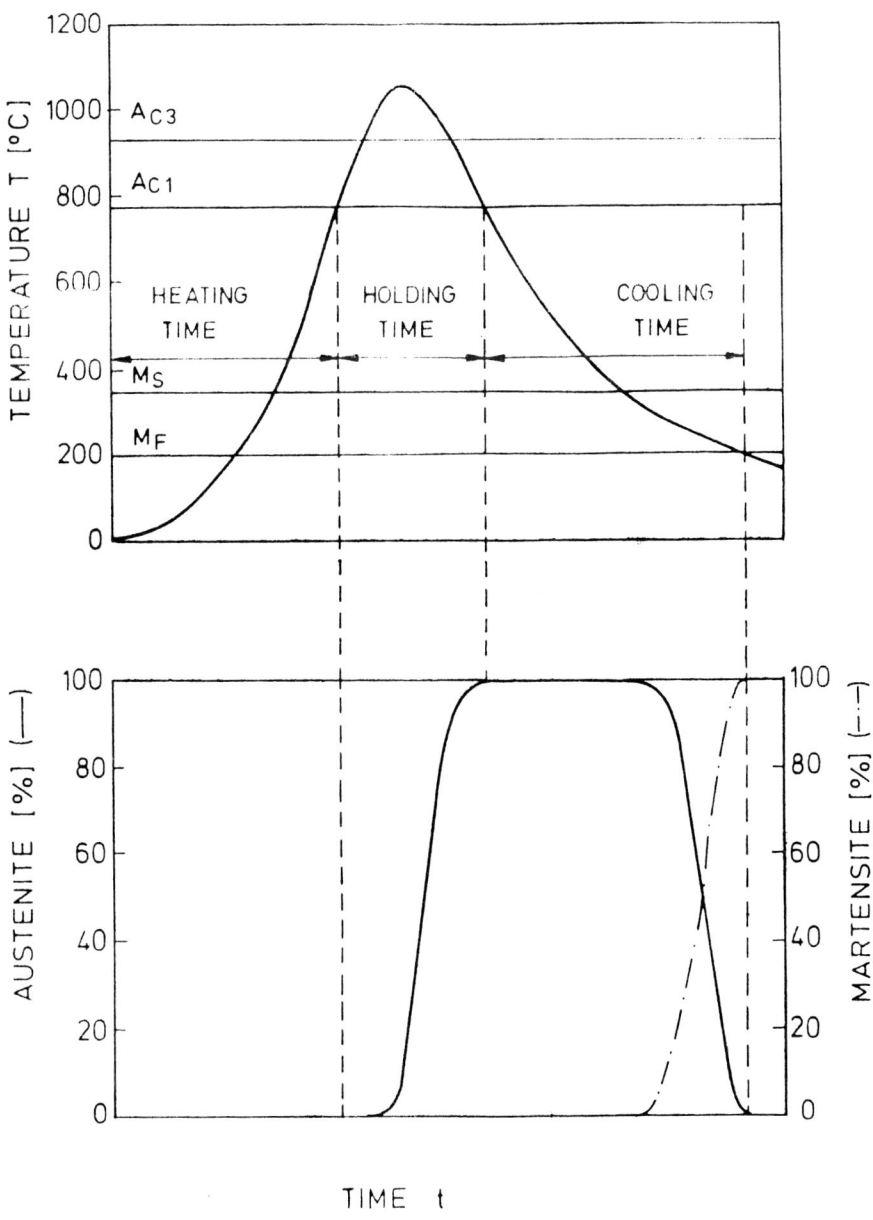

Fig. 1. Time temperature curve and corresponding schematic representation of austenite martensite transformation for a typical iron base alloy.

martensitic transformation (Fig. 1), in which the carbide is retained in solution in bcc(α)-Fe characterized by enhanced hardness [12]. The transformation occurs by nucleation and growth of the new phase in the matrix of the old phase. In slow heating as for conventional hardening (Fig. 2), the process will start at A_{C1} in a carbon steel and will be complete at the A_{C3} line. However, when the heating rate is high as for laser hardening (Fig. 2), the system is far from equilibrium conditions and the A_{C1} and A_{C3} line will tend to be displaced upward to higher temperatures. Thus, although the temperature may be sufficiently high to form austenite under condition of slow heating, the same temperature level may be insufficient even to initiate austenitization under high heating rates. Laser hardening parameters are, therefore, usually designed to give peak temperatures well above those employed in conventional hardening to ensure austenitization (Fig. 2).

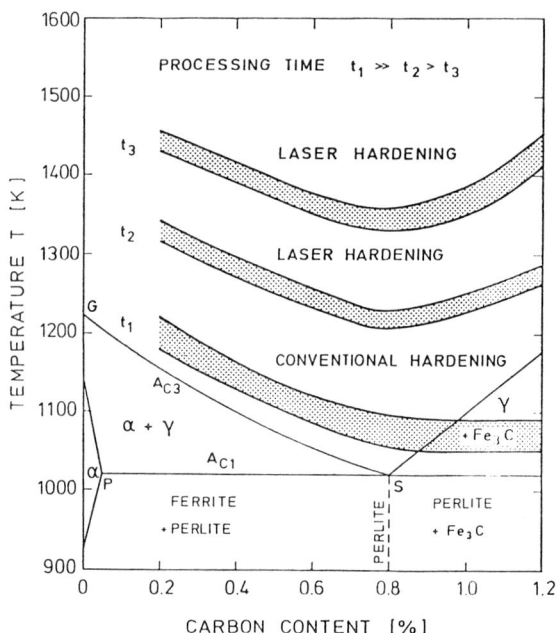

Fig. 2. Fe-C-phase diagram with related processing times and temperatures for conventional and laser transformation hardening.

For a pearlitic primary structure the transformation in terms of a grain by grain metallurgy starts from the lamellar segregation of carbon in a grain of pearlite, which occurs as sheets of ferrite, or pure iron, bounded by sheets of cementite, or iron carbide (Fig. 2) with the dimensions of individual grains depending upon the thermal history of the ferrous alloy. If the pearlite grain is heated to temperatures above the phase transition (Fig. 1) carbon in the cementite diffuses rapidly into the surrounding ferrite and the grain becomes a homogeneous distribution of iron and carbon called austenite. Upon cooling, the austenite will drop quenching to a temperature less than several hundred degrees celsius (Fig. 1) and enter a temperature region at M_S (start of martensite formation), in which martensite forms. From there, the transformation begins, thereby producing martensite finished (Fig. 1) at M_F (finish of martensite formation).

Following this outline transformation hardening of iron base alloys claims three requirements for the thermal cycle, with the maximum temperature, heating, cooling- and holding temperature determined (Fig. 3) from the temperature distributions within the surface region. Firstly, the heating of a thin skin layer to transformation temperature must take place at high energy density and heating times frequently shorter than the cooling times to minimize heat-losses due to the high thermal bulk conductivity. Secondly, the maximum temperature T_{MAX} must be higher than that of the material's transformation point A_{C1}, and also the surface temperature of the metal must be lower than the melting point T_M, because, if it is higher than the melting point, the surface will melt and become uneven, thus off-setting the advantages of laser quenching. Thirdly, cooling must occur from a temperature higher than that of transformation point A_{C1}, at a rate faster than the critical cooling rate, in order to avoid the crossing of the CCT (continuous cooling transformation)-curve and the TTT (time temperature transformation)-curve with reforming of pearlite by having the carbon diffuse back into layers of cementite and ferrite.

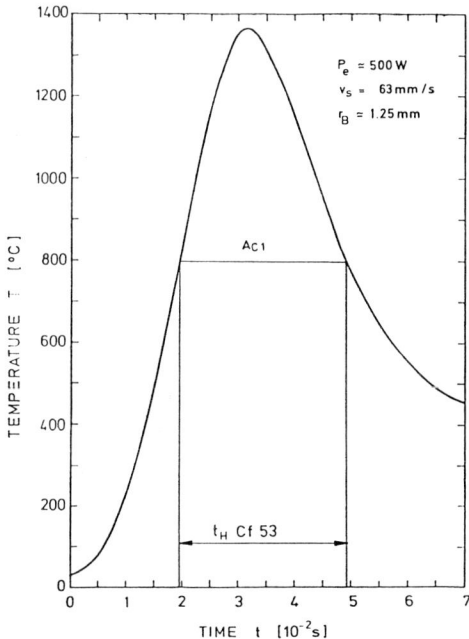

Fig. 3. Time surface temeprature curve for carbon steel Cf 53 as obtained from three-dimensional heat flow calculations.

The formation of composite austenite as well as the redistribution
of carbon, necessary to form a homogeneous γFe-C solid solution,
are governed by diffusion processes that require small but finite
time intervals [13-15]. However, this does not mean that the
carbon content is the same in all the composite crystals formed.
This ideal homogeneity of the austenite namely is dependent upon
the distribution of carbon in the original structure and upon
the temperature and duration of the austenizing process. The
kinetics of these processes under very high heating rates are
still somewhat uncertain, making the design of a laser surface
hardening process more difficult than that of a conventional
through hardening procedure. Thus, the minimum holding times
have been evaluated from experimental investigation [16] of
transformation hardening to electron beam in combination with
metallographic investigations of the heat affected zone (HAZ)
and numerical analysis [16] of the three dimensional heat conduction equation (see equation 2).

Because of the high metallic reflectivity, only few % of the
incident intensity of the laser is absorbed compared to ~ 70 %
for the electron beam, where the major loss is from backscattered
electrons. The dependence of absorptivity on morphology and
composition of the surface is less pronounced in addition to
the ease of alignment and variation of scanning speed with
incident power. The time-temperature curves have been calculated for every parameter set from electron-beam hardening and
metallography yielding with the transformation temperature the
necessary holding times (Fig. 3) for austenizing. As to be
seen from Fig. 4 the surface hardness increases for holding
times in the 10^{-2} s range. Proper adjustment of the processing
times allows for complete austenization with equalization of
carbon-concentration achieving a homogeneous martensite condition. It has been found, however, that in short-duration
hardening, irrespective of the process involved, appropriate
increase of the austenizing (hardening) temperature can accelerate diffusion to such an extent that homogeneous hardening
remains possible in spite of the abbreviated austenizing and
does not involve the formation of brittle coarse grains.

Processing variables

In transformation hardening lasers may be regarded as directed
energy sources with unique temporal and spatial characteristics
allowing the deposition of a great amount of heat into a selected
region of a material. Their main properties are: high incident
beam power, small spot size, capability of heating in a variety
of atmospheres, no x-ray hazard, and shallow absorption depth.
The laser beam raises the surface temperature, and removal of
the energy source results in self-quenching by conduction of
heat into the bulk. The laser properties have important implications with respect to heat flow and suggest that the laser
may be required to attain the highest cooling rates to produce
a rapidly cooled surface layer [17].

There are several independent process variables, which affect
single track laser transformation hardening [1,18,19]:

    - total incident laser power P_L |W|

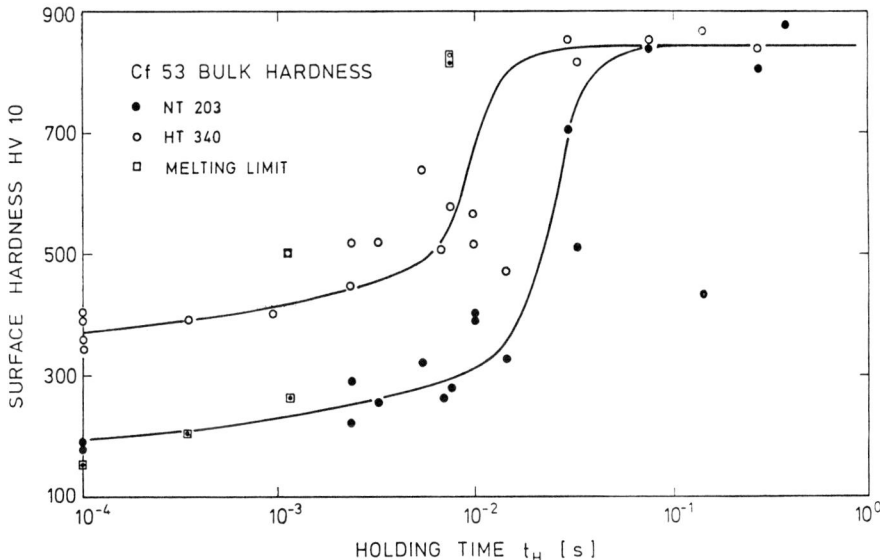

Fig. 4. Surface hardness versus holding time for non(NT)- and heat(HT)-treated carbon steel Cf 53 during electron beam transformation hardening.

- laser beam diameter at the surface r_B (mm)
- processing speed across the surface v (mm/s)
- thermophysical and metallurgical properties of the bulk.

The dependent process variables are considered to be [1,18,19]:
- depth of hardening z_H (μm)
- geometry of the HAZ
- microstructure and metallurgy of heat-affected region.

The investigation of wear performance of the hardened tracks give additional information on the response of a common engineering iron alloy.

It is convenient to characterize the process in terms of the spatial and temporal intensity distribution $I(r,t)$ and the interaction time $t_L \cdot I(r,t)$, which is determined by P_L and the focal area of the optical system, governs the energy density

$$\varepsilon = \frac{I(r,t)}{v} = \frac{I(r,t)}{l} t_L, \qquad (1)$$

where l is the length of the heated region in scan direction. The transformation properties and the depth of the material raised above a given temperature strongly govern the energy density. Low intensities and long interaction times result in longer heating, holding, and cooling times than those achieved with high intensity and short interaction time. In combination with the metallurgical properties, thus, it is possible to produce in a controlled manner surface-hardened zones of suitable iron base alloys.

The surface heat source is defined by r_B and $I(r,t)$, which depend on the mode-configuration ([20], Fig. 5), focussing optics, and method of beam manipulation [21], which are being used to produce suitable beam patterns at the workpiece surface. The method of applying the energy becomes a critical part for large area transformation hardening, since the laser beam diameter determines the coverage rate. Unlike other laser materials processing, a wider beam with uniform intensity distribution is preferred for laser hardening as opposed to a tightly focused high-peak-power low-order-mode TEM_{00} beam (Fig. 5). If a sharply focused laser beam is used, then the width of the hardened zone becomes quite narrow. It is more usual, though, to use as the most unsophisticated way of controlling the heat treated area a defocused beam to increase the area over which the laser energy is spread. This is because the uniform intensity distribution generates uniform case depth. There are different methods of beam manipulation [21] in modern work stations that can be used to obtain a broad beam with uniform intensity distribution as oscillator optics with scanning mirrors and integrator optics with segmented mirrors. Great care has to be taken to avoid beam overlapping, which often leads to low hardness at the overlapped zone [1,22] due to tempering effects. For hardening of cylindrical bodies a better alternative for appropriate shaping is to use a doughnut-mode $TEM_{01}*$ (Fig. 5) beam with toric mirrors [4,20].

The tools of customary machines normally are not affected by materials processing, except for some more or less controllable wear-out. The system parameters of the laser tool, however, may be changed during transformation hardening by optical feedback from the workpiece [23]. The laser and workpiece represent a system of coupled resonators with strong interactions. The workpiece controls the laser radiation depending on its reflectivity and location with respect to the focal plane. Tilting and moving of the workpiece and changes of the materials properties during processing can stimulate relaxation oscillations, mode instabilities, or chaotic temporal behaviour resulting altogether in stochastic fluctuations of $I(r,t)$, A controlled continuous operation may change into an uncontrolled pulse operation [24]. Optical feedback is less pronounced during laser transformation hardening because of the beam manipulation techniqies used (see above). In any case, laser-beam diagnostic and monitoring has to be performed during hardening with well-known procedures [25].

COATINGS FOR LASER TRANSFORMATION HARDENING

Laser-solid-interaction and energy coupling

Energy deposition and heating in an irradiated sample [26] are a consequence of the balance between the deposited energy, governed by optical parameters of the sample and characteristics of the laser radiation, and the heat diffusion, determined by thermal parameters and the interaction time. The hierarchy of relaxation time [26] governs the temporal evolution of carrier density, lattice heating, and the phase transitions possibly encountered in the processing.

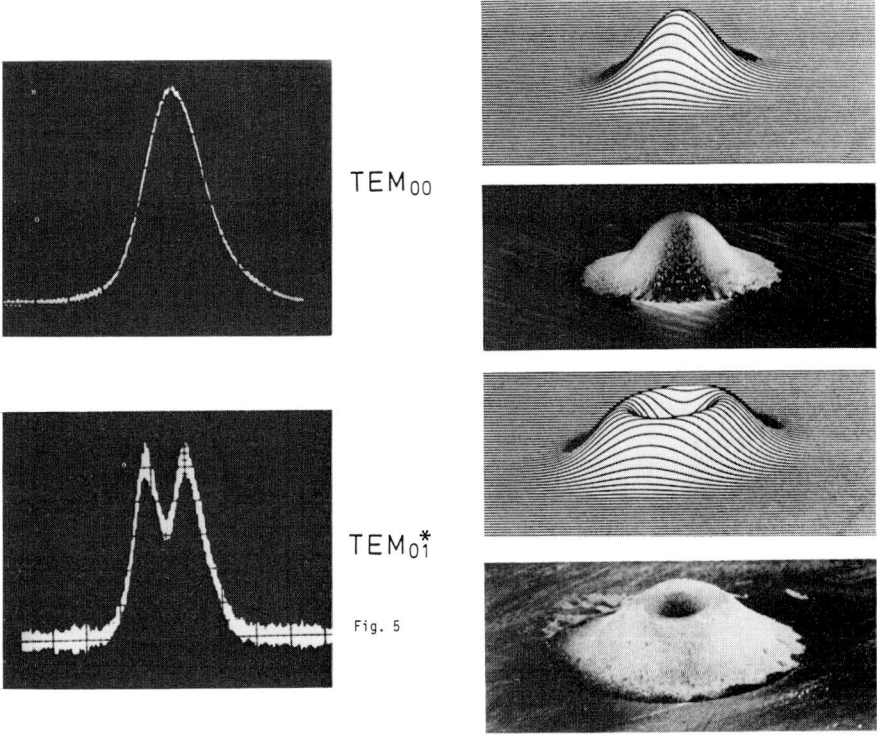

Fig. 5. Measured (cross section perpendicular to the propagating laser beam [25]), calculated, and PMMA burn in intensity distributions for low order modes.

The coupling of optical energy changes the energy distribution
of the carrier system by generation and relaxation processes.
The excitation mechanisms by inverse Bremsstrahlung [27], field
ionization [28] and photon ionization [29] involve optical free
carrier absorption, carrier avalanche multiplication, impurity
ionization, and effects of electron-hole plasma. For $I<I_M$ the
absorption is achieved by inverse Bremsstrahlung, i.e. laser
photons interact with free electrons, which raises the energy of
the electrons within the conduction band. The loss mechanisms
involve carrier-lattice collisions and carrier recombination.
The lattice heating is determined by carrier-phonon scattering
and phonon-phonon scattering. Depending on lattice temperature
phonon-assisted transitions, thermal free carrier absorption,
and thermal changes in band structure have to be considered.

For $t_L<\tau_e$ the optical energy is stored in the carrier system.
The lattice remains cool with considerable differences between
carrier and lattice temperature. The relaxation time τ_e, which
describes the conversion of electronic energy of thermalized
carrier distributions in random atomic motion, is introduced as
a thermodynamic parameter that covers the whole time in which
the excess electronic energy is shared among the different
vibrational modes in thermal equilibrium. For $t_L>\tau_e$ the optical
energy is shared among the different vibrational modes with no
remarkable differences in carrier and lattice temperature. It
can be assumed that the absorbed light intensity is instan-
taneously converted to heat which can diffuse according to the
conventional heat diffusion equation, i.e. the interpretation
within the framework of a simple thermal model ([30,31]).

The laser parameters characterizing the hardening process (Section
"processing variables") are wavelength λ, polarization p, and
spatial and temporal intensity distribution $I(r,t)$, where the
intensity in the area of interaction is the most relevant parameter governing working speed and efficiency. The material
properties (Section "processing variables"), which relate to
hardening, are the absorptivity A and some thermophysical and
metallurgical properties.

The absorptivity is strongly dependent on wavelength, temperature,
presence of dopant and free carriers, crystal perfection, and
surface properties like composition and morphology. As application of coatings, variation of angle of incidence and changes
of surface morphology may enhance the absorbed power during laser
transformation hardening. The minimization of losses by reflection and heat flow and the metallurgical requirements are main
criteria for the determination of the appropriate intensity.

At anything other than normal incidence the reflection coefficients on metal surfaces differ for laser light polarized in a
plane parallel ($R_{||}$) and perpendicular ($R_\perp$) to the plane of in-
cidence depending on the angle of incidence α. As to be seen
from the angular dependence $R_\perp$ is high for all angles and R
becomes very small at close to grazing incidence. The absorption
for incident light polarized parallel to the metal surface is
larger under these conditions and enhanced coupling occurs. This
higher absorptivity at large α suffers from the strong influence
on the spatial intensity distribution claiming additionally precise alignment of the workpiece and high degree of accuracy in

beam handling. Thus, the enhancement of absorptivity observed is much lower promising from $R_{11}=R_{11}(\alpha)$ and very difficult to utilize in laser processing [32,33].

In the interaction of plane electromagnetic waves with smooth ideal surfaces there will be refracted and reflected surfaces, both of which will have $\vec{k}_i$ as the parallel components of their wavevectors, where $\vec{k}_i$ is the component in the surface/interface of the incident beam. Rough surfaces, which are described by the profile function $S(x,y)$, characterized by the root mean square roughness δ and the correlation length σ, produce diffusely scattered radiation in all directions into the ambient, the surface and the bulk in addition to the specularly reflected beam. No correlation ($\sigma \to 0$) means that no intensity is scattered out of the specular reflected beam. For high correlation ($\sigma \to \infty$) diffraction dominates. An enhancement of the absorptivity is observed, if $\delta \cong \lambda$ and the deviations of surface irregularities about the mean are given by a distribution function. The enhancement of the scattering along the surface/interface during photon-solid-interaction requires, that $\sigma \cong |\vec{k}_i|$ and the deviations of correlation lengths about the mean are given by a distribution function. By both these effects the energy coupling into the surface selvedge becomes more probable, that the laser radiation is more strongly absorbed. On the other hand, a surface of microscopically columnar structures with dimensions smaller than the relevant laser wavelengths creates an average graded composition-depth profile, which smoothly changes the effective optical constants from unity of the ambient of those to the bulk. A reduction of reflectivity is observed [34] due to multiple internal reflections.

COATINGS PROPERTIES

If coatings are not used, the absorbed power is reflected and conducted away at a rate that does not permit to reach the desired transformation temperatures at the surface. Insulators and semiconductors in an amorphous state as paints, metals, oxides, metal phosphates, molybdenum disulfide, and colloidal graphite are suitable for applications. As main assumption coatings have to be easily applicable to the surface by dipping, spraying, or plating in controllable thickness and reproducible morphology without distortions of the production line.

The imperative properties, which govern the technical selection [35] of coatings, are
- absorptivity in the infrared portion of the electromagnetic spectrum
- affinity and adherence to the surface of the alloy with possibly easy removal after hardening
- thermal conductivity for heat conduction to the coating alloy interface
- energy coupling from coating to alloy for effective heat transfer
- thermal stability at high intensities and high temperatures
- compatibility with metallurgy of hardening
- chemical reactions induced during processing
- economical costs of material, surface preparation, application, and cleaning.

According to this selection outline the use of coatings ensures reproducibility of absorptivity by careful preparation and therefore of case depth and hardness. The more efficient coupling of the beam results in a shorter treatment cycle thereby minimizing distortion since the conduction of unnecessary heat into the bulk is reduced. Geometric selectively applied coatings can be used to produce hardness in an intricate pattern. Furthermore, the dependence of absorptivity on the angle of incidence is reduced and hence critical alignment of the surface to the laser beam is not necessary.

Optimization of coating selection

The absorptivity, the thermal stability, and the thermal conductivity of surface coatings have to be matched to the metallurgical and thermophysical properties of the substrate for an optimization of heat transfer during laser transformation hardening as a function of the processing parameters in combination with coating handling and economics of pre or post heat treatment procedures [7]. Effective and reproducible coatings necessarily imply the matching of the following properties [7]
- constant absorption within the range of processing variables
- improved thermal stability at transformation intensities and temperatures
- efficient energy coupling to the substrate

that the same processing conditions and arrangement will always create the same geometric and metallurgical properties of HAZ.

TRANSPORT PHENOMENA IN LASER TRANSFORMATION HARDENING

Heat Transfer

Heat conduction equation

Different methods for solving the heat conduction equations under various conditions have been described methodically by other authors [36-38]. Analytical mathematical solutions only are available for certain symmetrical boundary conditions. However, considerable physical insight can often be obtained from such approximate solutions expressed in analytic form. Detailed analysis of almost any practical laser transformation hardening problem requires that a numerical approach can be adopted. Numerical models [9,16,39,40] can be more general and can be made to allow for non-linear surface events as surface heat losses by radiation or convection. They can also be made to accomodate variations in the pwoer distribution across the laser beam (mode structure) as well as variations in the substrate optical and thermophysical properties due to temperature dependence as well as latent heat effects.

The response of a material after the absorption of laser radiation, which is equivalent to a source of heat in or on the solid, can be calculated by solving the three-dimensional heat transfer equation

$$\rho(T) \, C(T) \, \frac{\partial T}{\partial t} = \text{div} \, (K(T) \, \text{grad} \, T) + A(x,y,z,t), \qquad (2)$$

where ρ is the sample density, C the heat capacity, and K the thermal conductivity. The form of the three-dimensional temperature distribution $T(x,y,z,t)$ as solution of equation (2) is determined by the initial and boundary conditions pertaining to the problem and the source distribution function $A(x,y,z,t)$ giving the heat, which is supplied per unit volume in unit time. The amplitude of the solution is controlled by the magnitude and the temperature-dependence of the thermophysical constants.

During laser transformation hardening with characteristics interaction times $10^{-2} < t_L |s| < 1$ and processing intensities $10^3 < I_p \, |W \, cm^{-2}| < 10^5$ the optical penetration depth $4 \times 10^{-6} < \delta_{ph} \, |cm| < 8 \times 10^{-6}$ is orders of magnitude lower than the thermal penetration depth $5 \times 10^{-2} < \delta_{th} \, |cm| < 0.5$ for iron base alloys in the temperature range $20 < T \, |°C| < 1500$. Accordingly, the heat is supplied at the surface. For a homogeneous and isotropic body equation (2) follows

$$\frac{\partial T}{\partial t} = \kappa \, \Delta T, \qquad (3)$$

where $\kappa = K/\rho C$ is the thermal diffusivity determining the rate at which thermal equilibrium is reached in response to the sudden application of a heat source. Equations (2) and (3) are solved taking the initial condition $T(x,y,z,0) = 20 \, °C$ and the boundary condition (Fourier law) at the surface

$$\Phi_L = A \, I = -K \, \text{grad} \, T \qquad (4)$$

neglecting losses by radiation and convection, which are far below the heat flow Φ_L generated by the laser source. The laser beam of spatially and temporally constant intensity distribution is assumed to incident at $t = 0$ parallel to the z-axis normal to the surface of the workpiece in the x-y-plane at $z = 0$, which is moving parallel to the x-axis with processing velocity v.

Three-dimensional heat flow for flat semi-infinite body with temperature-averaged thermophysical properties

Using a Gaussian heat source (TEM_{00} mode)

$$w(x,y) = \frac{2AP_L}{\pi r_B^2} \exp\left(-\frac{2(x^2+y^2)}{r_B^2}\right) \qquad (5)$$

the expressions for the temperature distribution are as follows [16]

$$T(x,y,z,t) = \frac{1}{4\rho C (\pi \kappa)^{3/2}} \cdot$$

$$\int_0^t \frac{dt'}{\sqrt{t'^3}} \int_{-\infty}^{+\infty}\int_{-\infty}^{+\infty} w(x',y') \exp\left(-\frac{(x-x'+vt')^2 + (y-y')^2 + z^2}{4\kappa t'}\right) dx' dy' \qquad (6)$$

respectively after integration in space

$$T(x,y,z,t) = \frac{AP_L}{4\rho C(\pi\kappa)^{3/2}} \cdot$$

$$\int_0^t \frac{dt'}{\sqrt{t'}(\frac{r_B^2}{8\kappa}+t')} \exp\left(-\frac{(x+vt')^2+y^2}{4\kappa(\frac{r_B^2}{8\kappa}+t')} - \frac{z^2}{4\kappa t'}\right) \quad (7)$$

using temperature-averaged thermophysical properties (section "one-dimensional heat flow for flat semi-infinite body with temperature-dependent thermophysical properties"). For a stationary source (v=0) at x = y = z = 0 equation (7) reduces to

$$T(0,0,0,t) = \frac{AP_L\sqrt{2}}{\rho C \kappa \pi^{3/2} r_B} \arctan\left(\frac{8\kappa t}{r_B^2}\right)^{1/2} \quad (8)$$

giving the surface temperature at the centre of the laser beam.

According to the analytic solution by equation (6) the time-temperature curves (Fig. 3) representing heating and cooling cycles have been calculated revealing the experimental findings, which are expected from the intensity distribution. The formalism also describes in reasonable agreement the geometry of the hardened zone (Fig. 6) as derived from combined metallurgical investigations. The hardened case depth increases with scanning path yielding a constant z_H at prolonged processing due to the balance of heat generation and heat losses. Figure 6 indicates the uniformity of transformation hardening, which can be achieved in a single pass.

Following equation (8) the temperature in the centre of the heat source at the surface

$$T(0,0,0,t) \sim \frac{2P_L}{\pi r_B} = Ixr_B \quad ( )$$

is governed by the intensity and the beam radius. The absorbed laser power density, which is required to heat the surface to transformation temperatures, may be calculated as a function of pulse duration by the formalism derived. For short pulse durations, there is no time for transverse thermal conduction, and the hardening intensity is independent of the beam radius. For longer pulse durations transverse thermal conduction becomes important, and heat is conducted out of the interacting area more rapidly because of the higher thermal gradients. According to these considerations the scaling parameter Ixr_B governs the thermal losses by three-dimensional heat conduction in agreement with the experimental observations for different alloys (Fig. 7) and mode structures (Fig. 8). The hardened depth increases with Ixr_B showing a saturation region at high values of Ixr_B. At constant t_L and P_L the heat losses become higher with decreasing r_B, that smaller beam radii r_B require higher intensities for heating to the hardening temperature. The heat losses consequently increase with Ixr_B wince the intensity increases with

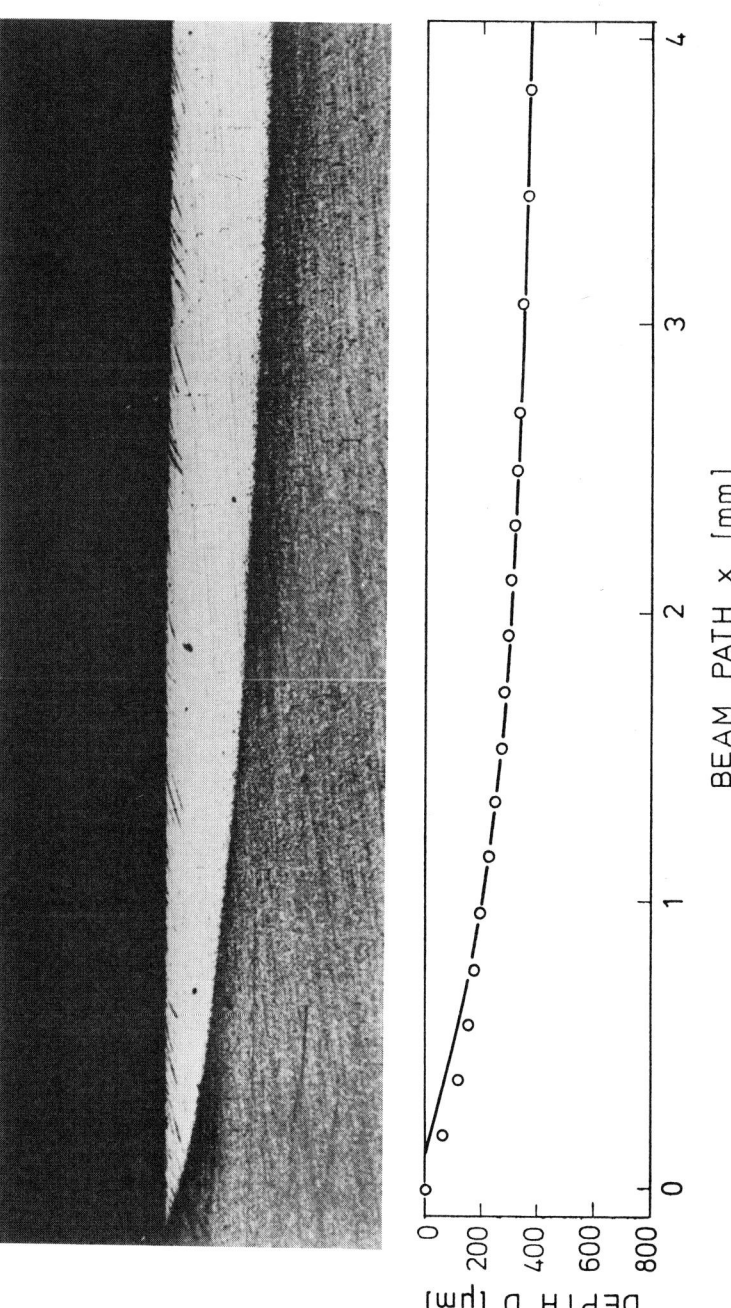

Fig. 6. Macrograph and hardened case depth (o) versus beam path for MoS$_2$ coated carbon steel Cf 53 compared to three-dimensional heat flow calculations (———).

decreasing r_B resulting altogether in a saturating z_H (Fig. 7 and 8).

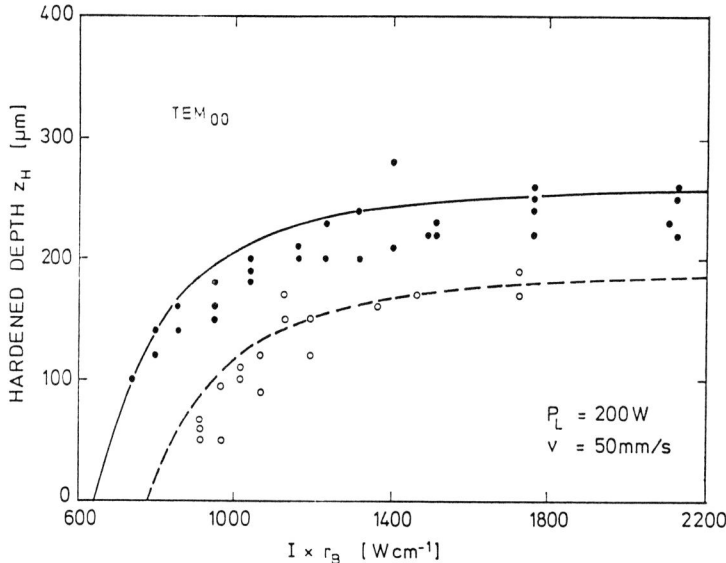

Fig. 7. Hardened depth for TEM_{00} mode as a function of scaling parameter Ixr_B for MoS_2 carbon steel Cf 53 (•) and alloy steel 100 Cr 6 (o).

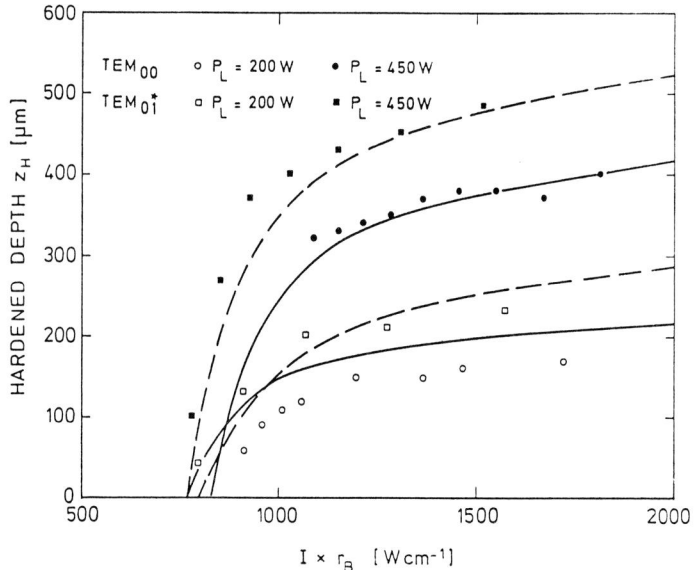

Fig. 8. Hardened depth for different modes as a function of scaling parameter Ixr_B for MoS_2 coated carbon steel Cf 53.

One-dimensional heat flow for flat semi-infinite body with temperature-averaged thermophysical properties

Uniformity of hardened case depth indicates that the heat flow can be regarded as predominantly one-dimensional, normal to the material surface. Three-dimensional heat flow results in non-uniform hardened case depths with enhancement of depth at the track centreline. For

$$r_B \gg \sqrt{8\kappa t} = \sqrt{2}\delta_{th} \qquad (10)$$

the one-dimensional heat flow dominates, whereas for

$$r_B \ll \sqrt{2}\delta_{th} \qquad (11)$$

the three-dimensional heat flow is dominant. As evaluated from equation (6) for a non-moving (v=0) heat source the analytical solution for the temperature distribution by one-dimensional heat flow is given by

$$T(z,t) = \frac{2AP_L}{\pi r_B^2 \rho C \sqrt{\pi \kappa}} \int_0^t \frac{1}{\sqrt{t'}} \exp\left(-\frac{z^2}{4\kappa t'}\right) dt' \qquad (12)$$

respectively after integration in time

$$T(z,t) = \frac{4AP_L}{\pi r_B^2 \rho C} \left(\frac{t}{\kappa}\right)^{1/2} \text{ierfc}\left(\frac{z}{\sqrt{4\kappa t}}\right) \qquad (13)$$

using also temperature-averaged thermophysical properties (see next section).

One-dimensional heat flow for flat semi-infinite body with temperature-dependent thermophysical properties

The influence of the temperature dependence of thermophysical properties on the temperature distribution in the workpiece is examined with a numerical method removing some of the limitations that apply to the analytical methods. In spite of the inherent advantages in numerical methods, only a few numerical models for heat flow in laser processing have been developed so far [9,16, 39,40].

The heat flow is represented by a one-dimensional heat conduction equation

$$\rho(T)C(T)\frac{\partial T}{\partial t} = \frac{\partial}{\partial z}\left(K(T)\frac{\partial T}{\partial z}\right), \qquad (14)$$

where ρ, C, and K are dependent on temperature. Boundary conditions are the usual ones: there are no heat losses from the surface and the workpiece is a perfect heat sink. Due to the nonlinearity of equation (14) the method of finite differences is used dividing the sample in space into slices of thickness

Δz and in time into slices of interval Δt. The time intervals are marked by n and (n+1), the space thickness by k. The time level (n+1) is weighted with α, the time level n with (1-α). Equation (14) becomes a system of equations

$$\rho_k^{n+\frac{1}{2}} C^{n+\frac{1}{2}} \left(\frac{T_k^{n+1} - T_k^n}{\Delta t} \right) = \lambda \left(\alpha \left(K_{k+\frac{1}{2}}^{n+1} T_{k+1}^{n+1} - K_{k+\frac{1}{2}}^{n+1} T_k^{n+1} - K_{n-\frac{1}{2}}^{n+1} T_k^{n+1} \right. \right.$$
$$\left. + K_{k-\frac{1}{2}}^{n+1} T_{k-1}^{n+1} \right) + (1-\alpha) \left(K_{k+\frac{1}{2}}^n T_{k+1}^n - K_{k+\frac{1}{2}}^n T_k^n - K_{k-\frac{1}{2}}^n T_k^n + K_{k-\frac{1}{2}}^n T_{k-1}^n \right) \right) \quad (15)$$

with the convergence parameter

$$\lambda = \frac{\Delta t}{(\Delta z)^2} \quad (16)$$

determined by the slice parameters. Equations (15) take into account the contributions of heat to the slices due to direct absorption from the lsser energy and to conduction from adjacent slices, respectively. Details of the numerical analysis are reported elsewhere [41].

A comparitive plot of the temperature distributions resulting from calculations with temperature-dependent and temperature-independent thermophysical properties is shown in Fig. 9. The calculations are performed with the formalism for one-dimensional heat flow in semi-infinite flat body using the analytic solution (equation 12) and the numerical analysis (equation 15) described. In the former case the thermophysical properties are assumed to be constant either represented by the temperature-averaged values in the range 20 < T [°C] < 1200 or the values at T 20 °C. In the latter case the temperature-dependence of the thermophysical properties is considered according to the literature. The surface tmeperature versus time as well as the bulk temperature versus penetration depth as obtained by the various procedures exhibit for iron base alloys reasonable agreement within the errors reliable for applications of laser transformation hardening in order to estimate the required processing parameters with relative ease without going through trial-and-error experiments. For materials other than iron alloys the temperature distributions calculated with temperature-dependent and temperature-independent thermophysical properties, respectively, may show less agreement, thus, claiming approximate formula or more sophisticated numerical analysis for technical parameter estimation.

Mass Transfer

Mass transfer mainly defines the carbon diffusion. The metallurgy requires holding times and temperatures, which are sufficient to allow adequate diffusion. The upper limit of the process, i.e. how fast one could heat treat iron alloys, depends on the diffusion distances during processing. The lower limit of the process is determined by the cooling rate for martensite formation to avoid the reforming of pearlite during processing.

On the base of a pearlite grain with alternating layers of ferrite and cementite Gregson [4] reported a model for a one-

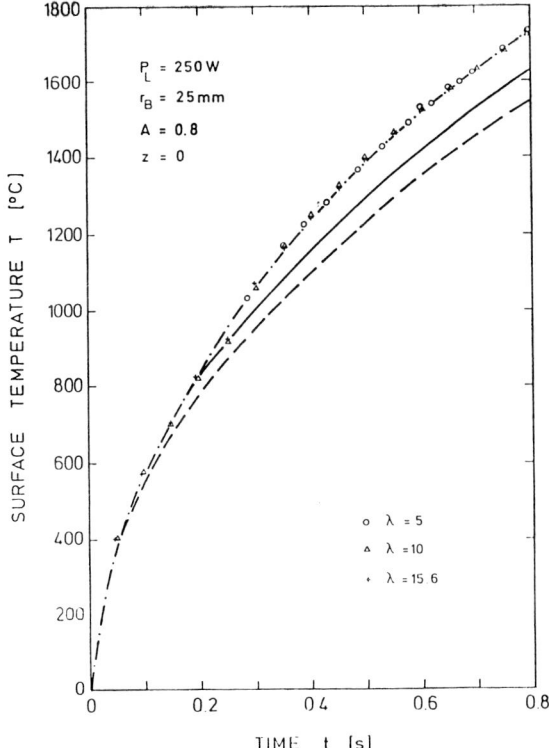

Fig. 9a. Calculated surface temperature versus time for carbon steel Cf 53 (——— one-dimensional heat flow with temperature-averaged thermophysical properties, ----- one-dimensional heat flow with constant thermophysical properties, -.-.- one-dimensional heat flow with temperature-dependent thermophysical properties.

dimensional calculation of carbon diffusion from a layer of cementite into an adjacent laser of ferrite. To permit the concentration of diffusion carbon to be determined approximately as a function of distance and time, the differential equation to be solved is

$$F(x,t) = \frac{C(x,t)}{C(0,t)} = 1 - \text{erf}\left(\frac{x}{2(Dt)^{\frac{1}{2}}}\right) \quad (18)$$

where $F(x,t)$ defines a dimensionless function, which is the ratio of the initial carbon concentration and the carbon concentration at the interface. The carbon ratio permits the calculation to avoid having to find the exact carbon percentages which will differ from alloy to alloy. At the start, cementite will have a constant carbon ratio value of 2 and ferrite a value of 0. As time progresses, carbon will diffuse across the boundary. If the carbon ratio at the midpoint of the ferrite layer has a value

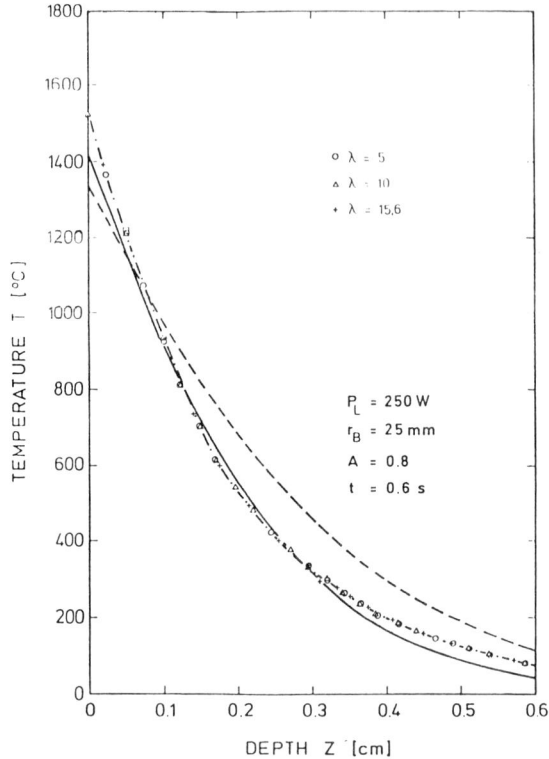

Fig. 9b. Calculated temperature versus depth for carbon steel Cf 53 (——— one-dimensional heat flow with temperature-averaged thermophysical properties, ----- one-dimensional heat flow with constant thermophysical properties, -.-.-. one-dimensional heat flow with temperature-dependent thermophysical properties.

of 0.25 then sufficient diffusion has occurred to produce martensite under adequate cooling rates [4].

A quick-and-simple approach would be to use an average value of D, but a better approach is to use temperature-dependent D or more complex diffusion models [42]. In fact, the best approach will be to solve a three-dimensional diffusion equation with temperature-dependent D, if extreme accuracy is needed. For iron alloys of eutectoid or near-eutectoid composition, where carbon is uniformly distributed in the starting microstructure, mass transport is not very critical since the time required for carbon diffusion is small. However, mass transport will be critical for materials where carbon is not uniformly distributed and where relatively higher soaking times are needed for carbon diffusion. For the characteristic interaction times of laser transformation hardening sufficient carbon diffusion has occurred for all the practical purposes of iron alloys in agreement with extended model calculations [16,41].

RESULTS AND DISCUSSION[2]

Absorptivity

The materials under investigation are iron base alloys subjected to conditions appropriate to surface transformation hardening. The surfaces are coated with paints, graphite, molybdenum disulfide, and iron oxide by spraying, plating, and oxidation in controllable thickness. Assuming no transmission the absorptivity of the coated surfaces is determined indirectly by static measurements of the incident laser power, the diffusely reflected power, and the specularly reflected power by an experimental arrangement described [43,44]. A 10 W, cw, 10.6 µm TEA CO_2 laser under inclination angle of 30° was focused into the center of the interaction region probing on-line the optical properties of the coated materials before, during, and after processing by common modulation techniques [7].

For $I \cong 10^2$ W/cm² $< I_p$ (I_p processing intensity) with intensity-independent absorptivity and thermophysical properties the ordinary reflectivity (specular and diffuse) at the wavelength of the CO_2 laser radiation (λ_L = 10.6 µm) increases in the sequence paints, molybdenum disulfide, graphite, i.e. the absorption coefficient is higher for paints and lower for graphite (Fig. 10). Thus, paints, which have high pigment and low organics, are superior to oxides and sulfides [4,7,43]. The pigments and fillers should be titanium oxide, silicon dioxide, and carbon black, which are all good absorbers for λ_L = 10.6 µm. The organics will only burn away, creating a laser induced plasma absorbing too much power [6].

The absorptivity strongly depends on the coating thickness (Fig. 11) owing to multiple interference effects within the coating and the formation of true solids on the substrate. The influence of refractive index and absorption coefficient on reflectivity and transmission of thin surface layers has been demonstrated [45] for various heterojunction systems by appropriate thin film optics expressions. The influence of composition and mixture on the electronic structure of various surface oxides has been shown [41] as function of temperature and conditions during growth as well as chemical reactions during processing (Fig. 12). In addition, the surface roughness governs the absorptivity [10,43,44] as supported by the reflectivity of surfaces with different finish (Fig. 9) by machining, sandblasting, grinding, or polishing. The surface morphology [35], which originates from the coating growth during deposition, also influences strongly the absorptivity. As to be seen from scanning electron micrographs [35] vapor-deposited tungsten coatings appear as a maze of wavelength-size columnar structures absorbing the incident radiation by multiple reflection [34]. Cupric oxide coatings exhibits a random array of loosely packed needles of about a quarter of the wavelength also with high absorptivity. Manganese-phosphate forms flat platelets with diameter larger than the wavelength resulting as tiny reflectors with lower

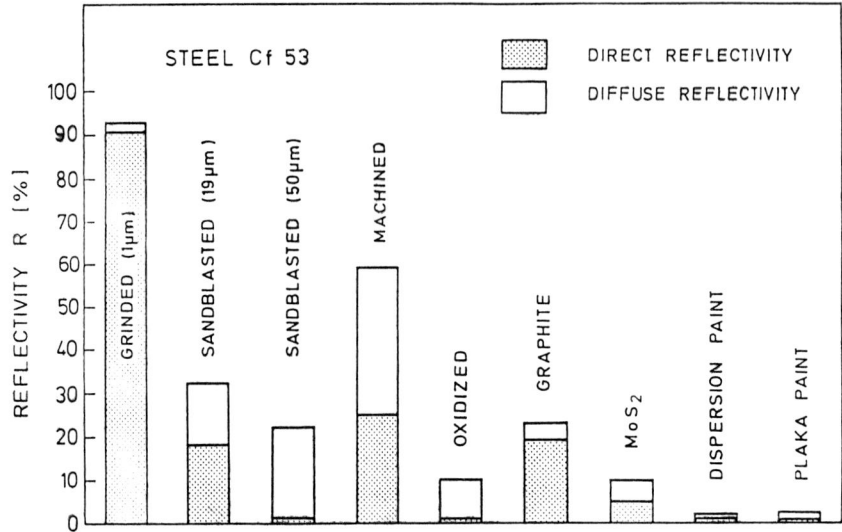

Fig. 10. Reflectivity of uncoated (different preparation) and coated surfaces of carbon steel Cf 53.

energy coupling. For optimum energy coupling the coatings necessarily should be highly absorptive for infrared electromagnetic radiation.

Stability

For $I = I_p > 10^3$ W/cm² with intensity-dependent absorptivity and thermophysical properties the direct reflectivity $R_D=R_p/R_O$ [7,43], where R_O and R_p are the time averaged reflectivities before and during processing, decreases for thin coatings as a function of Ixr_B and remains constant for thick coatings (Fig. 11) as derived from dynamic measurements (Fig. 12). If the coating is too thin, it will be destroyed via evaporation during laser heat treatment with subsequent power reflection at the underlying material surface yielding higher direct reflectivities (Fig. 11). For thick coatings the intensity is in any case too low to evaporate completely the coating, that the ordinary reflectivity of the coated material is measured independent of Ixr_B within the investigated range (Fig. 11). Because the heat transfer to the substrate is enhanced for low coating thickness D (see next section) and the coating absorption becomes higher with D, due to multiple reflection [45] within the coating, the direct reflectivity decreases for low values of Ixr_B with coating thickness (Fig. 11). According to these considerations R_D generally decreases with coating thickness (Fig. 11).

Photomicrographs before and after processing show, that for low values of Ixr_B and relatively large r_B with low intensities $I=P_L/\pi r_B^2$ the coating surfaces remain flat as prepared (Fig. 11),

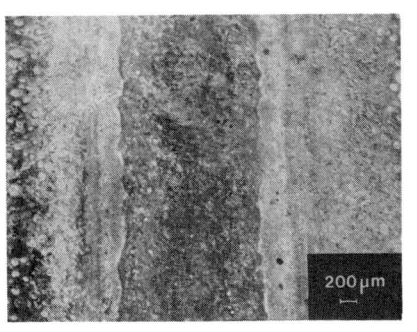

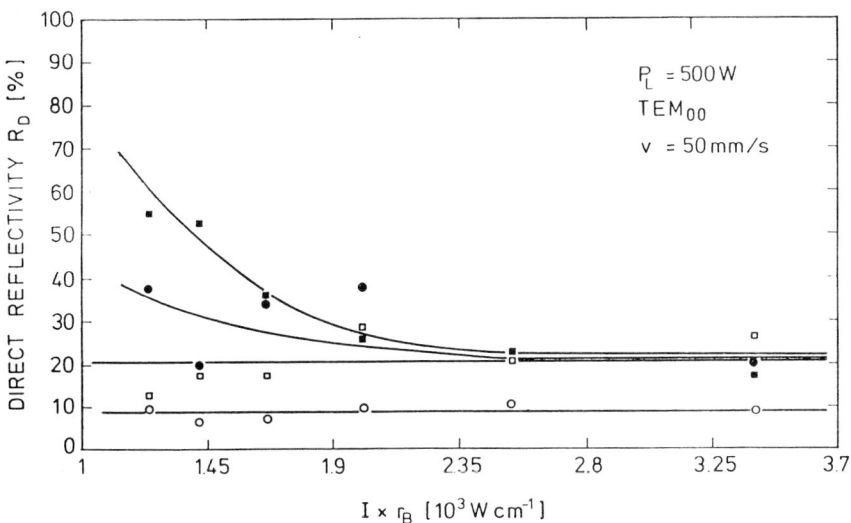

Fig. 11. Direct reflectivity as a function of scaling parameter for carbon steel Cf 53 coated with dispersion paint of different thickness (■ 11, ● 22, □ 33, o 54 μm) and optical micrographs at $1 \lessgtr I_p$.

whereas for high values of Ixr_B and relatively small r_B with high intensities the coating surfaces are roughened (Fig. 11). The direct reflectivity is high due to mainly specular scattering of the probing laser light at the flat surfaces, the direct reflectivity is low for diffuse scattering at the roughened surface (Fig. 11). Thin coatings are less stable than thick ones during prolonged processing, i.e. coatings should be able thermally to withstand high power densities and high temperatures.

The temperatures in the surface region are the result of an energy balance between the amount of incoming laser energy and the amount of thermal energy being conducted away from the surface. If the rate of incoming laser energy is high relative to the rate at which thermal energy is conducted away then temperature gradients are large with high temperatures concentrated near the surface. If the rate of incoming energy is nearly equal to the rate at which thermal energy is conducted away, the temperature gradients are small with lower temperatures, but a much deeper HAZ. Following the experimental observations large temperature gradients assumingly are generated within the coatings especially for high thickness withstanding high intensities and temperatures for prolonged processing. During laser heat treatment it seems reasonable that the coating surface will reach melting or boiling temperature, whereas the coating substrate interface is at considerably lower temperature. Evaporation of the coating, laser-induced chemical reactions with the ambient in the coating, cracking by different dilatation, incipient and complete surface melting change the surface topography of the coatings during transformation hardening depending on intensity and interaction time, at least destroying the coating (Fig. 11 and 12). As to be seen from typical track topographies (Fig. 11) the surface has undergone incipient melting resulting in the production of isolated globules, which form as the result of surface-tension effects. Increasing interaction times produced a progressive increase in both the size and number of globules until eventually they joined together to create a continuous melt [1,22].

Figure 12 shows the dependence of the direct reflectivity as a function of time indicating processing-induced changes of the coating properties. The coatings are roughened immediately after the beginning of processing, the diffuse reflectivity becomes enhanced resulting in a simultaneous decrease of specular reflectivity. The coating additionally is obliterated with processing time, the specular reflectivity subsequently enlarges due to pronouncing power reflection at the alloy surface. During further processing the surface is oxidized, the specular reflectivity measured represents the intensity-independent values of absorptivity (Fig. 10), which is higher for oxide covered alloy surfaces than for non-oxidized ones. During the growth of surface oxides the specular reflectivity decreases further according to the thickness dependence of the reflectivity [45] of the oxide formed or to the formation of different oxides by laser-induced chemical reactions [41]. The resulting surface oxide determines the saturation values of specular reflectivity after prolonged processing (Fig. 12). These informations enable the control of transformation hardening with respect to complexity of hardened patterns by control of coating properties with intensity and interaction time.

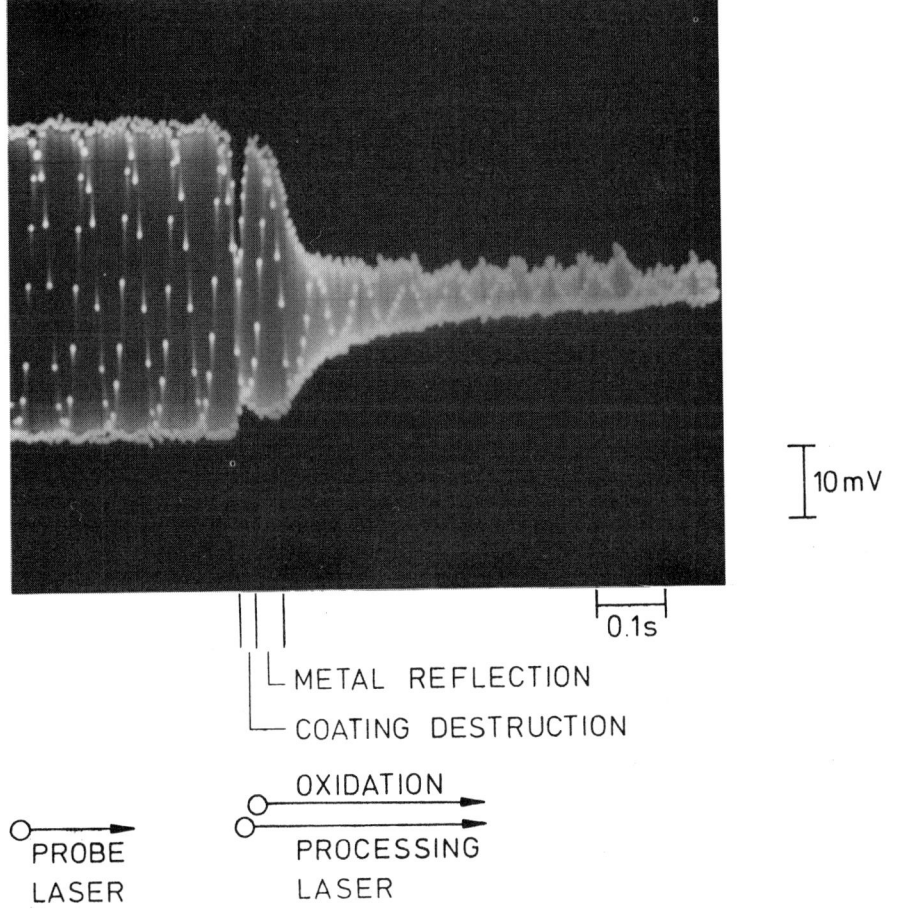

Fig. 12. Reflectivity of dispersion paint coated carbon steel Cf 53 versus time.

Energy coupling

The portion of the optical energy absorbed in the coating, which is transferred into the substrate, was determined by comparison of calculated temperature distributions in the alloy and metallographic examinations like microstructure and hardness. As for all practical purposes, the depth of the hardened zone was assumed to be very near the depth of the transformation temperature A_{c1} (transition temperature isotherm). Including the important processing variables the three-dimensional temperature distributions have been calculated as a function of time by solution of the heat diffusion equation in the approximation of flat semi-infinite body under conditions of temperature-independent averaged thermophysical properties and different low order beam profiles. In the limit of low intensities I<I_p the calculated transition (Fig. 1 and 2) temperature isotherms reveal the geometry of the hardened zone, if the absorptivity A represents the energy coupling coefficient A_{EC} (Fig. 10 and 13). In the limit of high intensities I≅I_p the calculated transition isotherms only reveal the geometry of the hardened zone, if A_{EC}<A is used for the calculations (Fig. 10 and 13). The energy coupling is higher for I<I_p. This is simply a consequence of less thermal stability of the coating at I≅I_p with all the drawbacks to the energy coupling.

The laser power coupled into the material is nearly constant at low values of the scaling parameter Ixr_B (Fig. 14 and 15). Depending on the coating A_{EC} decreases with increasing Ixr_B showing a saturation at high Ixr_B. The high absorptivity of the coating originates for low Ixr_B and relatively large r_B with low intensities in a high energy coupling, whereas for high values of Ixr_B and relatively small r_B with high intensities the coating is destroyed with subsequent power reflection at the metallic substrate (Fig. 11 and 14) as seen from micrographs of the heat-treated surface [7]. This effect is more pronounced for a TEM_{00} mode than for a TEM_{01}* mode because of the higher intensity at the beam centre (Fig 15). For low Ixr_B the hardened depth increases with Ixr_B and for high Ixr_B the counter effect of coating destruction at least originates in the observed saturation unless in a decrease with Ixr_B (Fig. 14). At low Ixr_B more energy is coupled into the substrate for e.g. MoS_2 coated alloys than for graphite coated ones yielding in deeper hardening. At high Ixr_B graphite withstands more effectively high intensities and temperatures, that z_H becomes larger than in the case of MoS_2.

The heat transfer from the coating into the substrate may be quantitatively described as a function of thermophysical properties of coating and substrate as well as laser processing parameters. For appropriate beam handling and forming with reasonably uniform distribution of power over the central region of the beam path the temperature distribution with depth during the temporal duration of the irradiation can be represented by the equations derived for simple, but idealized one-dimensional heat transfer assuming temperature averaged thermophysical properties. The use of high intensities and short interaction times favours the creation of HAZ cross-sections, where the heat flow can be regarded as predominantly one-dimensional and the analysis will accurately predict the temperatures in the heated material. The edges of the cross section are regions where the problem is two-dimensional and the simple picture will not accurately predict the induced temperatures. One must use a more complex description to predict edge effects. Low intensities and long interaction times favour the creation of HAZ cross-sections with three-dimensional heat flow and analysis (see equations (5) to (9)).

Following the model calculations the relative temperature T_B/T_S of temperature T_B at the boundary coating/substrate and temperature T_S at the coating surface decreases with coating thickness independent of the coating type (Fig. 16), i.e. less heat is transferred to the substrate at high D. The heat transfer becomes higher in the sequence molybdenum disulfide, iron oxide, graphite.

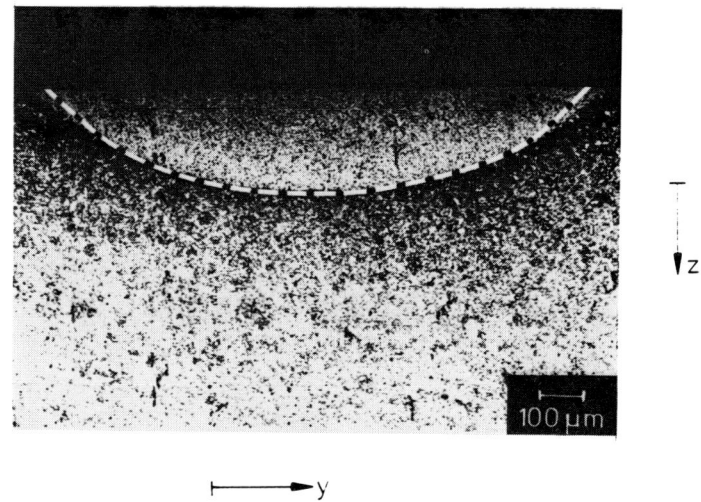

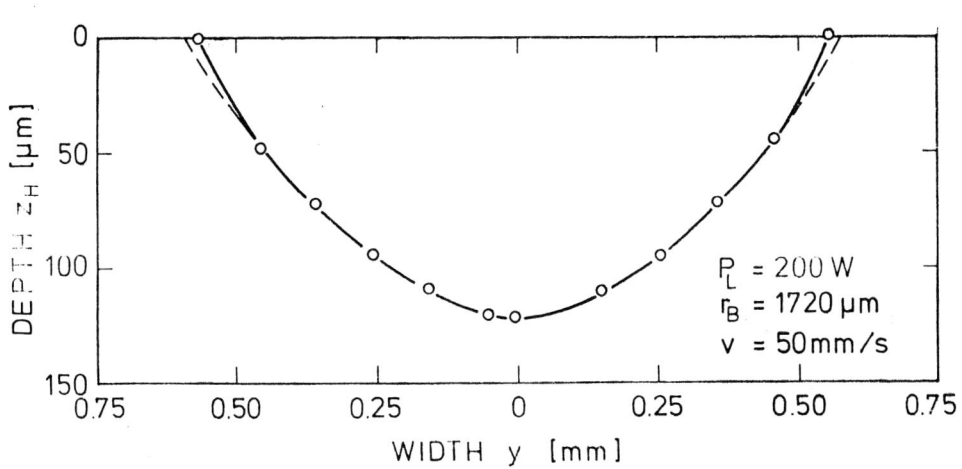

Fig. 13a. Macrograph and hardened depth (o) as a function of hardened width for MoS$_2$ coated carbon steel Cf 53 at I<I$_D$ compared to three-dimensional heat flow calculations (---) with intensity-independent absorptivity A = 0.86.

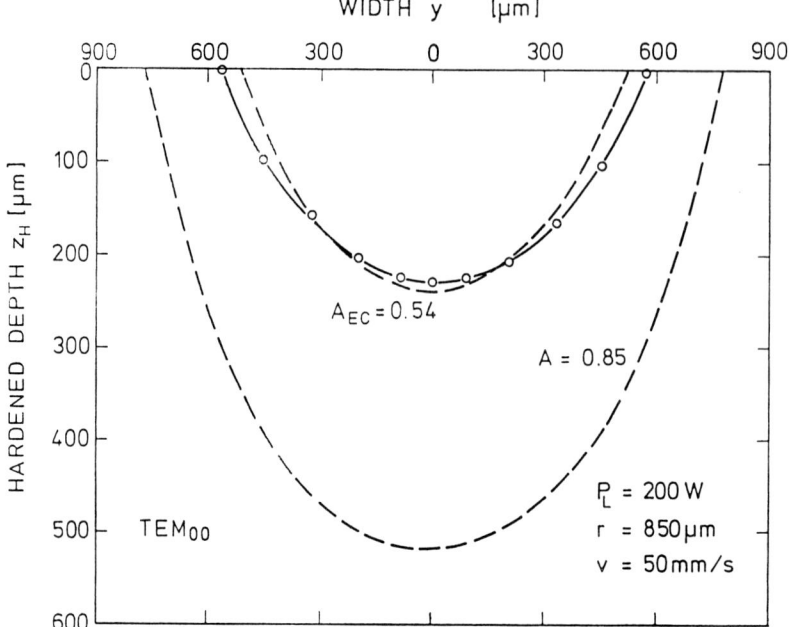

Fig. 13b. Hardened depth (o) as a function of hardened width for MoS_2 coated carbon steel Cf 53 at $I < I_p$ compared to three-dimensional heat flow calculations (---) with intensity-independent (A = 0.85) and intensity-dependent (A = 0.54) absorptivity.

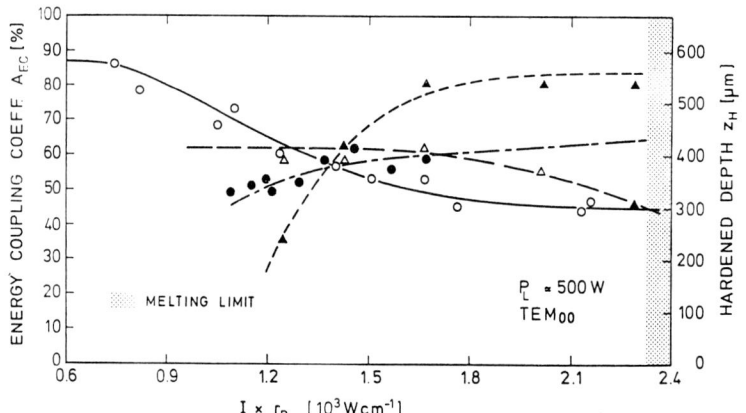

Fig. 14. Energy coupling coefficient (o △) and hardened depth (● ▲) versus scaling parameter $I \times r_B$ for carbon steel Cf 53 coated with graphite (△ ▲) or MoS_2 (o ●).

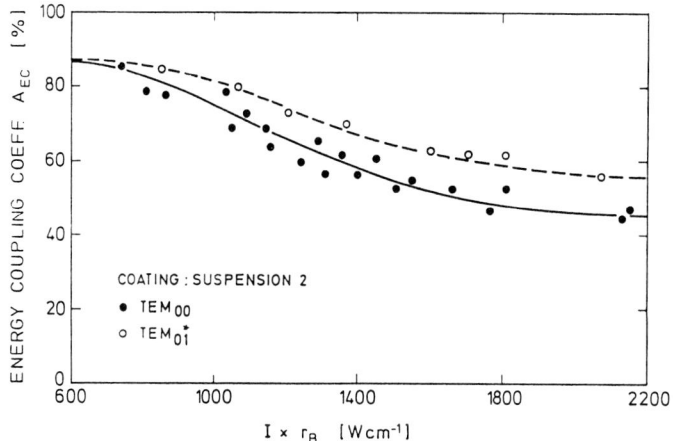

Fig. 15. Energy coupling coefficient for different low order modes as a function of scaling parameter $I \times r_B$ for MoS_2 coated carbon steel Cf 53.

The dissipation of laser energy by three-dimensional heat flow and the absorption of the laser radiation within the coating pronounces with increasing coating thickness that the heat transfer into the material decreases. The part of the laser power coupled into the coating obviously becomes higher with increasing D and the laser power A_{EC} coupled from the coating into the processing material decreases with D that HAZ and z_H decrease with coating thickness. Despite of high thermal stability coatings should be good thermal conductors to conduct the heat with minimum losses into the processing material.

Molybdenum disulfide exhibits a high absorptivity (Fig. 10) but a low heat transfer into substrate (Fig. 16). For paints with high absorptivity (Fig. 10) but low boiling temperature the incoupled energy has a tendency to decrease with increasing laser power (Fig. 14) because of the remarkable destruction of the coating due to evaporation and plasma formation.

SUMMARY

The independent process variables, which affect laser transformation hardening, and incident laser power, beam energy distribution at the surface, interaction time, surface absorptivity as well as thermophysical and metallurgical properties of the material. The dependent process variables are the depth of hardness, the microstructure and the geometry of heat-affected zone, and the metallurgical properties of the transformed material. The use of coatings is inevitably necessary because the coatings start the absorption of the laser radiation and transfer the absorbed energy into the material at wavelengths, where normally most of the incident energy is reflected.

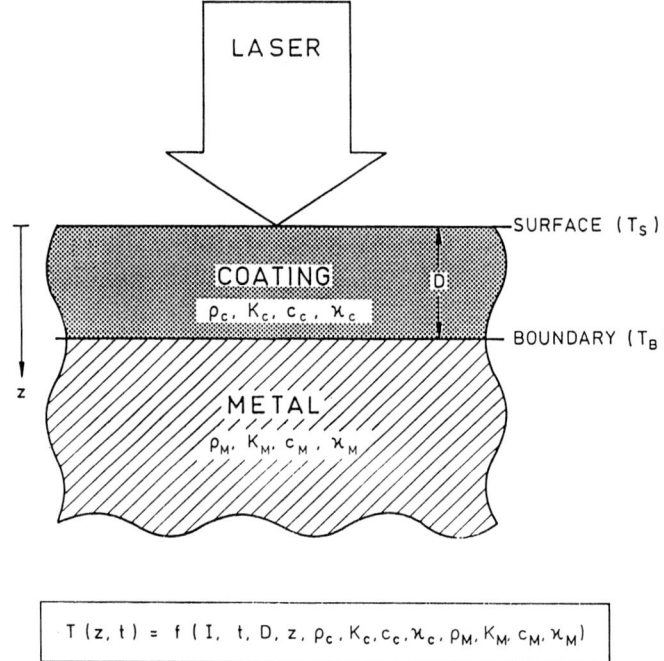
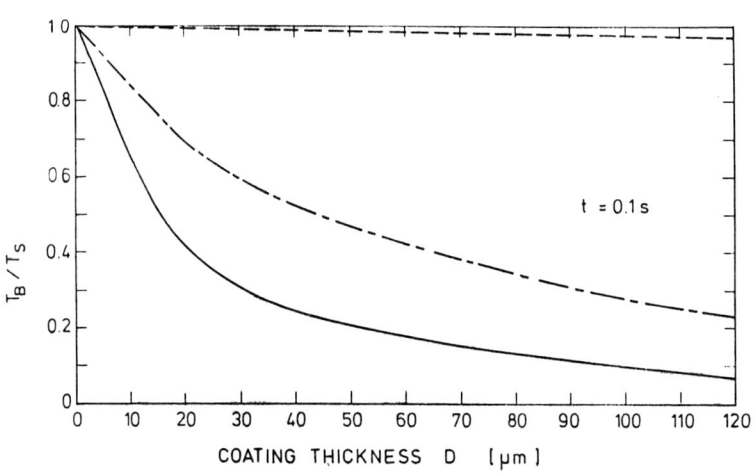

Fig. 16. Schematic representation of coating-metal-arrangement with corresponding thermo-physical properties and relative temperature of carbon steel Cf 53 versus coating thickness (——— MoS$_2$ —·— graphite - - - Fe$_3$O$_4$) obtained by one-dimensional heat flow calculations.

Efficient coatings for reproducible hardening imperatively imply the optimization of the properties constant absorption within the range of processing variables, improved thermal stability at transformation intensities and temperatures, and optimum coupling to the substrate, that the conditions and arrangement of processing will always create the same geometric and metallurgical properties of the heat-affected zone. For $I<I_p$ with the highest absorption achieved for paints the physical and chemical properties of the coatings remain unchanged and the laser power coupled into the substrate is given by the absorptivity and thermophysical properties of the coating. For $I<I_p$ the coating properties are strongly dependent on the interaction time. The intensity, the spot size, and the thermal stability govern the energy coupled into the substrate.

The mathematical models for one-dimensional and three-dimensional heat flow in the approximation of flat semi-infinite body under conditions of temperature-dependent or temperature-independent averaged thermophysical properties and different beam profiles allow to calculate temperature distributions, heating and cooling rates at the surface and at distances below the surface for a given beam profile, laser power, and processing speed. The mathematical models provide an adequate prediction of the hardened case depths facilitating rapid process optimization. Full modelling of laser transformation hardening should add more details of the metallurgical response accounting for the transformation kinetics and the diffusion mechanism involved.

ACKNOWLEDGMENT

The authors appreciate the stimulating interest and active support of Prof. Dr. G. Herziger, Fraunhofer-Institut für Lasertechnik Aachen. They are very indebted to K. Behler and A. Gillner, Fraunhofer-Institut für Lasertechnik Aachen, for making available some experimental results prior to publication. They are much obliged to Mrs. H. Salow for the cooperation in typing the manuscript. They are very grateful to Mrs. H. Mosna for her skilful assistance in drawing the diagrams.

REFERENCES

[1] D. N. H. Trafford, T. Bell, J. H. P. C. Megaw and A. S. Bransden, Metals Technology, 10, 69 (1983) and references therein.
[2] H. Kawasumi, Technocrat, 11, 11 (1978).
[3] D. M. Roessler and V. G. Gregson, Appl. Opt, 17, 992 (1978).
[4] V. G. Gregson, in Laser Materials, Processing M. Bass, Ed. (North-Holland Publ., Amsterdam, New York, Oxford 1983) 203.
[5] G. Herziger, Feinwerktechnik & Meßtechnik 91, 156 (1983).
[6] E. Beyer, L. Bakowsky, R. Poprawe and G. Herziger, in Optoelectronic in Engineering (Springer-Verlag, Berlin, Heidelberg, New York, Tokyo 1984) 367.
[7] K. Behler, A. Gillner, G. Herziger, E. W. Kreutz and K. Wissenbach, in Induced Defects in Insulators (Les Editions des Physique, Les Ulis 1984) 47.
[8] D. S. Gnanamuthu, C. B. Shaw, Jr., W. E. Lawrence and M. R. Mitchell, American Institute of Physics Conf. Proc. 50, 173 (1979).

[9] P. Henry, T. Chande, K. Lipscombe, J. Mazumder and W. M. Steen, Laser Institute of America Proc. 31, 25 (1982).
[10] Y. Arata and I. Miyamoto, Department of Welding Engineering Osaka University II W Doc. IV-50-71 (1971).
[11] J. H. P. C. Megaw, Surfacing J. 11, 6 (1980).
[12] H. Schumann, Metallographie (VEB Deutscher Verlag für Grundstoffindustrie, Leipzig 1980).
[13] K. J. Albutt and S. Garber, J. of the Iron and Steel Institute 12, 1217 (1966).
[14] H. Schlicht, Härterei Technische Mitteilungen 29, 184 (1974).
[15] G. Stähli, Härterei Technische Mitteilungen 34, 55 (1979).
[16] K. Wissenbach, thesis, TH Darmstadt (1985).
[17] S. M. Copley, Laser Institute of America Proc. 31, 1 (1982).
[18] C. Courtney and W. M. Steen, Proc. Int. Conf. on Advances in Surface Coating Technology 1, 219 (1978).
[19] J. Mazumder, J. of Metals 35, 18 (1983).
[20] G. Ripper and G. Herziger, Feinwerktechnik & Meßtechnik 92, 6 (1984).
[21] L. Bonello and M. A. H. Howes, Heat Treating 12, 32 (1980).
[22] E. Beyer, P. Daab, E. W. Kreutz and K. Wissenbach, to be published.
[23] H. Eichler and G. Herziger, Zeitschrift für Angewandte Physik 23, 297 (1967).
[24] P. Loosen, L. Bakowsky, G. Herziger and F. Rühl, in Optoelectronics in Engineering (Springer-Verlag, Berlin, Heidelberg, New York, Tokyo 1984) 247.
[25] P. Loosen, L. Bakowsky and G. Herziger, Feinwerktechnik & Messtechnik 92, 1 (1984) and references therein.
[26] H. Kurz, L. A. Lompré and J. M. Liu, J. de Physique Colloque C5, 23 (1983).
[27] P. Mulser, Zeitschrift für Naturforschung 25a, 282 (1970).
[28] L. V. Keldish, Sov. J. Exp. Theor. Phys. 20, 1307 (1965).
[29] B. A. Tozer, Phys Rev. 137A, 1665 (1965).
[30] P. Baeri, U. Campisano, G. Foti and E. Rimini, J. Appl. Phys. 50, 788 (1979).
[31] R. F. Wood and G. E. Giles, Phys. Rev. B23, 2923 (1981).
[32] F. O. Olsen, Deutscher Verband für Schweißtechnik Berichte 63, 197 (1980).
[33] E. Beyer, K. Wissenbach and G. Herziger, Feinwerktechnik & Meßtechnik 92, 141 (1984).
[34] H. G. Craighead, R. E. Howard and D. M. Tennant, Appl. Phys. Lett. 37, 653 (1980).
[35] S. L. Engel, Society of Manufacturing Engineers Dearborn USA MR76, 857 (1976).
[36] H. S. Carslaw and J. C. Jaeger, Conduction of Heat in Solids (Oxford University Press, London, New York 1959).
[37] J. F. Ready, Effects of High Power Laser Radiation (Academic Press, New York 1971).
[38] W. W. Duley, CO_2 Lasers: Effects and Applications (Academic Press, New York 1976).
[39] B. H. Kear, E. M. Breinan, L. E. Greenwald and C. M. Banas, Society of Manufacturing Engineers Dearborn USA MR76, 867 (1976).
[40] S. Kou, S. C. Hsu and R. Mehrabian, Met. Trans. 12B, 33 (1981).
[41] K. Wissenbach et al., to be published.
[42] L. C. Brown, J. Appl. Phys. 47, 449 (1976).
[43] K. Wissenbach, L. Bakowsky and G. Herziger, Feinwerktechnik

& Meßtechnik 91, 327 (1983).
[44] K. Wissenbach, L. Bakowsky, H. G. Treusch and G. Herziger in Optoelectronic in Engineering (Springer-Verlag, Berlin, Heidelberg, New York, Tokyo 1984) 312.
[45] O. S. Heavens, Optical Properties of Thin Solid Films (Butterworths, London 1955).

Chapter 4

LASER MACHINING

LASER CUTTING
Michel Querry
Manufacturing Processes Research Laboratory (L.E.P.F.) and C.A.L.F.E.T.M.AT. INSA — Lyon, 20 Av. A. Einstein, 69621 Villeurbanne, France

For the last twenty years, the use of lasers for the forming of materials has continued to increase, in a particularly rapid manner, by the multiplication of the fields of application, the increase in the reliability of still-young technologies, and the development of different techniques.

In the field of forming by the removal of material, the laser would appear to be a multi-purpose instrument, bringing original solutions to problems of technical feasibility, the quality of the manufactured product, and economic competitiveness. The association of the laser with computer systems controlling the complete production installation and, particularly in the case of cutting to shape, the use of digitally-controlled movements have resulted in the concept of the "laser-tool", whose careful use and control improve the efficiency of production lines and reduce their costs.

Among the properties by which lasers, as light sources, are distinguished, are:
- power and/or energy emitted,
- mono-chromaticity,
- coherence in time and space,
- directional control,

the specific applications in the field of materials make use essentially of the very high brilliance of the source, associated with the mono-chromaticity and the low divergence of the resultant beam. These properties allow the beam to be focused on a very limited area of the part to be machined. Thus, the principal characteristic of the laser with respect to the forming of materials lies in its capacity to concentrate a high energy on a very small working surface, and thus produce particularly a high light flux (or power density), generally of the order of 10^4 to 10^8 $W.cm^{-2}$.

The industrial applications have been abundantly illustrated by the particular studies published in technical literature, and therefore we will not be compiling here a review of the different technologies; we shall present, in the following pages, a summary of the principal characteristics of the process of cutting by laser, and the principal significant features.

I - DESCRIPTION OF THE PROCESS

Cutting by laser results from the association of a laser source of high power with a system controlling the movement of the beam, its focusing on the part, the relative movement of beam and part, and the shielding gas in the vicinity of the point of impact.

The choice of the source is dependent on the power density required (10^5 to 10^7 W/cm², Fig. 1) to achieve the temperature of fusion (or of vaporisation in certain cases) of the material to be cut.

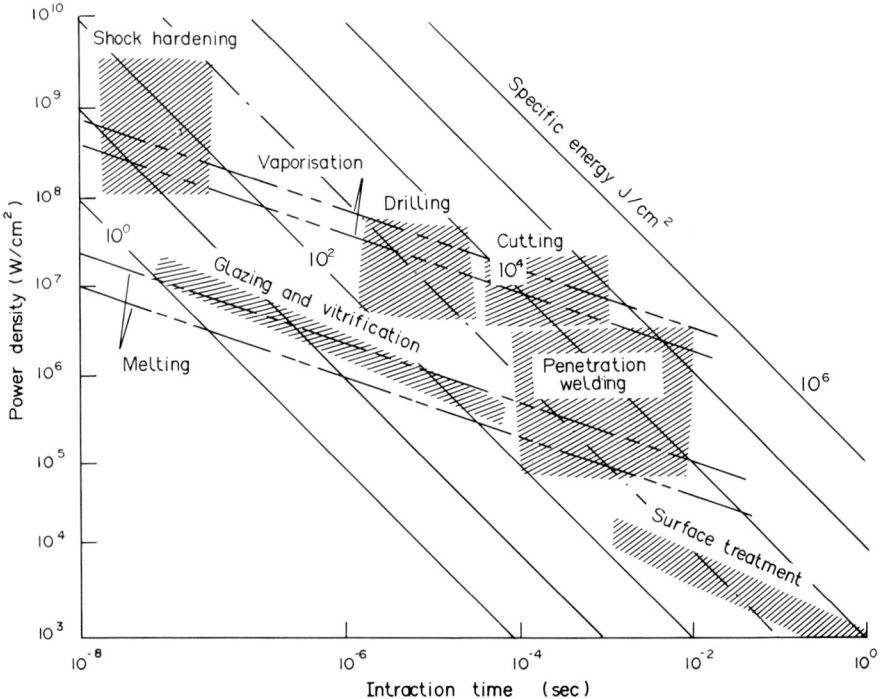

Fig. 1. Diagram of the operating ranges of lasers and the different groups of applications.

Since no source is capable of providing such a light flux, the use of the beam necessitates the introduction of an optical focusing system, concentrating the energy on a focal spot of small area (Figs. 2 and 3). This amplification, of several orders of magnitude, is nevertheless limited by theoretical

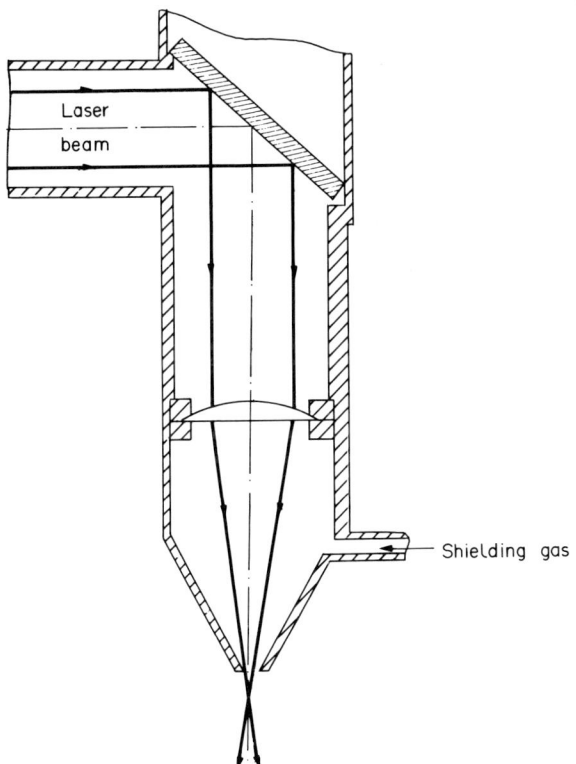

Fig. 2. Lens focusing system.

factors (diffraction, the existence of aberrations), and by technological ones (resistance of the optical components to a very high light flux); sources which are suitable for the cutting of materials, particularly metals, must thus furnish high powers, of the order of 10^2 to 10^4 watts. Among the very large number of sources developed, only two types have, at the moment, the characteristics both of output power and of reliability which are required by the industrial laser-tool (Fig. 4):

* CO_2 source (gas laser),
  generally continuously-emitting
  wave length λ = 10,6 µm (middle of the infrared range);

output power up to several kilowatts for a reliability of industrial standards, and up to several tens of kilowatts if a lower reliability is accepted (laboratory or research and development **applications**);
energy efficiency in practice of the order of 10%.

* YAG Source (solid laser),
  generally pulsed emissions (from 0,1 to several tens of milliseconds);

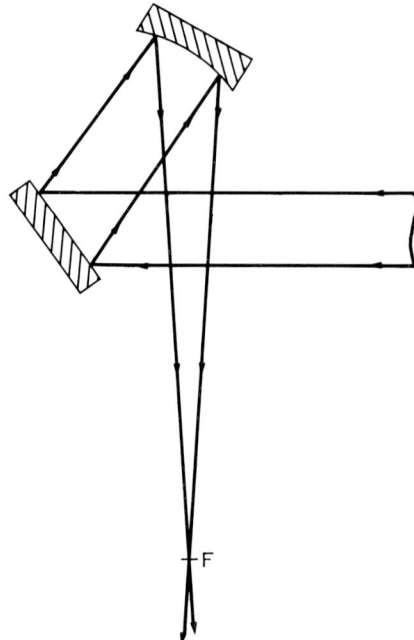

Fig. 3. Mirror focusing system.

wave length λ = 1,06 µm (close to infrared);
mean power up to 400 watts, but maximum power can be very much higher, dependent on the energy and the duration of the pulses; overall efficiency of the order of several per-cent.

In fact, experience has shown that, for drilling the pulsed YAG laser is more suitable, while cutting is associated essentially with the continuous CO_2 laser (Fig. 5), which will be the only one considered in the remainder of this study.

The focusing of the beam, referred to above, is achieved in two different ways:

* by a dioptrical system, essentially a convergent lens; the wavelength of the CO_2 laser enforces the use of particular materials for the making of these optical devices (essentially zinc selenurium ZnSe, with an anti-reflecting treatment), of high cost (e.g., 700 to 1500 $ for a 30 mm diameter lens, which accepts a beam of 1 to 2 kW), and with a limited life (a few hundred hours); beyond several kilowatts, the use of such lenses becomes problematical;

* by a catadioptric system, using **suitable metal mirrors** (copper, molybdenum), usually with a surface treatment; of a lower cost than the lenses, these mirrors have the disadvantage of a higher light absorption; their use is limited to the focusing of beams of high power, or to the shaping of the

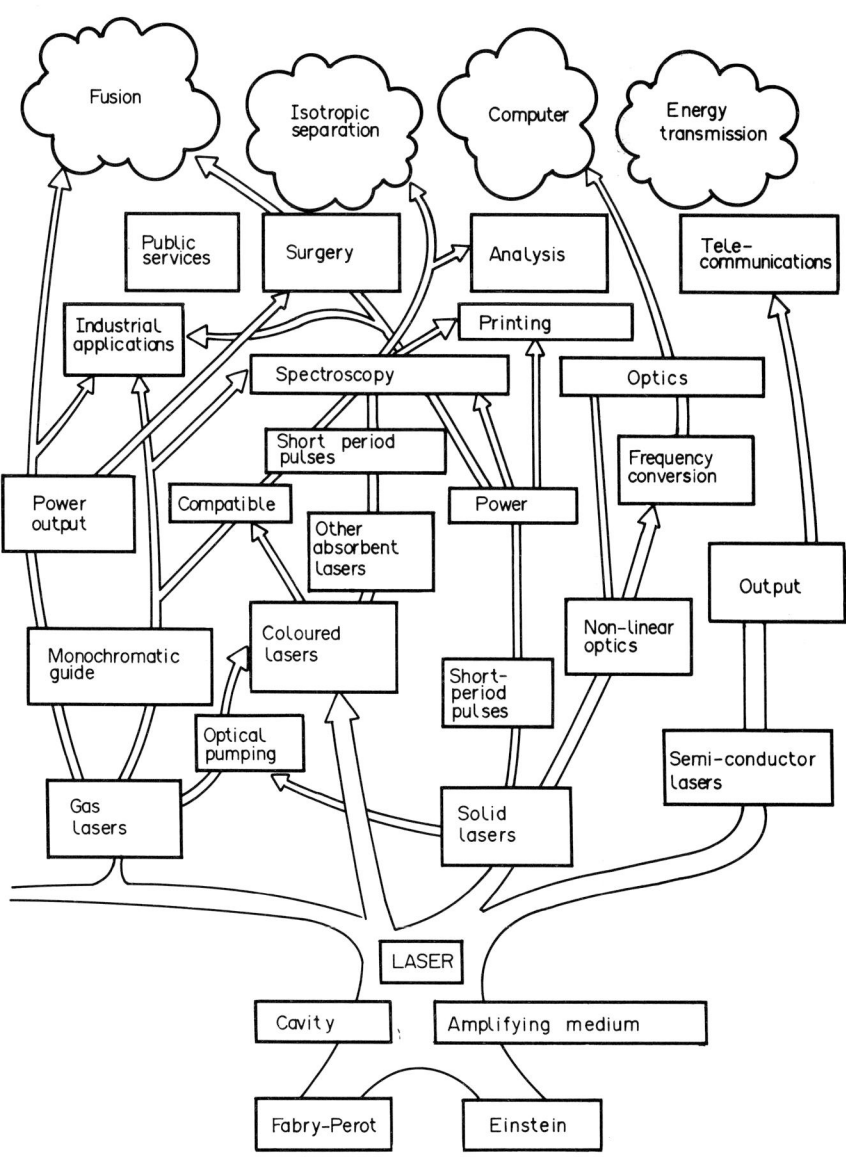

Fig. 4. The laser family tree (by B. Decomps).

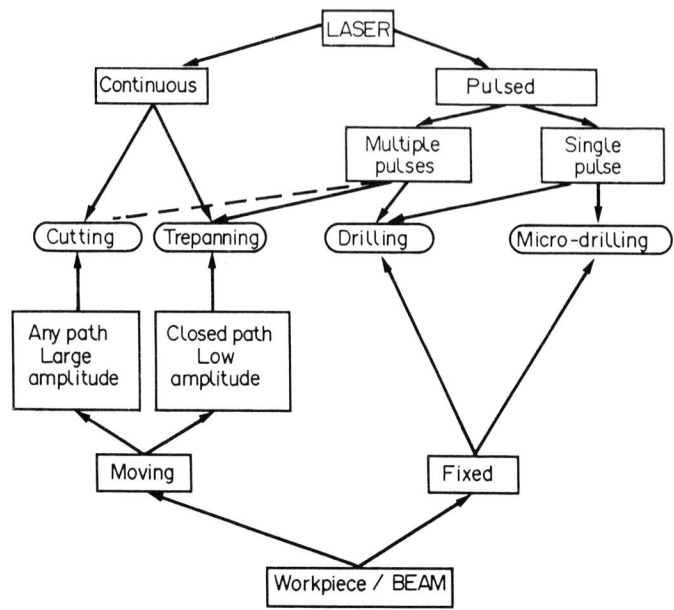

Fig. 5. Classification of the applications.

beam (change of the energy distribution), which is outside the field of cutting (heat treatment applications).

It should be noted that the quality of the focusing, represented essentially by the maximum power density in the vicinity of the focus, is dependent on the distribution of the energy within the beam emitted, and also on its natural divergence; the ideal case is that where the energy distribution is gaussian (mode indicated by TEM_{00}), which is often used as an assumption for calculations. Figure 6 shows, in tabular form, the principal optical parameters, and several orders of magnitude. The CO_2 lasers, whose power is less than 1 or 2 kW, frequently operate in this mode, while for higher powers, the technology available provides mainly multi-mode sources, not necessarily resulting in the productivity being proportional to the power employed.

The system for the transmission of the beam provides for the transfer of the light energy from the outlet window of the source to the work position(s) equipped with a focusing system. The first requirement to be complied with in industrial applications concerns the safety of nearby workers. The beam must be totally inaccessible - it is to be enclosed along its complete length. The source may thus be relatively remote from the work position; the limitation is associated with the natural divergence of the beam and its compatibility with the opening of the optical systems located downstream: for a CO_2 laser, a divergence of 1 milliradian is an acceptable order of magnitude; a beam diameter of 20 mm at the outlet from the source will be doubled after 10 m travel.

Fig. 6. : Optical parameters

Assumptions : TEM_{00} mode beam

- Gaussian energy distribution
- Divergence : limited by diffraction
  1/2 angle : $\theta = 2/\pi \cdot \frac{\lambda}{d}$
- Convergent lens aperture d - Focal length f - No aberrations

Limiting case :

- Focusing of the beam on a spot of radius r
  if $f \cong d$, $r \cong \lambda$
- Power density at the centre of the spot :
  $$I = \frac{\pi}{2} P \frac{d^2}{\lambda^2 f^2}$$
→ f low
- Depth of field ($r \cong cste$)
  $$z = \frac{4}{\pi} \lambda \frac{f^2}{d^2}$$
→ f high

The concept of the transmission of the beam allows installations with multiple work positions, similar or otherwise, to be envisaged, supplied by a single laser source. With this arrangement, a high rate of use of the source can be achieved, by carrying out the ancillary operations (loading, positioning, unloading, etc) in non-productive time, or by taking advantage of the installation of a major power source.

The relative movements of the part and the beam, in the case of a continuous laser, define the duration of the interaction of beam and material. For this reason, the speed of the displacement of the beam constitutes an important parameter in the process.

For the displacement, very diverse configurations of machine are called upon, which can be summarised by classifying them in 3 groups :

- fixed beam, moving part, i.e. fixed cutting head and moving work table;
- fixed part, moving beam, by means of mirrors moving from side to side and/or rotating;
- mixed, generally when there is a requirement for more than 2 degrees of freedom (non-planar parts).

These elementary movements are achieved and synchronised on current cutting machines by digital automatic systems ; their complexity and sophistication can result in the cost of the associated machine being of the same order of magnitude as that of the laser source.

The introduction of cutting by laser requires the use of a shielding gas: a stream of pressurised gas, generally coaxial with the focusing system, provides an aerodynamic means for the evacuation of the molten pool which forms in the impact zone of the beam; this gas can:

*either have a purely mechanical action, as in the case of inert gases (helium or argon principally), which in addition provide chemical protection of the surface against the risk of oxidation or combustion during the cutting.

*or, most frequently during the cutting of metals, combine the mechanical action with a thermo-chemical oxidation process, if air or oxygen is used; in this case, the formation of oxides assists the absorption of the incident radiation by the material being cut, and the energy produced by the exothermal reaction accelerates the process.

In addition, the shielding gas, active or inert, protects the optical system against any vapour which might flow back, or even liquid splashes which, by intense absorption of the radiation, would result in the instantaneous destruction of the equipment.

II - EXAMINATION OF THE PARAMETERS

I - PRINCIPAL PARAMETERS

A study of the available publications will reveal the parameters which play an important role in the achievement of rapid, high-quality cutting.

Figure 7 shows the focusing of the beam on a part to be cut; the parameters which determine the cutting conditions, indicated on the diagram, are as follows:

- pressure, flow rate and temperature of the shielding gas
- geometry of the nozzle (diameter, length and shape of the orifice)
- focal length: f
- distance from the surface of the part to the focal point: D
- distance from the nozzle to the surface of the part: d
- thickness of the material: e
- speed of the laser: v
- power of the laser: P
- distribution of the energy in the beam, dependent on the mode of the source
- polarisation of the beam

It will be noted that the possibility of monitoring and of modifying the majority of these parameters depends on the design of the cutting head.

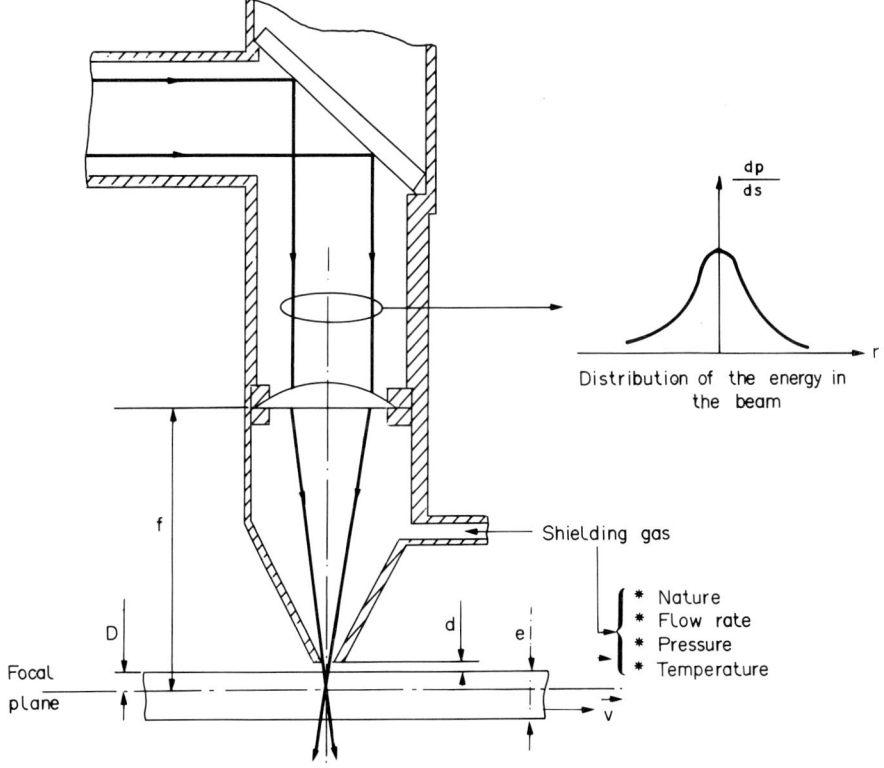

Fig. 7. Principal parameters.

2 - THICKNESS/SPEED/POWER RELATIONSHIPS

These three parameters are closely linked by the thermal effects associated with the interaction between the beam and the material:

- fusion and vaporisation of the material being worked, determined by their latent heats,
- heat transfer, by conduction through the part, by convection in the molten pool formed and in the shielding gas, and by radiation from the surfaces which have been heated to high temperatures.

Numerous models have been developed, mainly in the welding or heat treatment fields; the complexity of the **phenomena does not** make them very suitable in the case of cutting. Nevertheless, a very simple model can be used to obtain a first approximation of the relationship between the three technical parameters, thickness - speed - power:

$$v = \frac{Pabs}{Q.l.e}$$

Pabs = power absorbed by the part

l = width of the cut
Q = energy necessary to heat and melt a unit volume

$$\text{or } v = \frac{P \cdot \eta}{C_p \cdot T_f \cdot l \cdot e}$$

P = incident power
η = efficiency, dependent on the absorption of the radiation by the surface, on its roughness, and on the nature of the shielding gas
C_p = specific heat of the material
T_f = temperature of fusion

This model reveals 3 characteristics:

* proportionality between v and P
* inverse relationship to the thickness being cut
* influence of output terms, liable to alter the preceding conclusions profoundly, dependent on the conditions under which the process is applied.

A bibliographical summary is illustrated by the figures:

- 8 : maximum cutting speed, with respect to the thickness being cut, for medium power values (0,2 to 1,2 kW) and for unalloyed or low alloy steels;

- 9 : maximum cutting speed, with respect to the incident power, for different thicknesses;

If the properties deduced from the preceding model are qualitatively reflected by the experimental results, these nevertheless call for several remarks:

* the results are not immediately comparable to one another, because of differences in the operating conditions (focal length, injection of oxygen, undefined peculiarities, powers and thicknesses).

* a large dispersion of the maximum speeds will be noted, particularly for the smallest thicknesses (although in this area the data are the most numerous). This dispersion arrives mainly from the high number of secondary parameters.

Miyazaki et al. have suggested a mathematical model, obtained by producing a smooth curve from experimental results for the cutting of steels, which is expressed as:

$$v = 3{,}5 \cdot e^{-0.56} \cdot P^{0{,}5}$$

(v in m/min, e in mm, P in kW). This model is valid only for thicknesses of less than 6 mm, and we have used the data extracted from the bibliography consulted to produce a model of the same type **(power function with two variables); 69** sets of data, covering the range defined by 1 < e < 50 mm and 0,25 < P < 15 kW, for unalloyed or low alloy steels, are processed by a multiple-variable regression program; the smoothing is performed on the basis of the method of least squares, resulting in the model:

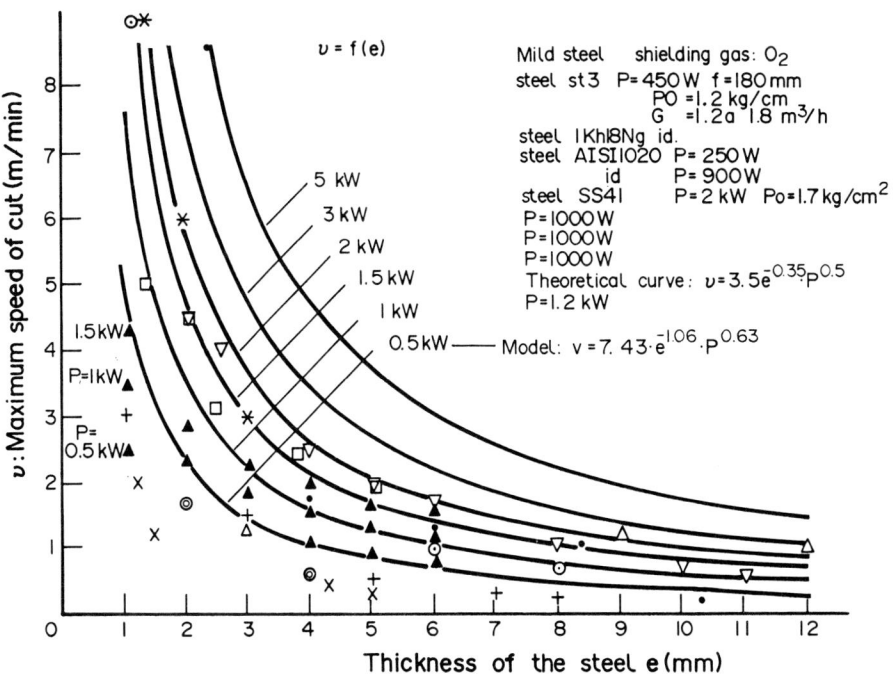

Fig. 8.

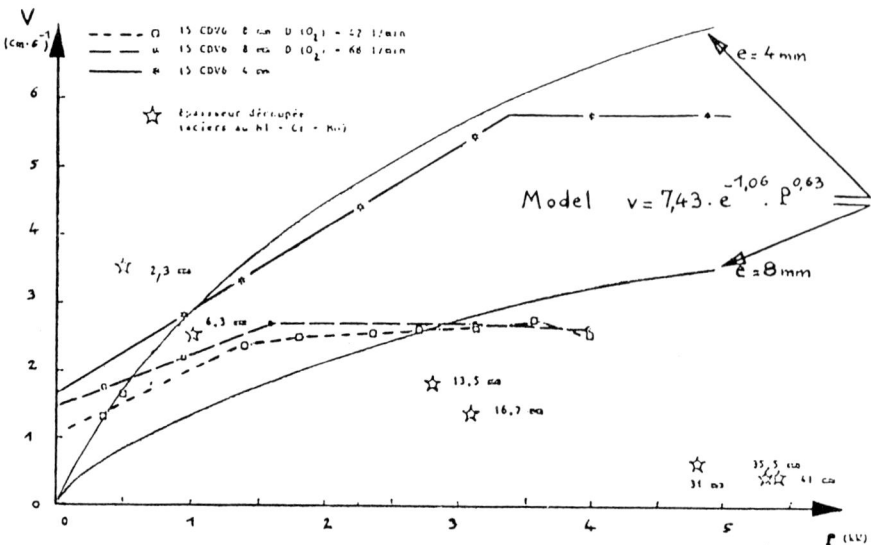

Fig. 9. Limiting speed of cut for a steel 15 CDV 6 (experimental results by Gerbet - ETCA).

$$v = 7,43 \cdot e^{-1,06} \cdot P^{0,63}$$

(v in m/min, e in mm, P in kW).

This model is represented graphically by the curves in Figures 8 and 9, and calls for several remarks :

* the model seems to be "optimistic", in other words, gives high speeds for the low thicknesses (e < 2 mm); beyond that, the agreement between the model and experimental results seems satisfactory.

* this model agrees fairly well with that proposed by Miyazaki for thicknesses of 4 to 6 mm.

* the speed varies roughly as a function of 1/e, which is very frequently to be found in the references.

* the model does not reveal the phenomenon of the "saturation" of the speed, when the power increases beyond 3 to 4 kW.

To conclude, this model remains imperfect, but is fairly representative of the variables e and P ; it can be used as a basis for estimates, bearing in mind that the operating conditions for the process (nature of the part, shielding gas, etc) can have a considerable influence on the results.

3 - THE ROLE OF OXYGEN

The use of a stream of gas in conjunction with the laser beam improves the cutting performance, whatever the material being cut; the inert gases (nitrogen, argon, helium) or compressed air are suitable for the cutting of non-metallic materials; conversely, for the cutting of metals, oxygen is most frequently used. As with standard oxygen cutting, the laser maintains the exothermal reaction of the oxidation of the metal by the stream of oxygen; the main difference lies in the dimensions of the area on which the incident thermal flow is concentrated (typical values for the diameter : 3 mm for an oxygen cutting flame, 100 µm for a CO_2 laser). The addition due to the exothermal reaction may be as much as 70% of the energy involved in the cutting, the remaining 30% being supplied by the laser source which initiates and maintains the phenomenon. Under these conditions, an excess of oxygen and/or too low a relative beam/part speed can result in a runaway of the reaction (self-burning), characterised by a very wide cut and unacceptably distorted surface conditions; conversely, an injection of cold oxygen at a high flow rate can reverse the thermal balance-sheet for the operation, resulting in too great a cooling of the lower layers of the part ; resolidifying of the metal (and/or oxidation) takes place behind the zone of activity of the beam, and cutting is therefore incomplete.

Analysis of the role of the oxygen has led to the introduction of the preheating of the cutting gas; the performance of the process is thereby improved, but at the cost of considerable complexity in the cutting head. Since the shielding gas has the role of protection and frequently cooling of the optical system, the preheating of the cutting gas demands the introduction of a second gas for the thermal protection of the lens; it would seem that this variant of the process, although interesting, has no industrial potential.

The consulted publications quote oxygen injection pressures of 1 to 5 bars; the upper limit would seem to be associated mainly with the strength of the lens which encloses the gas injection chamber. The flow rates are very rarely mentioned; the values which are published lie in the range 1 to 4 m^3/hr; the maximum speed of cut increases with the mass flow of the gas, but this effect seems to find a ceiling at high flow rates. The effectiveness of the stream of gas appears to be linked to its velocity rather than to the pressure existing in the cutting zone ; the ejection of the molten pool (metal + oxides) seems due essentially to a transfer of quantity of movement, rather than to pressure forces. From this point of view, the desirability of a supersonic flow is demonstrated, and this point is made by several authors. In addition, the establishment of such conditions is accompanied by a shock wave whose role could be significant; the existence and the stability of this shock wave are dependent on the geometry of the oxygen injection nozzle, coaxial with the beam ; the majority of designs are of the convergent/divergent type (or convergent/parallel/divergent), with a throat diameter of 1 to 2 mm, resulting from a compromise between the overall dimensions of the beam and the gas flow velocity. In the absence of a systematic study of this particular point, the published opinions are fairly divergent, particularly concerning the effect of the gas flow on the surface condition of the cut.

The distance from the nozzle to the part must be as short as possible, for reasons of gas economy; however, under these conditions, the flow rate varies considerably with the distance, being susceptible to fluctuations due to geometrical defects in the part. Studies of the mechanics of fluids relative to the impact of a jet on a surface indicate that this variation becomes a minimum when the nozzle/surface distance becomes greater than a quarter of the diameter of the nozzle outlet duct. Cutting under these conditions provides a satisfactory compromise between the speed of cut, gas economy, and the stability of the process.

4 - OPTICAL PARAMETERS

The focal length is the distance between the lens and the focusing point of the beam, The publications mention focal lengths of between 63,5 and 254 mm (2,5 to 10"). It is necessary to achieve a compromise between a short focal length, which results in the irregular cutting of large thicknesses, due to the rapid divergence of the beam on either side of the focal plane, and a long focal length, representative of a considerable diffraction of the beam and thus a reduction in the power density available in the focal spot.

The optimum position of the focus in relation to the surface of the part is indicated on Fig. 10, which shows the penetration of the beam into the material, independently of the ejection of the latter; the results for cutting indicate a relationship of the same type.

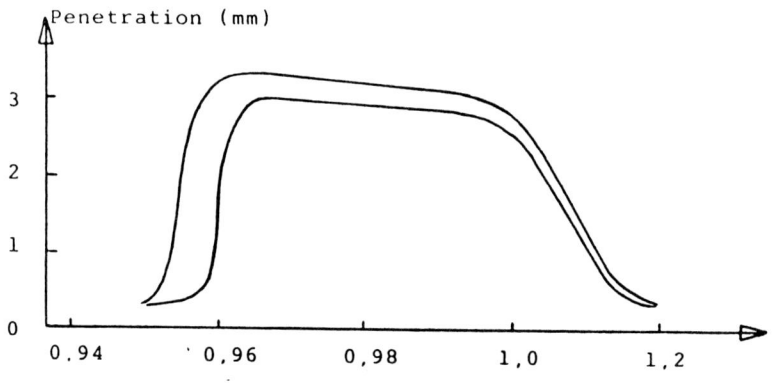

Fig. 10. Penetration of the molten zone.
Focal lengths from 50 to 150 mm - from ARATA.

5 - CLASSIFICATION OF THE CUTS

Observation of the macro-geometry of the cuts has led ARATA to suggest a classification (Fig. 11) which is an indication of its quality ; the five classes are interpreted in terms of the two

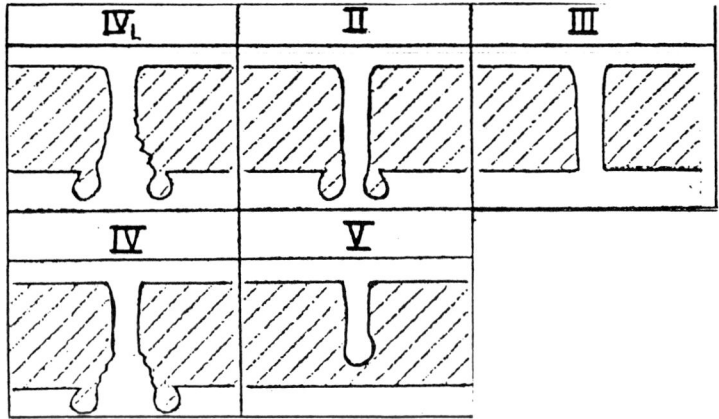

Fig. 11. Classification of the cuts (from ARATA).

parameters "speed of cut" and "thickness" : for the lowest speeds and thicknesses the cut is irregular, and contains holes whose diameter is greater than that of the focal spot (CLASS I). As the speed increases, the edges of the cut tend to become parallel, with oxides present on the outlet face (CLASS II). The latter then tend to disappear (CLASS III). Beyond this, an increase in the width at the bottom of the cut is to be noted (CLASS IV). For even higher speeds, cutting becomes impossible (CLASS V), due either to the incomplete penetration of the beam, or to a resolidification beyond the focal spot.

Figure 12 shows the ranges over which the different classes extend, for an unalloyed steel and for 2 levels of power; the usable range is, in general, that in Class III, and possibly classes II and III together, dependent on the quality required.

It will be noted that the notion of the maximum speed of cut, envisaged earlier, which constitutes a major part of the published results, is in fact the boundary between ranges III and IV; the determination of this boundary is sometimes difficult. This notion constitutes only a small part of what must be considered as the "identity card" of a material with regard to cutting by laser. Range III is generally a curved-sided triangle, one acute angle of which points towards the large thicknesses; beyond a maximum thickness, range III ceases to exist and the cut passes directly from type I to type IV. By increasing the power, range III can be extended towards the greater thicknesses.

III PROSPECTS

Developments of the laser-tool are dependent on foreseeable changes in technology, at any rate in the medium term. The principal change likely consists of the introduction on an industrial scale of increasingly powerful sources, either extrapolated from existing installations which have reached a sufficient level of reliability, or derived from sources in the

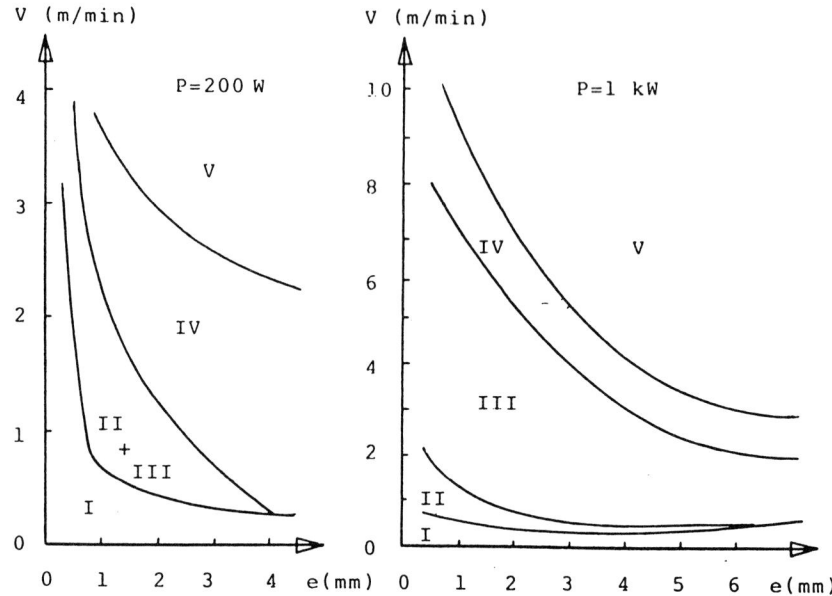

Fig. 12. Quality of the cuts (from ARATA).

in the course of development in the laboratory. Such could be the case for iodine or ammonia lasers, both with emissions in the infrared range, or for excimer lasers which, with short-wavelength emissions, could allow the difficult problem of the high reflectivity of metals to be resolved.

At present, many laser applications have been developed before the fundamental mechanisms of the interaction between the beam and the material have been understood, essentially because such understanding requires a multi-disciplinary base. The disciplines implied - optical absorption, thermodynamics and heat transfer among others - have so far tended to remain relatively shut-off from one another. It would therefore appear desirable and, given the setting-up of Research Centres aimed at applications of power lasers, probable, that effective synergies of different disciplines be generated, leading to some original, high-performance industrial applications.

IV ELEMENTARY BIBLIOGRAPHY

H. Maillet, The laser - Principles and Techniques of its application - Technique et Documentation Paris 1984.

Technical and economic study of the industrial applications of power lasers. Institut de Soudure and SERI - 1977.

Baujoin, J, Techniques of the engineer
Article 2725 : Machining by laser.

The applications of high-energy lasers to material forming - CALFETMAT
Conference summary. Centre d'Actualisation Scientifique et Technique INSA-LYON 1984 and 1985.

Machining Data Handbook - Vol. 2
Art. 12.7 - Metcut Research Associates
Inc. Cincinnati - 1980.

LASER CUTTING USING CNC TECHNIQUES

J. C. Beitialarrangoitia, G. E. Garcia de Vicuna and S. K. Ghosh

Department of Mechanical and Computer-Aided Engineering, North Staffordshire Polytechnic, Beaconside, Stafford ST18 0AD, UK

ABSTRACT

Laser cutting/processing has been accepted as standard cutting technique in a wide variety of industries due to the large number of advantages achieved by this technique in comparison with the conventional methods. Nowadays, there are many manufacturers of laser cutting equipment, and most of them provide their machines incorporated or supported with CNC cutting tables or robots. This integration of CNC systems has become somewhat compulsory to make good use of all the possibilities provided by laser cutting technology.

In this paper an introduction to the laser cutting technology using CNC is given. Furthermore, an overview of the different CNC systems developed for laser cutting/processing applications in the industry is outlined.

INTRODUCTION

Many efforts are being realised in order to improve the utilisation of the laser cutting technique as a cutting procedure for high productivity. Success in these attempts are due to the different investigations that are simultaneously being undertaken considering all the different factors involved in this cutting technique. Those factors can be enumerated as follows:

- physical mechanism of the cutting process,
- properties of the processed materials,
- achievement and control of the Laser Beam Generation system,
- control of the laser delivery system.

As a result of the many improvements obtained since the first applications of laser cutting in the 1960's, this technique has become well known and familiar for most users involved in the cutting industry.

A summary of the most significant advantages and disadvantages of laser processing have been widely discussed in the literature; see References (1-7).

Some of the advantages are listed below:

- significant increases in productivity
- substantial reductions in material processing costs
- reduction in manpower
- higher product quality kept within narrow limits (tolerances)
- reduction or elimination of scrap (material savings)
- reduced or eliminated rework due to its smooth finishes
- improved environmental conditions (reduced noise, vibrations, fumes, chips, lubricating and refrigerating fluids, etc)
- on-line processing (repetitive process parameters)
- fast response of fully automated production systems
- multi-process capability and dissimilar material processing
- adaptability to Flexible Manufacturing Systems (FMS)
- product design changes
- manufacturing system design changes
- flexibility controllability and ease of automation
- interfacing with CADCAM
- does not require (in order to apply its energy) the motion of mechanical means of sufficient mass and rigidity to ensure precision and efficiency to the process
- minimal material distortion
- minimal heat-affected zones (uniform mechanical property)
- non-interference from any electric or magnetic field created by previous processing of the part
- full integration with automated or manual part loading systems.

There are, however, some disadvantages which are given below:

- Laser equipment cost (large capital investment)
- Difficult to install (skilled personnel)
- Maintenance of the equipment (High Down Times) required about every 500 hours.

The main disadvantage is obviously the high initial costs of the laser equipment in comparison with other techniques. However, the trend to lower laser costs is evident and more accentuated than in other cutting technologies, for most new industrial laser devices. As a result of this trend the application of laser cutting equipments is already showing insignificant increases.

A brief and simple explanation of the process developed for the generation of the laser beam and the main components of the laser system have been given by Bannister (Ref. 8). In terms of basic requirements, the laser has three prerequisites, see Fig. 1. These are:

(i)   the laser medium or metastable state;
(ii)   the excitation source; and
(iii)   the optical resonator.

These three ingredients are necessary for the generation of every type of laser beam. Putting these requisites in more practical

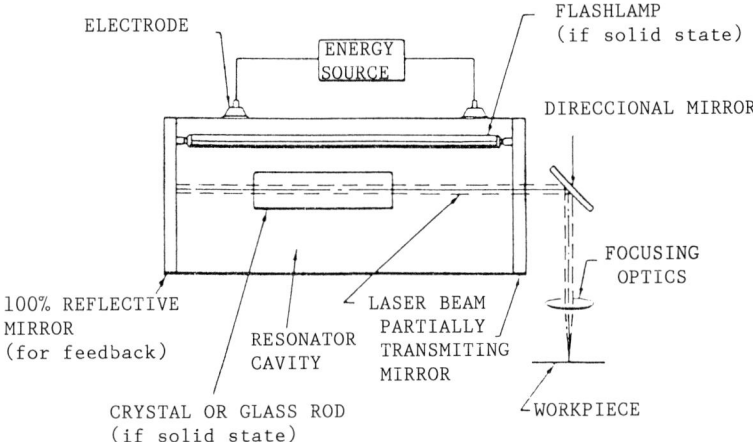

Fig. 1. Basic elements of an industrial laser (Ref. 8).

terms, the figure further indicates the basic function of an industrial laser. In this particular case, a solid state laser is demonstrated in contrast to a gas generated laser.

Figure 2 demonstrates the heart of the laser delivery system. Basically, it features a tube, bending mirror, extension tube, focusing lens, nozzle and finally a workpiece. The laser to the left is the energy source and the extension and focussing lens, the delivery system.

Charschan in Ref. (19) summarises the process of laser facility for material removal or cutting. This consists essentially of:

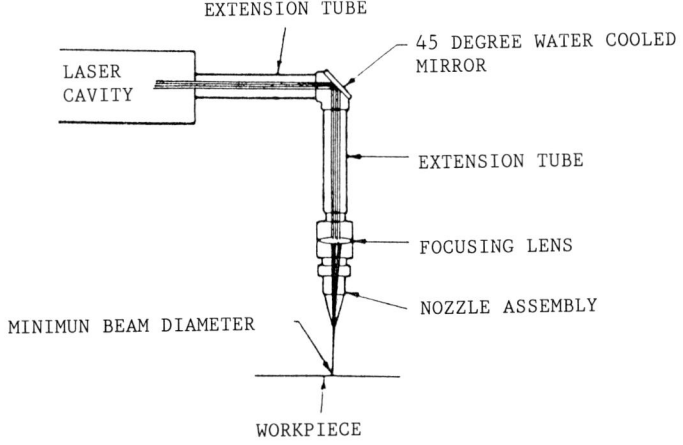

Fig. 2. Laser beam delivery system (Ref. 8).

- generation of a focused high intensity laser pulse;
- absorption of the energy by material;
- surface temperature rise to vapourisation point;
- minimal heat conduction from irradiated spot;
- material removed by vapourisation and particle ejection; and
- plume **generation of vapour and particulate matter**

Laser beams are known to be two main forms - brief pulses or a continuous stream, (wave) of light. Pulsed lasers are often used for drilling or welding small components while continuous-wave lasers are most suited to cutting and shaping of materials and treating of surfaces (Ref. 26).

One of the most important advantages of laser cutting technique is its capability to guide the laser beam easily by refracting it properly and precisely through an appropriate delivery system to the workpiece area desired to be cut. This beam can be reduced into a small spot (decimal fractions of a mm) producing a very narrow and precise kerf, assuring the required tolerances and desired results. To utilise these characteristics fully the necessity to implement laser technology with precise and exact motion delivery systems became obvious CNC systems have been the main factor to contribute to that necessity.

Nowadays, laser beam delivery systems implemented with a CNC table or adapted to industrial robots can cut easily complex two or three dimensional components with high speed and accuracy. An analysis of the role that CNC systems have played in laser cutting technology is given in the section below.

CNC IN LASER CUTTING TECHNOLOGY

Laser motion delivery systems can be classified into one of the following types:

- motion of the laser head;
- motion of the workpiece;
- motion of the optics (laser head and workpiece static)
- motion using robots.

MOTION OF THE LASER HEAD

This method is particularly suitable for lasers with light weight output powers of up to 500 W, since devices can be constructed from light weight materials and connected to separate power suppliers by a suitable "umbilical" cord. Some manufacturers have used twin cross-carriages to support lasers of 1 kW output; however, this can impose a "dead" area of 2 to 3 mm on the table because of the size of the resonator cabinet. Also, since a higher power can usually improve cutting speeds for a given material, inertia problems can arise from the increased laser mass (Ref. 9).

An advantage of this method is given in Ref. (10). This describes that the lightweight laser head provides a good example of the use of existing machine tool technology, thus avoiding some of the development costs of producing new laser cutting machines.

MOTION OF THE WORKPIECE

This method is suitable for small and light workpieces such as usual cuts from sheet metal working. The limits have been estimated by several authors (Ref. 9, 10) oscillating between cuts over an area up to 1 m² and up to 3 m x 1 m in exceptional cases. The obvious disadvantage to increase these limits is the floor space needed to relocate such tables.

MOTION OF THE OPTICS

This method develops as it has been recognised by Leece (Ref. 10) one of the fundamental benefits of laser light, which is its ability to be positioned under computer control using 'moving' optical components. It is suitable when either the workpiece or the laser presents characteristics which make them inappropriate to be moved. These systems draw less from the existing machine tools, making them therefore more expensive than those using the conventional systems of motion.

The basic functional layout of these systems is shown in Fig. 3. Note both mirrors are set in position accurately by the CNC controller.

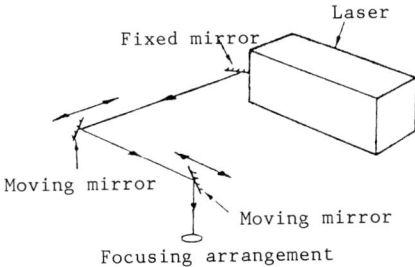

Fig. 3. Diagram of moving optics system (Ref. 10).

MOTION USING ROBOTS

This method maximises the advantage mentioned in the previous case, utilising the benefits of the laser technology combined with industrial robots, as described by Bannister (Ref. 8).

The laser beam can be articulated by adapting the delivery system with an industrial robot arm, as it is being done now in many flexible manufacturing systems. In fact, the development of so-called "laser robotics technology" has been very rapid due to its integration with FMS systems or cells. For recent developments on these topics the reader is referred to Refs. (6, 11-17, 24).

The role played by CNC systems in all these different motion systems mentioned above has been varied to-date. This has meant however that a variety of versatile microprocessors can be fitted to these machine-tools, providing the following characteristics.

- with CNC positioning systems the laser cutting can be used with high precision, giving profiles in almost unlimited form;

- all functions of the laser cutting tool such as shut down sequence, intensity power control etc, are under control of the computer, and synchronised with the rest of the process;

- off-line self-programming (or teach and play) is possible, using simple CAS systems which generate all the required datum points from a drawing to be recorded in a microprocessor and transform them in precise order to the machine tool;

- available software such as that required for nesting systems can be implemented easily with CNC in order to improve the productivity of the system.

Figure 4 shows as example of a laser cutting system for metallic sheets, controlled by computer.

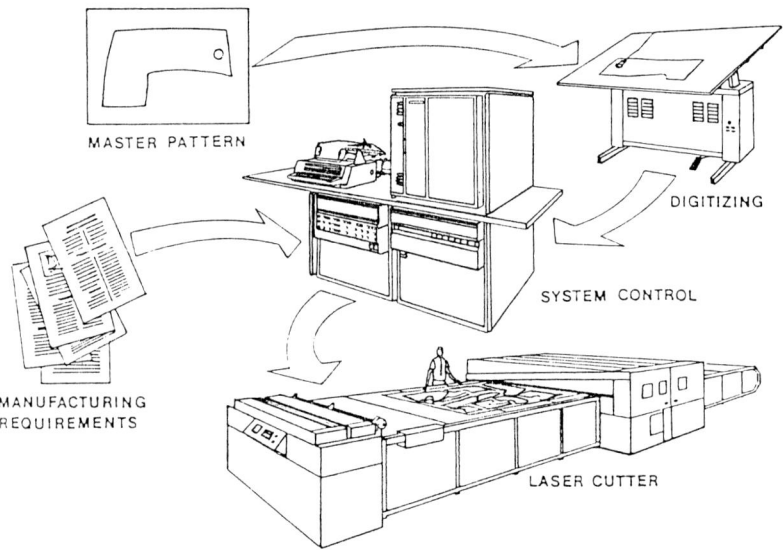

Fig. 4. Computer controlled lasercutting system (Ref. 18).

APPLICATIONS

Most manufacturers of laser profile cutting machine tools have developed machines involving simple motion systems: 3 or 4 axis CNC controlled, moving laser head, moving table, or mix of both. One example of these devices (moving the laser head) has been the FALCON S profile cutting machine in operation with Ferranti MF 400

Laser produced by Hancock Cutting Machines Ltd. This machine is
controlled by the GN1H control system. Control of the cutting
system may be automatic or semi-automatic depending on the shut-
ting mode. The output of the laser is about 400-450 W, with a
power density of 5 MW/cm².

The same laser, MF 400 has been incorporated by Laser-Kombinations
together with two S systems GmbH (Ref. 29) in the PLS (Precision
Laser Saw), available in two forms, depending on the control:
PLS 100 KS/NC or PLS 200 KS/NC, both with the movable X-Y carriage.
The PLS 100 KS/NC has been successfully applied in many companies
in Europe and U.S.A. (Ref. 23) as a laser die-cutting system.
The advantages of the application of laser techniques for die-
cutting are presented by Ulmer (Ref. 21) and Forbes (Ref. 23).

A general example of 'mix motion' system can be seen at Figure 5.
In this case the 'x' transverse is proportioned by the laser head.
In fact, most laser profile cutting machines have combined movements between the laser head and the table; this can be seen, for
example   Ferranti-Asquith 4 axis CNC universal laser
cutting machine, Fig. 6, developed by those two companies and
equipped with the MF continuous wave CO_2 laser generator of 1.2
kW. This machine incorporates the Allen Bradley 8.200 CNC system.

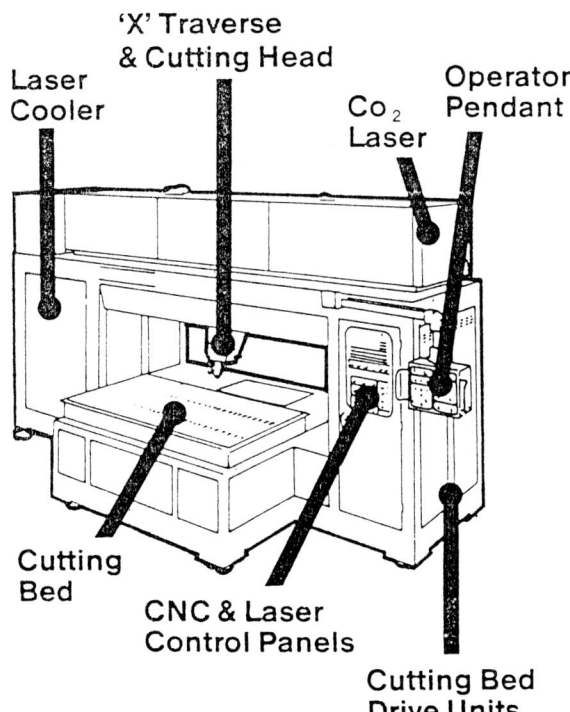

Fig. 5. LMC Laser cutting centre. (Courtest of Laser Scientist Services L.S.S.).

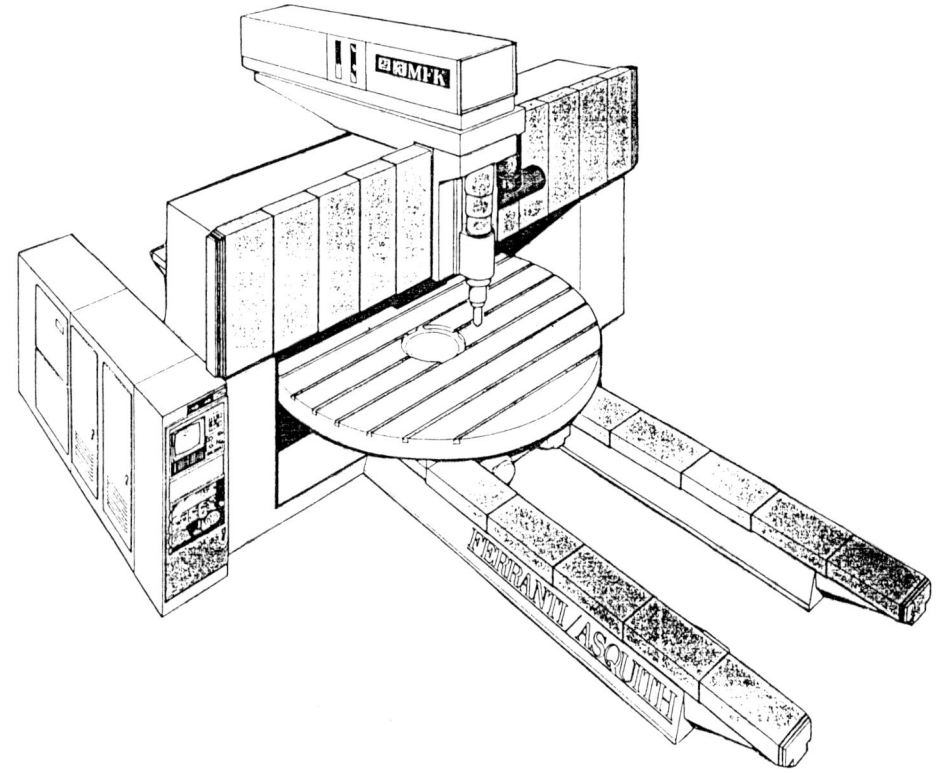

Fig. 6. FERRANTI-ASQUITH 4 axis universal cutting machine (Courtesy of FERRANTI LTD.).

The machine has two **working** areas; an area of 3 by 2 m is for flat plates and is of similar design to a conventional profiling table. However, at the end of the machine, a 'manipulator' consists of a large-diameter rotary table mounted with its pivot axis horizontal and supported by a saddle which has 2 m of vertical travel on a column. Curved workpieces can be mounted to this table, which will rotate 360°, and can be worked on by the laser head because of its cantilevered construction. This machine has been adopted by GEC Turbine Generator at Manchester to profile mild and stainless steel ranging from 6 to 10 mm thickness (Ref. 27).

Similar CNC system, the Allen Bradley 7320, has been incorporated in the Coherent-Everlase 525-W CO_2 laser, which cuts in continuous and pulsed modes and was manufactured by Sciaky Brothers, Chicago, for Globe Engineering Co., a sheet metal fabricator specialising in aircraft components (Ref. 30).

Figure 7 shows an example, see (Ref. 9), of a system of moving optics developed for shipbuilding. Laser of 2 kW output power

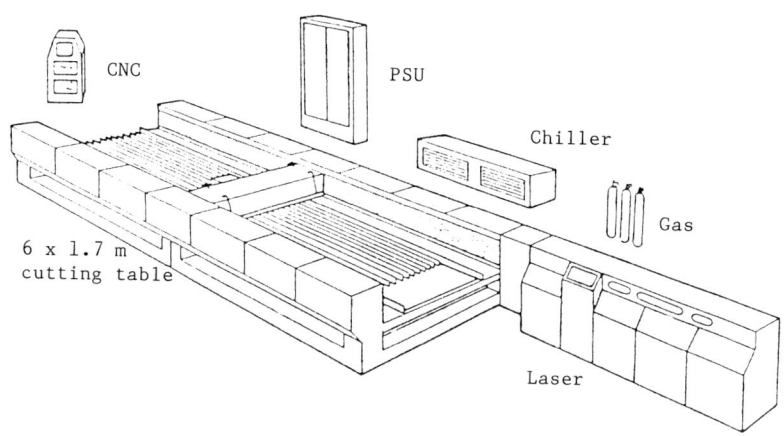

Fig. 7. 2 KW CNC moving optics laser cutting systems for shipyard (Ref. 9).

is able to cut steel plates of up to 10 mm thick using this system.

The most successful examples of moving optics cutting systems have relied on machine bed construction techniques developed for high speed CNC spindle-routing (Ref. 9).

Many other systems have been developed to cut through 2-dimensional sheet metal contours or plastic components using CO_2 lasers. One such system is the Behrens 5 axis laser cutting system (Ref. 16) which is shown in Figure 8. This system has incorporated the RS 1000-1kW CO_2 laser and the IBH CNC-MODUS controller, and it can cut steel plate of up to 8 mm thickness with assistance of oxygen gas. The five axis CNC has spline interpolation capability and can be programmed via DNC-Interface.

Examples of multi-station systems are described in Ref. 6. This work presents a scheme of single laser, multi-station laser system using "beam hopping" for station at head change; see Figure 9. System of similar configuration have been installed at the Mirafiori Plant and built by COMAU with the cooperation of FIAT research centre (CRF) and Fiat Auto Mechanical Production Technologies. The two stations are powered by a single Spectra Physics 973 Laser (Ref. 6).

Figure 10 shows an example of a laser processing system constructed by Raycon Co. and developed in conjunction with Automotive Exhaust Pipe to cut preshaped lengths of exhaust manifold pipe into segments. The sharing of the laser beam among three workstations provides maximum utilisation of laser-time.

With regard to the robots applications in laser cutting systems, probably most successful implementation up to-date has been the FLS COBRA used by many companies in Europe as well as in the U.S.A. This system incorporates the ASEA IRB6 or IRB6/2 and is

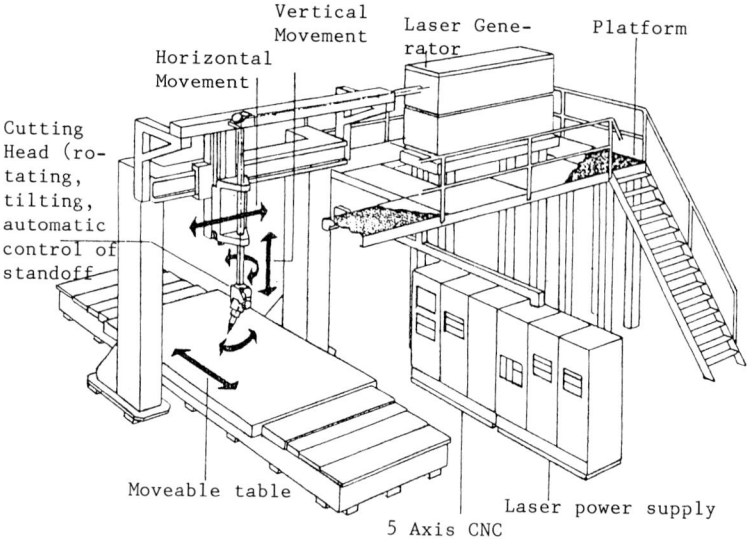

Fig. 8. Three-dimensional 5 axis laser cutting system (Ref. 16).

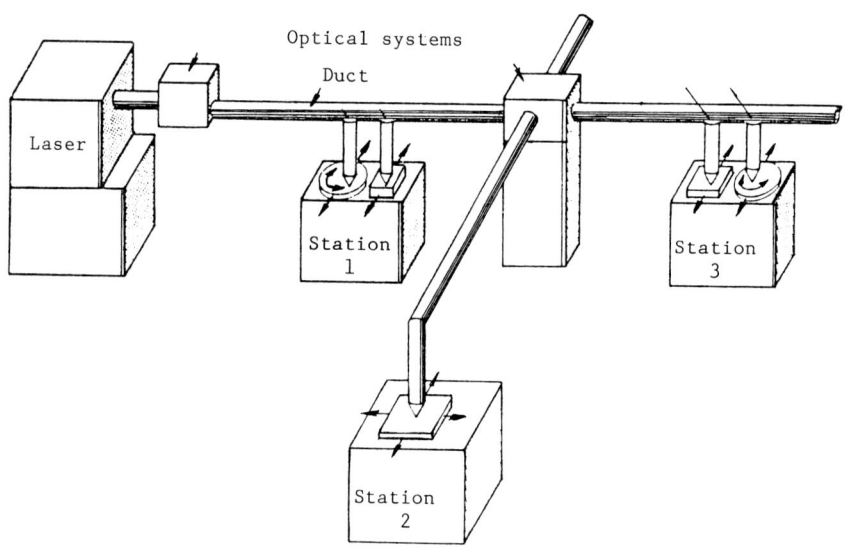

Fig. 9. Schematic of single laser, multi-station, (each single or multi-head) laser system using "beam-hopping" for station head change. The workpieces are moved relative to laser heads (Ref. 6).

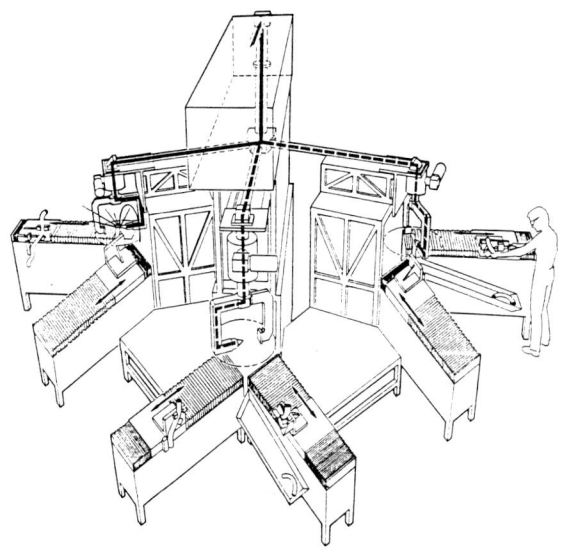

Fig. 10. Raycon Special CO_2 Laser Machining System for Automotive Exhaust Pipe (Courtesy of Raycon Corporation).

manufactured by FERRANTI INDUSTRIAL LASERS LTD. Its output power is 1.0 kW and many of its characteristics and applications are given in Refs. 24, 26.

Seelig (Ref. 20) shows that there are significant contributors to high power lasers from the large research centres that are results of their endeavour to develop lasers for nuclear fusion purposes. Some of these systems like Novette (a Nd-laser at Livermore Laboratory), Antares (a CO_2 laser at Los Alamos) or Asterix (an I-laser at Garching) are capable of peak power outputs in the 10 TW range at a pulse duration in the ns-range and with excellent beam quality. Through these tremendous output powers and the respective technology by far exceeds the requirements of current industrial demands, many of the research results associated with the development of these lasers are of practical interest to industrial laser systems, such as the design of high power optical components, unstable resonator, amplifier stages, as well as the development of methods for frequency conversion, the use of non-linear optical effects, etc (Ref. 20).

The ability of laser technology to machine very small structures has been demonstrated, dealing with microdimensions such as those involved in semiconductors (Ref. 22).

CONCLUSIONS

A brief explanation about laser cutting system, and its advantages and disadvantages are presented. The different implications between CNC systems and laser cutting techniques has been outlined. Also practical applications of the several types of summary systems are discussed. Finally, a tabular summary of different cutting parameters relating to processing of different materials by laser cutting is given in Table 1 (Ref. 15).

Table 1 : Cutting Parameters (Ref. 15)

Material	Thickness (mm)	Speed (m/min)	Power (Kw)
Non-metallic			
ABS	4.0	4.0	0.5
PVC	0.3	3.0	0.1
LEXAN	0.75	12.6	0.38
TEFLON	0.04	15.0	0.38
QUARTZ	2.0	1.0	0.40
PAPER	0.00	1600.0	0.40
ALUMINA	0.06	2.5	0.5
WOOD	25.00	1.5	8.00
Composites			
BORON/ALUMINIUM	3.0	1.2	1.0
FIBRE-GLASS/EPOXY	12.5	4.5	20.0
BORON/EPOXY	8.0	1.65	15.0
Metallic			
MILD STEEL	3.0	0.5	0.5
STAINLESS STEEL	3.0	5.0	1.0
HIGH SPEED STEEL	5.5	0.6	0.5
ALUMINIUM	12.5	2.3	15.0
TITANIUM	9.5	2.5	10.0

REFERENCES

1. Jefferson, T. B. "Laser cutting is a job shop speciality" Welding Design and Fabrication, March 1980, 109-111.
2. La Rocca, A. V. "Laser applications in manufacturing" Scientific American, March 1982, 80-87.
3. La Rocca, A. V. "Laser applications in manufacturing", Proceedings IVA's Beijersymposium, The Factory of the Future, Stockholm, September 1983.
4. La Rocca, A. V. "Laser application in manufacturing". The Factory of the Future, Symposium - The Royal Swedish Academy of Engineering Sciences, September 1984, Stockholm, Sweden.
5. Belforte, D. A. "High Power CO_2 Lasers in US Manufacturing Operation", Belforte Associates, Studbridge, Mass, U.S.A. - Proceedings VDI Internationaler Workshop, March 1984, Dusseldorf, Germany.

6. La Rocca, A. V., Pera, L. "Some considerations on the definition of laser robot systems" Laserobotics 1, April 1985, SME Technical Paper. Vol. MS85-491 (1985), 21 p.
7. Belforte, D. A. "Economic justification of industrial laser applications" SPIE vol 27. Applications of High Power Lasers (1985), 18-27.
8. Bannister, R. D. "Lasers and Robots - A Tale of Two Technologies", Laserobts 8, Conf. Procee. Appl. Today, Vol. 1 (June 1984), 25 p.
9. Martyr, D. R. "Laser cutting - a new tool for shipbuilding". Laser Welding Inst. Cambridge, Engl. (1984), 28-32.
10. Leece, J. "An analysis of machine tool systems suitable for laser profiling". Laser Welding, Inst. Cambridge, Engl. (1984), 13-17.
11. Delle Piane, A. "Five Axis Laser Robot for Cutting and Welding". Laserobotics 1, April 1985, SME Tech. Pap. Vol. MS85-492 (1985), 11 p.
12. Plankenhorn, D. "An approach to high power laser robotics using the Unimatic 6000". Laserobotics 1, April 1985, SME Tech. Pap. Vol. MS85-493 (1985), 6 p.
13. Ream, S. L. "Rectilinear robotics for high power laser materials processing" Laserobotics 1, April 1985, SME Tech. Pap. Vol. MS85-494 (1985), 16 p.
14. Sibayama, K. "Three dimensional laser cutting machine", Laserobotics 1, April 1985, SME Tech. Pap. Vol. MS85-495 (1985), 6 p.
15. Desforges, C. D. "Laser application", Technical file no. 46, Engineering, October 1977, 1-6.
16. Tamaschke, W. "Behrens 5 axis laser cutting system", Laserobotics 1, April 1985, SME Tech. Pap. Vol. MS85-496 (1985), 11 p.
17. Tight, T. "Review of Laserobotics. Challenges remaining to fully develop the technology. "Laserobotics 1, April 1985, SME Tech. Pap. Vol. MS85-497 (1985), 95.
18. Salyer, R. A. "Applications for Industrial Lasercutter Systems", SPIE, Vol. 86. Industrial Applications of High Power Laser Technology (1976), 50-59.
19. Charschan, S. S. "Laser cutting or surgery". LIA vol. 32, ICALEO (1982), 22-29.
20. Seelig, W. "High power gas lasers". SPIE, Vol. 455, Industrial Applications of High Power Lasers (1984), 2-9.
21. Ulmer, W. "The application of CO_2 lasers for diecutting : A report on 10 years of experience on the introduction of lasers into the folding carton industry". L.I.A. Vol. 31. ICALEO (1982), 129-130.
22. Arthurs, E. G. "Precision laser micro machining for semiconductors" L.I.A. Vol. 33, ICALEO (1982), 62-68.
23. Forbes, N. "Die board cutting by CO_2 laser", SPIE Vol. 247. Advances in Laser Engineering and Applications (1980), 8-17.
24. Anon. "FLMS Systems - a new generation of machine tool?" The Production Engineer (December 1983), 16-17.
25. Anon. "Lasers set to lead the trend in sheet metal cutting". The Production Engineer (October 1983), 21-22.
26. Spalding, I. "Lasers: a cut above the rest": Link up, January-March (1986), 26-28.
27. Astrop, A. "CNC Lasers cuts platework costs". Machinery and Production Engineering, July 1985, 40-41.
28. Vaccary, J. A. "The lasers edge in metalworking", American

Machinist, August 1984, 100-114.
29. Anon. "A compact workshop laser". Welding and Metal Fabrication (January 1974), 13-14.
30. Anon. "Lasers, one punch press boost sheet-part production", Welding Design and Fabrication, July 1985, 40-41.

DEFECTS ARISING FROM LASER MACHINING OF MATERIALS

G. E. Garcia de Vicuna, J. C. Beitialarrangoitia and S. K. Ghosh

Department of Mechanical and Computer-Aided Engineering, North Staffordshire Polytechnic, Beaconside, Stafford ST18 0AD, UK

ABSTRACT

In laser cutting/processing of material there are many variable and qualitative factors that are difficult to be controlled the optimal utilisation of the resources due to the complexities should achieve correct combination of all these factors relevant to each application. Hence, many trials are necessary on the real work to obtain the required results. Several defects may damage the work and these should be avoided to reduce cost of development and wastage in production.

In this paper, a review is presented of the most common effects/defects that occur in the manufacturing processes of laser cutting and drilling applied to different materials. Discussion is also given on how to reduce or eliminate, if possible, these effects according to requirements laid down in the job specification.

INTRODUCTION

In any laser machining process one can distinguish between two kinds of factors as a result of interaction of the laser beam with the workpiece. On one hand, the proper quality and behaviour of the laser beam has influence on the final result of the work. On the other hand, the interaction of the beam with the workpiece generates some effects that have to be controlled to obtain an acceptable finished product.

The discussion below is divided into two main parts, dealing first with the proper quality of the beam to process the workpiece properly, and secondly, the effects/defects generated when the beam has processed the workpiece material.

LASER BEAM QUALITY

It is necessary to pay attention in the process of production to a good quality laser beam to obtain the necessary results. Therefore, there are different elements that should be taken into account before the beam is used to process any material.

Cooling of the laser site

It is generally necessary to provide substantial cooling capacity to the laser site, due to the laser's relative inefficiency (Ref. 4) in power conversion.

Traditional gas lasers employ water cooling jackets, where 5 to 50 gallons/minute of water flow is required, depending on the laser size and design. The quality of the cooling water is important. Some lasers use the same water to cool both the heat exchanger and the electrodes. If this water contains a high amount of minerals or other impurities, it may produce corrosion, deposits in the cooling system and imbalance in the electrical characteristics of the laser.

Recently developed systems include convectively cooled systems in which the laser gas itself is circulated and the temperature reduced by passing this through a heat-exchanger. A small fraction of gas is removed continuously since breakdown of carbon dioxide to carbon monoxide and contamination occurs which reduces laser efficiency (Ref. 2). Fresh gas is added to compensate for the loss.

With Nd:YAG lasers the technique of cooling the cavity, the rod and the lamp has two significant drawbacks. One is the loss of efficiency due to absorption of pump light in the water and the other is reduced quality imaging of the lamp in the rod due to turbulence in the water. To resolve these problems (Ref. 3), the cavity is often cooled by a flow of water around the outside surface while at the same time the rod and lamp are cooled by flowing cooling water through an annular region between the rod and a transparent cooling jacket.

Optics maintenance

The dust accumulated on mirrors or lenses increases the surface absorption of the laser beam and causes heating of the optic; local damage (Ref. 1) of the optic can be caused when the dust particle burns.

The problem is that the optic is usually close to the work, and the weld spatter can easily damage the entire coating on transmission optics.

Optics with upward-facing surfaces are most suitable to collect dust. One way to reduce the rate of dust accumulation is to provide a transverse air flow across horizontal optics, but it still occurs. To assure consistent performance of the beam path perhaps the best way is a periodic optics cleaning. This cleaning technique depends of course on the type of optic.

Ventilation

Most of the laser material processing activities generate some form of effluent; possible are particulates, condensates, recombinations and gases of unknown chemical form. Good vented system of the laser interaction zone and discharge in a safe manner of the exhaust are imperative. In some processes it may be necessary to use filter bags and even scrubbers.

Laser Characteristics (Refs. 4,5)

Divergence: the beam divergence can be reduced by expanding and collimating the beam by a factor inversely proportional to the diameter expanded beam. Plane mirrors are more suitable because spherical mirrors and output windows result in a higher divergence of the laser beam, see Fig. 1.

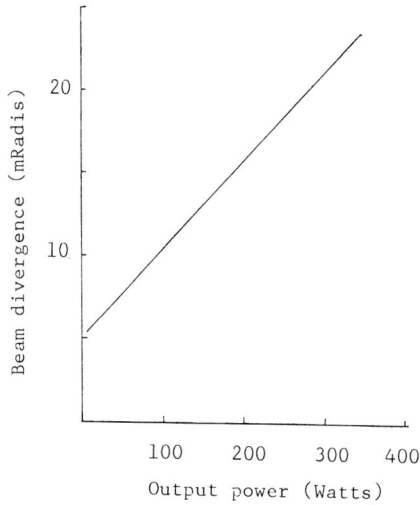

Fig. 1. Variation of beam divergence with output power, (Ref. 29).

Mode structure: The mode of the laser beam has to be good, and the focal spot has to be as small as possible, so that the width of the cut kerf can be minimised. Sometimes it is desirable that the output should be of a single spatial mode so that the diffraction-limited focused beam diameter can be achieved.

Factors that affect the mode structure include the geometry of the laser cavity and optics, the gain of the cavity, inhomogeneities in the laser medium and the pumping power.

Mode selection is possible by varying the mirror curvature and by including apertures with or without additional lenses, resonant reflectors or saturable absorbers within the optical cavity.

Coherence: The degree of temporal coherence is often described by the coherence length which is the distance over which the output intensity retains a measurable correlation of phase. The degree to which the frequency of the longitudinal mode is stable is the limiting factor which governs the coherence length. The stability of the longitudinal mode is limited by fluctuations in mirror geometry, thermal variation in the laser medium, and changes of the mirror separation.

Polarization: The polarization of the laser beam must be plane, and in contour cutting the plane of polarization must be turned, so that the plane of polarization is always parallel to the actual cutting direction, Fig. 2. A polarized beam can be achieved from an unpolarized laser beam by using suitable polarizing materials, however, a reduction in intensity of the light output occurs.

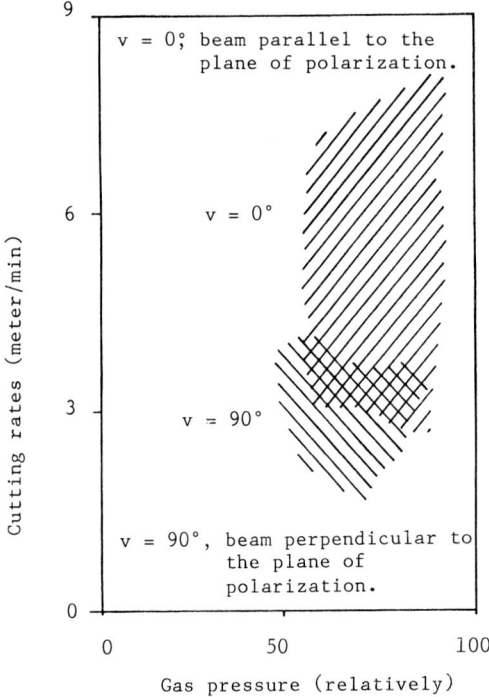

Fig. 2. Cutting rates versus the angle v. (Ref. 5).

PROCESSING EFFECTS

Material removal factors

The basic factors to be determined in an analysis of the power

balance in laser-metal interactions are: (i) what fraction of
incoming laser beam power is absorbed; (ii) how much material
is ejected from the metal specimen as a consequence of the absorbed power; and (iii) what fraction of this material is removed
as molten material (Ref. 6).

Material removal begins with vapourisation. As the vapourisation
proceeds, the absorption of the laser beam increases rapidly as
the depth of the crater increases rapidly the temperature of the
crater's inner wall. At the same time, the thermal diffusion
process begins as the local temperature rises.

The behaviour of material removed in the liquid phase depends on
characteristics of the pulse, as well as material constants such
as the thermal conductivity, the difference between vapourisation
and melting temperatures, and the ratio of the heats of vapourisation and liquifaction.

An increase in metal removal rate could be achieved by the use of
untrasonic vibration during laser cutting. The vibration would
either shake away molten material or cause a better penetration
of unreacted oxygen to the hot surface (Ref. 7). Ultrasonic
vibration of a workpiece increases the mass or volumetric rate.
It has an optimum value at a cutting velocity just below that
which will still pierce a plate, see Table 1 in this context.

Table 1. Different velocities (Ref. 7).

Oxygen Pressure kN/m^2 (psi)	Cutting Velocity mm/sec		Plate Thickness mm
	Vibration	No vibration	
14 (2)	10	13	1.65
34 (5)	19	28	1.65
69 (10)	32	45	1.65
103 (15)	40	46	1.65
14 (2)	6	8	2.9
34 (5)	13	17	2.9
69 (10)	28	30	2.9
103 (15)	28.5	31	2.9

Heat Loss (Ref. 7)

There are several causes of heat loss while the beam cuts a hole,
part of the beam energy is lost. As workpiece velocity is increased the fraction of beam energy lost this way is reduced until
the point is reached when the plate is no longer pierced. However,
heat is lost by conduction through the workpiece; once as a result

the power per unit area of metal surface reduces with increasing speed, the desired temperature is not attained.

This effect is exaggerated with the infrared beam of the CO_2 laser because the absorption coefficient also reduces with surface temperature. With assistance provided by oxygen the situation is further complicated.

The higher velocity due to the high pressure would reduce the thickness of the diffusion layer and increase the reaction rate.

Effects due to Nozzle Design

At a given supply pressure there is an optimum nozzle diameter. This result is explained by mapping out the stagnation pressure distributions from the various nozzles as a function of height above the workpiece and distance from the centre of the nozzle, Fig. 3. The maximum cutting speed for a given nozzle will be proportional to the distance between the pressure axis and the point of intersection obtained from the horizontal line through the minimum pressure required to remove molten material from the kerf in the time available and the nozzle pressure distribution curve.

The configuration of the nozzle may influence the quantity of oxygen transported down the cutting kerf and how efficiently this is done.

In the pressure range up to 4 bar the maximum cutting rates for high quality cuts increase with increasing pressure. Over 4 bar a further increase in pressure will allow no (or very little) increase in cutting speed. At a pressure higher than 5 bar a new

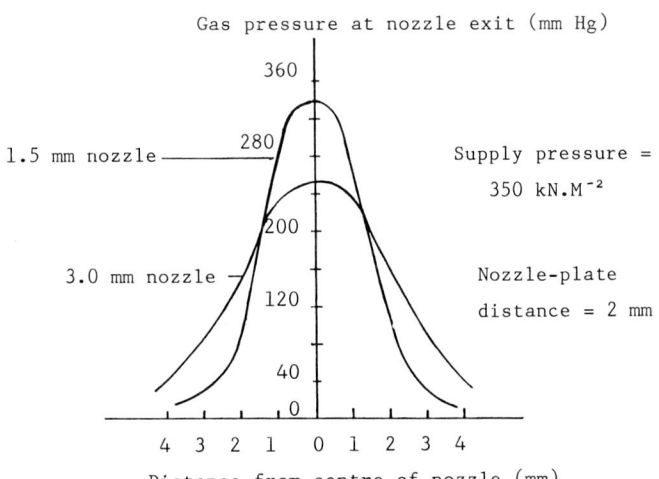

Fig. 3. Pressure distribution for nozzles of different diameter at a constant supply pressure, (Ref. 3).

phenomenon is observed. This is self-burning marks in the kerf, but very small and regular marks compared to those seen at low speed, (Ref. 8).

Gas Flow Effects (Ref. 3)

An increase in the gas flow rate increases the cutting speed up to a maximum after which further increase in gas flow rate causes a fall in cutting speed, Fig. 4.

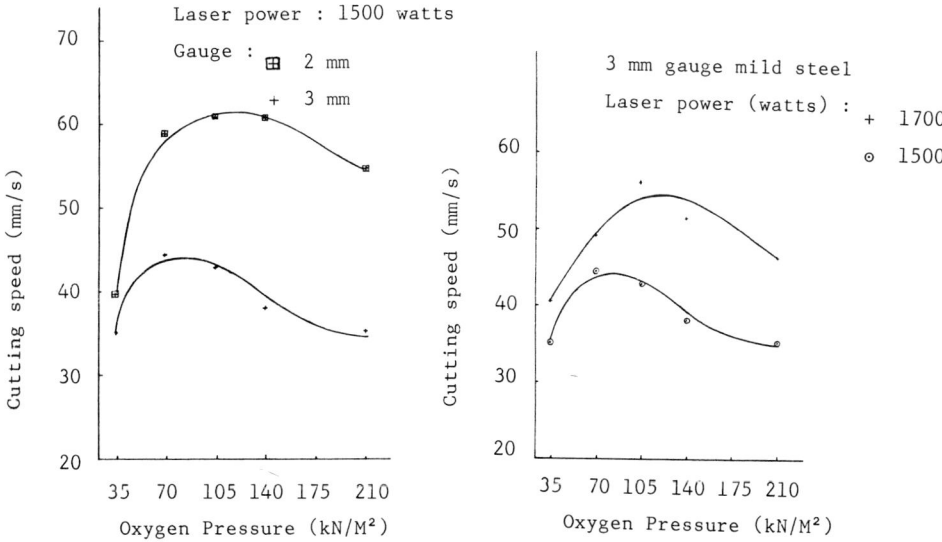

Fig. 4. Variation of cutting speed with oxygen pressure, (Ref. 3).

At high gas pressures a density gradient field (DGF) is formed above the workpiece and lies within the region covered by the gas jet; this is shown in Fig. 5. The shape and size of the DGF depends on: (a) gas pressure; as the gas pressure is raised the DGF covers a wider area and the strength of the DGF increases; (b) distance between the nozzle and the workpiece; for >3mm the strength of the DGF decreases with increasing h; and (c) nozzle diameter; increasing nozzle diameter increases the size of the DGF and the gas consumption.

Fig. 5. Presence of a density gradient field, (Ref. 3).

Beam scattering of the DGF will give rise to an increased incident beam diameter on the surface of the workpiece and will result in decreased cutting performance.

If the jet is too close to the plate severe back pressures on the lens will be created apart from mechanical jamming against splattered dross particles. If the jet is too far away there is an unnecessary loss of kinetic energy.

The type of gas used in the jet affects how much heat is added to the cutting action. There is a large difference between using oxygen and using argon in cutting any metal. However, with some materials oxygen is too reactive and causes a ragged edge.

Energy coupling (Ref. 2)

There are two different mechanisms to consider. Firstly, there is the initial coupling when the focused beam strikes the flat surface of the workpiece and there is the later coupling of the laser power within the kerf. The first is a problem of reflectivities, initial plasma formation and surface profiles under thermal stress. The second is a highly complex situation involving multiple reflections off a reacting rippled molten surface.

Thick material (Ref. 9)

One of the reasons that the laser is not a cost-competitive cutter for thick-section material is that jet-assisted parameters cannot be scaled with laser power. Another factor is consequence of the narrow cut width achieved. Narrow cones introduce two primary problems. First, a small-diameter assist jet must be utilised to match the width of the cut. The coherence length for an overexpanded free jet is typically of the order of a few jet-orifice diameters. This leads to jet expansion in a short distance. Secondly, the thin layer of liquid metal in the narrow-cut zone is strongly bound to the solid by surface tension forces. Clean removal of the liquid material is difficult.

Pulse Effects (Ref. 10)

For a very short illumination duration the temperature rise approaches a linear dependence on time. For longer light pulses, a larger portion of the laser energy is lost to the substrate by heat conduction during the pulse.

When the pulse duration is short enough. The machining of thin films substrates is most efficiently obtained due to the temperature rise linearly proportional to the pulse duration.

Machining could be obtained either by front illumination or by back illumination, the efficiency is approximately equal. In many cases back illumination is advantageous since the vapour from the surface is directed away from the focusing and deflecting optical components.

Varying the laser pulse intensity, the machining spot area varies
nearly linearly with the pulse intensity over a certain range
some threshold value, Fig. 6.

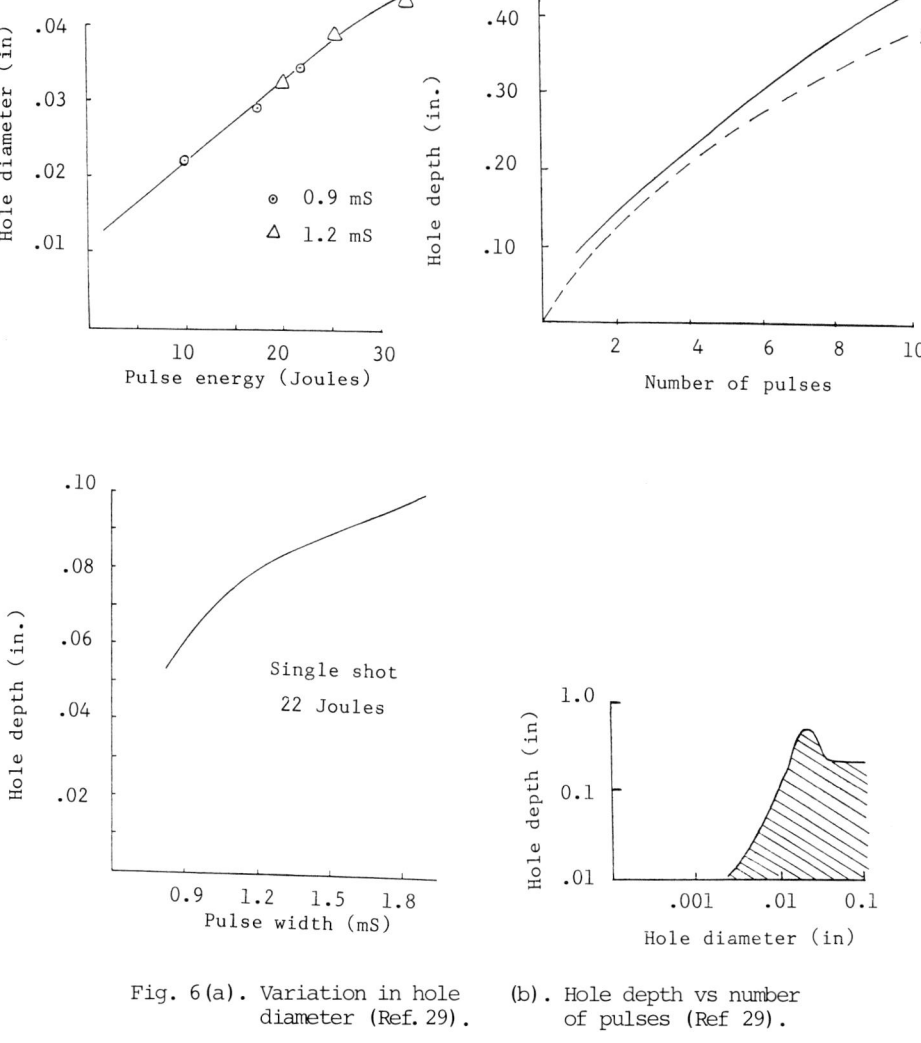

Fig. 6(a). Variation in hole diameter (Ref. 29).
(b). Hole depth vs number of pulses (Ref 29).
(c). Variation of hole width (Ref. 29).
(d). Depth/diameter envelope in ferrous alloys (Ref. 29).

Different improvements (Refs. 3, 11)

(a) Use of faster jets in particular supersonic; the results
have been disappointing due to excessive cooling of the
molten product and shock waves interfering with the fine

focus on the laser beam.

(b) Use of multiple jets and concentric jets that can increase the jet potential core and so give a higher stagnation pressure above the cut slot, thus improving slot velocities. Other variations include the simultaneous use of jets above and below, cross blowing beneath to remove the under-bead, and cross blowing over a venturi to introduce suction from below.

(c) Changes in gas composition; chlorine can react exothermically with most metals. It is also more dense than oxygen and thus has more drag on the molten product. The problem is that it is lethal. An alternative method of introducing more energy is by adding an electric arc to the laser interaction zone.

(d) Combination of laser types; in circumstances where surface tension becomes important (cutting or drilling with very small beam size) the combination of cw with pulsed laser would bare a greater advantage over simple cw irradiation. The penetration of thin plates can be speeded up significantly by the application of a trailing pulse to a cw beam. Here the advantage remains in the elimination of the need to liquify all the metal and a single final pulse is adequate.

DRILLING

Drilling takes place generally by repetitive pulsing using from one to a number of pulses per hole. The material inside the hole melts and is removed by vapourisation, which makes the process violent and therefore difficult to control (Ref. 12).

Optimal results of metal drilling efficiency of a Nd : YAG lasers can be expected only in a quite narrow intensity region - for most metals roughly from 5 to 50 MW/cm^2. For lower intensities heat conduction and reflection losses are dominant. For higher intensities, effects such as beam defocusing in the vapour cloud or even induced air-breakdown severely degrade the drilling efficiency and the reproducibility of the drilling process.

For expulsion of metal in liquid form, a mechanism involving radial liquid movement caused by the evaporation pressure is considered. The expulsed liquid must come from the bottom of the holes and not from the walls. The expulsed liquid jet forms the envelope of a cone (Ref. 13).

To increase the depth of laser drilling the laser pulse duration must be increased, but an increase in the pulse duration results in a larger volume of the molten material leading to poorer accuracy. In this case we should determine the optimum ranges of pulse duration and power density at which the volume of the molten material is minimised and to increase the depth of drilling one should use the multipulse technique (Ref. 14).

Hole requirements

Laser variables having an effect on one or more of the hole requirements - diameter, depth, taper, recast, microcracking and angle - are pulse energy, pulse duration, number of pulses, focal length of lens, beam diameter at the lens and beam quality (Ref. 15).

Figure 7 shows the faults most likely to occur in laser drilling.

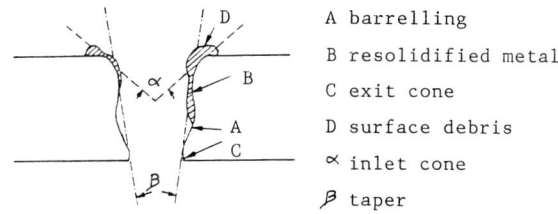

A barrelling
B resolidified metal
C exit cone
D surface debris
α inlet cone
β taper

Fig. 7. Faults in a laser drilled hole (Ref. 16).

Figure 8 shows how a typical laser drilled hole might appear in cross section in a metal sample. Uncontrollable diameter variations exist, starting with the funnel-shaped entrance to the hole. A thin layer of recast clings to the side walls of the hole. This material has been melted and resolidified in such a short time that the frozen metal has a different structure than that of the parent material (Ref. 18).

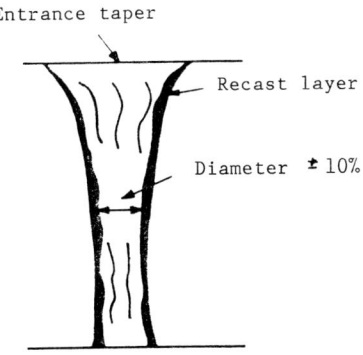

Fig. 8. Typical cross section of laser drilled hole, (Ref. 18).

Diameter: The diameter of the hole is dependent on the incident beam diameter. This diameter is determined by the laser cavity optics, the focusing optics and the laser wavelength. Passing the beam from the laser through an aperture has two benefic effects. Firstly, the spot size is independent of the pumping power, thus giving more reproducible results, and secondly, the power distribution across the heated spot is more uniform than would be the case with simple beam focusing.

Some specific techniques for improving the edge and rim of a hole depend on either blowing the material away or reducing the wettability of the hole lip. Other methods of improving edge quality have been reported using a gas blast effect from blowing on the underside with a strong gas jet or coating the hole exit side with epoxy resin (Ref. 3).

The difficulty in laser drilling holes smaller than 0.0005" (0.0127 mm) diameter is due to the inability to maintain adequate depth of focus at this spot size.

<u>Taper</u>: It is a result of erosion caused by the expulsion of molten and vapourized material from the hole. Degree of taper can be controlled by the number of laser pulses, pulse energy level and optical system design (Ref. 15). Taper is generally reduced in thicker samples, whilst inlet cone and barrelling depend strongly upon the position of the beam waist with respect to the surface plane of the workpiece. High quality drilling is achieved when the point of maximum intensity is placed some distance above the surface (Ref. 16). Figure 9 shows the relationship found between material thickness and hole taper.

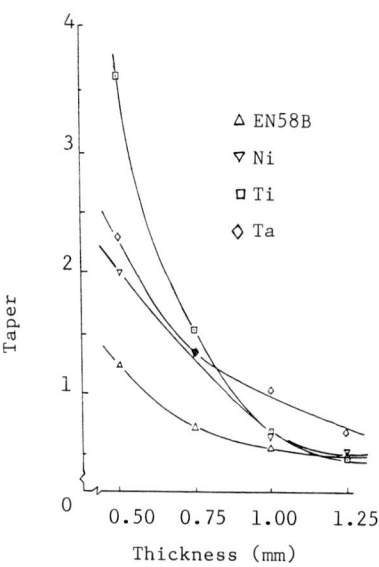

Fig. 9. Taper versus thickness (Ref. 16).

<u>Recast and microcracking</u>: Any molten or vapourised material that

is not completely expelled, but resolidifies inside and around the hole, is known as recast. Thermally induced stresses during the solidification process produce microcracking, which if propagated into the parent material at grain boundaries, weakens the final product. Recast and microcracking are usually reduced by selecting power densities that effectively expel molten and vapourized material, along with pulse widths short enough and repetition rates slow enough to minimise heating of the surrounding parent material. See Fig. 10 and Ref. 15 in this context.

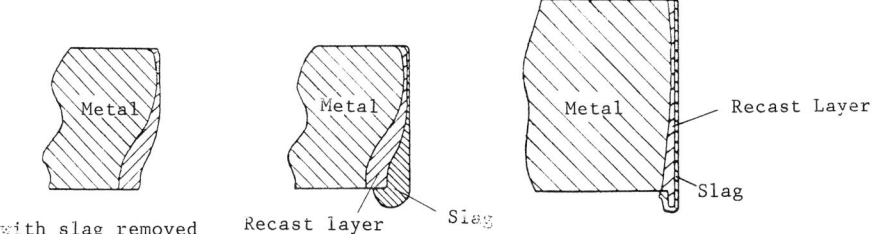

Fig. 10. Recast produced (Ref 23).

<u>Drilling angle</u>. Drilling perpendicular to the surface maximises the effective surface power density and permits the widest range of focal lengths and focusing angles. However, at this angle the maximum amount of expelled debris is directed back to the focusing lens, so it is necessary to protect the lens. The gas stream appears to have little if any effect on the drilling process other than to protect the lens.

As the drilling angle with respect to the surface becomes smaller, the volume of debris directed toward the lens decreases, but the focusing geometry becomes more complicated. Because of the angle between surface plane and focal plane, the effective surface power density is reduced, making it necessary to increase laser power and/or improve laser beam quality (Ref. 15).

<u>Pulse energy</u>. At high energy levels there is a tendency to cause more deformation on the top surface.

<u>Pulse duration</u>. Generally, shorter pulse durations (0.4 milliseconds) will produce higher quality holes, but will require more pulses to drill through. Longer pulse durations (to 3 milliseconds) will remove more material per pulse, but will produce holes of lower quality.

<u>Focal length of lens</u>. Shorter focal lengths are typical for small diameter and/or shallow holes drilled at a near normal incidence.

Different methods

The nitrogen-carbon dioxide laser is operated in the continuous wave mode. Transitions between vibrational states of molecules are less energetic than are the conic processes occurring in solid state devices. The width of the beam in the focal region of the focusing optics is proportional to the wavelength of the radiation so the intensity varies as the inverse square of the wavelength. The intensity achieved at the workpiece also depends on the form of the laser beam and the optimal quality and **behaviour** of the various elements transmitting and reflecting the beam.

Restrictions found are based on the ability of the laser to generate a sufficiently high power intensity within the workpiece that non-conduction limited mechanisms are generated. (Ref. 16).

Substantial savings in energy and time can be obtained by first heating the metal by continuous laser radiation and after sufficient heating, suddenly applying a sharp pulse of radiation, pushing the metal plug by the explosive pressure created by the pulse (Ref. 17).

Trepanning

This operation involves the laser beam being focused more tightly so that its diameter is much smaller than the hole to be produced. The beam is then used to cut around the profile of the hole. This involved more pulses than the direct drilling technique, which meant that drilling times could be reduced substantially. The energy transmitted to the workpiece is reduced and with it the likelihood of micro-cracking because it used a small spot size. By this way any shape hole could be produced. A co-axial assist gas is required to remove debris from the drilling cone (Ref. 12).

The burr on the hole exit edge can be minimised by applying a liquid laser coat on the bottom surface of the part prior to laser cutting. The laser coat dries rapidly to a powdery chalk-like substance that can be removed by water washing, (Ref. 18).

Various materials

The drilling of hard brittle-like materials by laser is attractive due to the extremely difficult task for current tool technology. An accepted guide used is the limiting ratio of hole depth to hole diameter. The maximum hole depth achievable is limited by the amount of energy lost due to reflections from the hole wall and by the decrease in the aperture of the hole as a result of vapour from the bottom of the hole being cooled by, and depositing on, the wall of the hole (Ref. 19).

With metals, generally, when the laser pulse energy is kept constant, the largest hole depths are produced for low-melting point materials. Drilling with a superposition of pulses usually yields holes whose sides have less taper than those of holes drilled with one pulse of higher energy (Ref. 20).

Low-power drilling is accompanied by an extensive heat-affected region adjacent to the hole boundary and some of the melt is deposited on the surface of the steel. At high incident intensities drilling occurs in a few milli-seconds and little heat is transferred laterally into the sheet. The result is a clean hole with well-defined edges and no evidence of a residual melt region. Holes drilled in vacuum are much more uniform than those drilled in air.

CUTTING PROCEDURES

The amount of energy usefully employed for cutting by an incident light beam will depend on the optical and thermal properties of the material. Loss due to reflection from the surface can be very high, particularly for metals which at room temperature only absorb a few per cent of the radiation. However, as energy is absorbed and the temperature of the metals rises this reflection coefficient decreases rapidly. With non-metals, reflection losses are less severe and for certain materials infra-red absorption bands coincide with the laser wavelengths. Once the energy has been absorbed, losses can occur by conduction into the bulk material (Ref. 21).

Different types

Three different versions of laser cutting can be distinguished (Ref. 22) as described below:

(a) In laser sublimation cutting, the focused laser beam heats the material to its evaporation temperature. A jet of inert gas carries the vapour out of the cutting front. A narrow kerf with high quality surfaces on both sides can be generated. The heat affected zone is almost eliminated but the thermal effects that will accumulate from the presence of molten residue could be serious enough to crack in brittle material.

Pulsed cutting may have advantages over cw with heat sensitive materials since the evaporative mode of cutting leaves less heat in the workpiece.

(b) In laser fusion cutting, a stronger inert gas jet is used to blow the molten material out of the kerf. The material has to be heated only above its melting point. The required cutting energy per unit length is less, and higher cutting rates can be reached. The quality of the cut, however, is reduced due to striations and solidified droplets of residual melt clining at the lower cutting edge.

(c) In the case of laser gas cutting, instead of an inert gas oxygen is used reacting exothermically with the material as soon as the ignition temperature is reached. As shown in Fig. 11 high cutting rates are attainable combined with cutting surfaces of high quality.

Some advantages and disadvantages of gas jet assisted laser-cutting of metals can be summarised (Ref. 20) as follows:

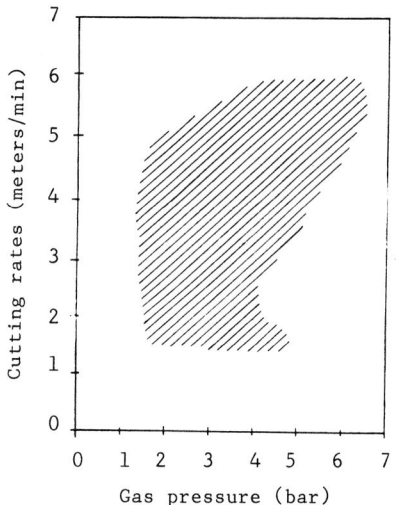

Fig. 11. Cutting rates versus gas pressure (Ref. 5).

- Oxidizable metals can be cut at high rates using an oxygen gas jet assist.

- Cut widths are usually smaller than those attainable with conventional thermal cutting methods.

- The edges of the cuts are square and a planar surface is produced.

- The heat-affected zone can usually be kept small but may be different from one side of the slab to the other.

- It is difficult to cut metals with a high melting point, high reflectivity at 10.6 μm and/or large thermal conductivity with the cw CO_2 lasers.

- The oxygen gas jet is most effective only for highly oxidizable metals.

Quality of Cut

The quality of the cut is judged mainly on the burrs where the best quality is a clean, smooth cut without any dross or burrs, small heat-affected zone [HAZ] sharp edges and fine striation. A quality with sharp edges but a small, just visible burr will however be acceptable.

The quality is shown as a function of cutting speed and oxygen pressure. At low cutting speed self burning appears. Around the parameter range where clean cuts are obtained there is a zone with formation of burrs, and at high speed the materials are not cut through, (Ref. 8).

The oxygen jet has two effects, it increases the **absorptivity** of the material contributing to the formation of the oxide film on the surface and removes the film and the melt from the cutting zone until the material is completely out. The quality of cutting is better at high gas flow rates and smaller distances between the nozzle and the material surface (Ref. 14).

Special care has to be taken that oxygen pressure is built up thoroughly before the optical shutter is released, otherwise the instant metal vapour formed within the expanded plasma plume penetrates the nozzle orifice and coats the focusing lens. An enhanced radiation absorption causes lens damage (Ref. 24).

The parameters which affect the quality of the cutting are the power of laser. The feed rate, metal thickness, nozzle design and the gas used in the jet (Ref. 23).

Properties of the cutting surface

The geometrical properties of a cutting surface are characterised by the surface roughness, inclination of the cutting surface, roundings at the upper and lower cutting edges, presence of squeezings, blowpipe bead and clinging dross.

In laser gas cutting at low cutting speeds, self burning of the material occurs characterised by an irregular flashing frame cone. This leads to an enlarged kerf and poor cutting surfaces. Increasing the cutting speed, the self burning is increasingly suppressed, and the quality of the cut improves until the optional cutting speed is reached.

Exceeding the optimal cutting speed, the quality of the cut now deteriorates rapidly. The erosion process becomes unstable inducing a strong rifling, and the increasingly incomplete oxidation of the material leads to blowpipe bead clinging at the rear surface of the plate (Ref. 24).

Kerf surface

There is an optimum kerf width for a given material thickness in order that the dross should clear easily. Where the thickness of a metal sheet over-rides the kerf width by more than a factor often, one carefully has to trim oxygen pressure and cutting velocity either to achieve best results (Ref. 24).

The kerf width depends on the cutting speed, oxygen pressure, polarization and the focal spot. The spot size depends on the mode quality, the focal length of the lens and the position of the specimen relative to the focal point (Ref. 8). The optimum spot size is a function of material thickness to allow adequate material removal from the kerf.

The main problem in producing fine cuts lies in the proper extraction of molten material through the narrow kerf width. If the physical size of the slag forms tiny spheres and the momentum vector of it lies in parallel. The optical energy flux, laser cutting performs very well, but as soon as a considerable momentum transfer of slag is directed towards the kerf wall. The material flow will be unbalanced and an enhanced dross formation or even an overgrowing of the kerf is the result (Ref. 24).

Striations

The striations which occur on the edge of laser cut material are a function of the viscosity of the slag associated with each material. With low viscosity slag the perturbations in the laser output are evident while with high velocity slag other coarser striations occur which are associated with the cooling process of the molten material (Ref. 23).

The formation of striation is caused not only by blowing out the droplets from the front of the cut, also by differences in the velocity of the moving laser beam amd the velocity of the oxidation front. The striations can be formed on the sides of the cut kerf as waves in the thin film of molten material behind the moving laser beam.

At low pressure (about 2 bar) and medium cutting rates, the striations typically are straight lines, sometimes with a line parallel to the cutting direction, Fig. 12.

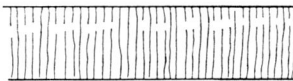

Fig. 12. Low pressure, medium speed (Ref. 5).

At medium gas pressure (4 bar) and higher cutting rates the striations are curved and often there are lines starting from the top of the cut kerf running to the bottom of the kerf, Fig. 13.

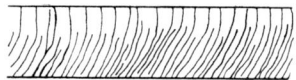

Fig. 13. Medium pressure, high speed (Ref. 5).

When the pressure is in the range of 6 bar, and the cutting rates are high, cuts can be deserved,when there is a region in the middle of the cutting kerf, where no lines are observed, Fig. 14.

We have two types of observations:

- At some special parameter sets (cutting rate, pressure, working distance) the process is unstable (the striations suddenly change characteristics from fine striations to rough striations)

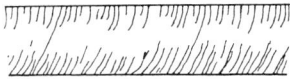

Fig. 14. High pressure, high speed (Ref. 5).

- Very often when examining the two opposite sides of the cut, different striations are observed on the two sides. Two different directions on the striations and two different striations frequencies can be observed (Ref. 5).

Improvement methods

Some promising results have been obtained by blowing slagging agents such as lime (calcium oxide) into the kerf with a view to lowering the dross melting point and reducing its viscosity to increase fluidity. However, this process also increases the thermal load. An exothermically reacting slagging agent would be more effective but probably too costly (Ref. 16).

By adding an arc to a laser there is an increase in speed or penetration without significant loss of quality. At low arc currents and high speeds when the arc would not be stable on its own, the arc is stabilised by the laser-generated hot spot because this is a region of increased electron concentration; see Figs. 15 and 16.

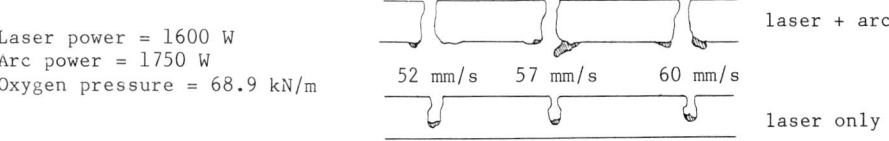

Laser power = 1600 W
Arc power = 1750 W
Oxygen pressure = 68.9 kN/m

Fig. 15. Variation in cut geometry with cutting speed for 4 mm mild steel (Ref. 27).

The arc is not of much help when it is used on the same side as the laser because it flickers from one side of the kerf to the other and so damages the cut edges. The arc is much more stable at the underside of the cut and initially keeps the dross more fluid and thus extends the velocity (Ref. 3).

Different materials (Refs. 25, 26).

4340 Steel Alloy

Cutting performance generally increased in proportion to laser power for both oxygen and inert-gas jet-assisted conditions for

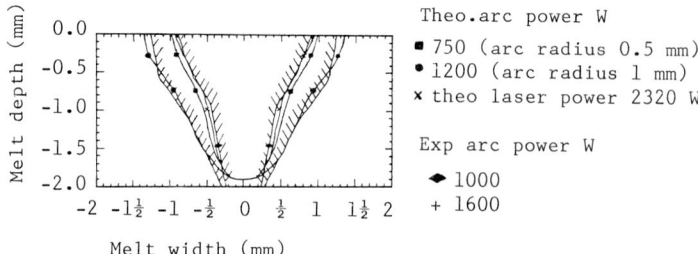

Fig. 16. Graph of cutting speed against input power. Input power up to 1870 W is only by laser; above 1870 W the extra power is supplied by the electric arc (Ref. 30).

oxygen assisted cutting. The material can be severed at substantially higher speeds than those for best-cut quality, but a speed reduction and increased gas pressure is required to prevent tenacious dross at the base of the cut. In inert-gas-assisted cutting, optimum speeds are of the order of 50-60 percent of oxygen-assisted rates, and a generally smoother edge is obtained.

Titanium alloy

In oxygen assisted cutting the cut quality does not improve with increased power level. The higher incident power density leads to a violent reaction in the oxidation-prone titanium and to extremely rough cuts. Inert-gas-assisted cutting gives a generally smooth cut surface with less apparent heat damage and easily removal adherent slag at the lower lip of the cut. With helium assist, some of the irregularity in the cut surface is removed to give a relatively smooth edge. Surface irregularities are influenced by appropriate control of the fluid dynamic characteristics of the jet.

Non-inflammable materials

In cutting these materials the gas jet mostly carries out the cleaning of the cutting zone blowing off the vapour and molten drops and cooling the material adjacent to the cutting zone. Cooling is especially important for cutting di-electric materials without glazing and charring. The gas jet also protects the optical system from the disintegration products of the material.

Waspaloy

Since the material is resistant to oxidation, smaller gains in cutting speed are attained with oxygen. Oxygen-assisted cutting results in a smaller lower edge burr than for inert gas assist. The slab generated during inert gas-assisted cutting could readily by chipped from the edge, leaving a relatively smooth cut surface. Small increases in power substantially increase cutting capability for the thickness indicated.

Ceramics

As ceramics have poor heat conductivity and high Young's modulus the conventional gas-jet laser cutting does not often produce good results. This results in the super-heating of the material near the cut and its disintegration owing to high thermal stresses. To reduce the total power input the pulsed regime is typically used.

Wood

The amount of laser light absorbed by wood is a positive function of the power density, the coefficient of absorption, the conductivity of light, and the depth of light penetration into the material being cut. More power is required to cut wet wood than is required for dry wood if feed speed is held constant because its less overall absorption coefficient cuts are of good quality with some tendency for charring to occur on the side of the sheet not subjected to the gas assist. The sides of the cut are parallel and the kerf width can be adjusted by a change of laser power or gas flow there appears to be little gain in using inert gases such as nitrogen or helium in the gas assist. Wood is not easily cut without a gas jet assist.

Textiles

Many layers of material can be cut simultaneously and an early problem of lateral penetration of the gas jet between layers which produces discoloration and charring at the bottom of the stack, has apparently been overcome. One distinct advantage with synthetic textiles is that the edge of the cut is sealed by the heat from the laser, thus eliminating the frayed edges typical of knife cuts.

Paper and cardboard

Some attractive features of paper cutting with high-power CO_2 lasers are: not contact between the cutting tool and the paper, eliminating the replacement of tools; the paper is vapourised and not cut away, resulting in a superior edge quality; and no dust is generated by the cutting process.

Plastics

The penetration was found to be greater with a gas assist only at high incident laser powers. However, the quality of the cuts produced was significantly better with the gas assist even at low laser powers. An air gas assist was the best compromise.

CONCLUSIONS

In laser cutting/machining, to obtain a good quality finished product, there are many factors and characteristics to be taken into account. For each particular application a detailed preliminary study is necessary to explore the different values and to deal with the necessary adjustments. It is hoped that defects/effects described above will enable reader to consider various process parameters carefully.

REFERENCES

1. S. L. Ream. "Rectilinear Robotics for high-power laser materials processing", SME Technical Paper MS85-494, 1985, 16p.
2. M. J. Fletcher. "High power laser cutting of metals", Welding and Metal Fabrication, September 1973, pp 308-311.
3. M. Bass, "Laser materials processing", North-Holland Publishing Company 1983.
4. J. E. Harry. "Industrial lasers and their applications", McGraw-Hill 1974.
5. F. O. Olsen. "Investigations in optimizing the laser cutting process", Lasers in Materials Processing Conference Proceedings - American Society for Metals 1983 pp 64-80.
6. M. K. Chun and K. Rose. "Interaction of High-Intensity Laser Beams with Metals", Journal of Applied Physics, V41 N2, February 1970, pp 614-620.
7. S. J. Ebeid and C. N. Larsson. "Ultra-sonic assisted laser machining", Proceedings of the 18th International Machine Tool Design and Research Conference. London. 14-16 September 1977 pp 507-514.
8. Flemming Bach Thomassen, Flemming O. Olsen. "Experimental Studies in Nozzle Design of laser cutting", Proceedings SPIE Industrial Applications of High Power Lasers, 1984, pp 169-181.
9. Conrad M. Banas and Robert Webb. "Macro-materials Processing" Proceedings of the IEEE V70 N6 June 1986 pp 556-565.
10. D. Maydan, "Micromachining and Image Recording on Thin Films by Laser Beams", The Bell System Technical Journal V50 N6 July-August, 1971, pp 1761-1789.
11. L. C. Towle, J. A. McKay and J. T. Schriempf, "The penetration of thin metal plates by combined cw and pulsed-laser radiation", J. Applied Physics, Volume 50 N6, June 1979, pp 4391-4393.
12. ANON. "Why a laser is better than EDM for drilling", The Production Engineer, December 1983, pp 13-14.
13. M. von Allman, "Laser drilling velocity in metals", Journal of applied physics, Vol. 47, No. 12, December 1976 pp 5460-5463.
14. N. RyKalin, A. Uglov and A. Kokora. "Laser machining and welding", MIR Publishers, Moscow, 1978.
15. W. B. Tiffany. "Drilling, marking and other applications for industrial Nd YAG lasers", SPIE Vol. 527, Applications of High Power Lasers, 1985, pp 28-37.
16. Brain F. Scott. "Laser Machining and Fabrication - A Review" Proceedings of the 17th International Machine Tool Design and Research Conference, Birmingham, 20-24 September 1976, pp 335-339D.

17. B. Steverding, R. W. Conrad and H. P. Dudel, "Explosive puncturing of metal plates by lasers", J. Applied Physics, Vol. 50 N H November 1979 pp 6713-6718.
18. Steve Bolin, "Laser Drilling and Cutting", SME Technical Paper MR81-365, 1981, 10p.
19. Francis P. Gagliano, Robert M. Lumley and Laurence S. Watkins. "Lasers in Industry", Proceedings IEEE, Vol. 57, February 1969, pp 114-117.
20. W. W. Duley, "CO_2 lasers - Effects and Applications", Academic Press, 1976.
21. F. W. Lunan and E. W. Paine. "CO_2 Laser Cutting", Welding and Metal Fabrication, January 1969, pp 9-14.
22. I. Decker, J. Ruge and U. Atzert. "Physical models and technological aspects of laser gas cutting", Industrial Applications of High Power Lasers, SPIE, Vol. 455, 1983, pp 81-87.
23. N. Forbes. "The role of the gas nozzle in metal-cutting with CO_2 lasers", Laser 75 Opto-electronics Conference Proceedings, Munich, 24/27 June 1975, pp 93-95.
24. Erich H. Berloffa, J. Witzmann. "Laser materials cutting and related phenomena", Industrial Applications of High Power Lasers Proceedings SPIE, Vol. 455, 1983, pp 96-101.
25. J. Huber and M. Warren. "Production Laser Cutting", Applications of Laser in Materials Processing, 1979, pp 273-290.
26. Vladimir G. Barnekov, Charles W. McMilhn and Henry A. Huber. "Factors influencing laser cutting of wood", Forest Products Journal, V36 N1, 1986, pp 55-58.
27. J. Clarke and W. M. Steen. "Arc augmented laser cutting", Laser 79 OPTO - Electronics, Conference Porceedings, 1979, pp 247-253.
28. R. E. Wagner. "Laser drilling mechanics", J. Applied Physics, Vol. 45, No. 10, October 1974, pp 4631-4636.
29. S. R. Bolin. "The Effect of F # and Beam divergence on quality of holes drilled with pulsed Nd:YAG lasers", LIA, Vol. 31, ICALEO, 1982, pp 135-140.
30. William M. Steen. "Arc augmented laser processing of materials", J. Applied Physics, Vol. 51, No. 11, November 1980, pp 5636-5641.

SL 25, A NEW HIGH PERFORMANCE 2.5 KW INDUSTRIAL CO_2 LASER

V. Fantini and G. Incerti

SOITAAB s.a.a., Via F. Filzi 3/5–20052 Monza, Italy

INTRODUCTION

In order to meet the increasing demands of the industry for more reliable and economical laser material processing, SOITAAB has developed and put in the market SL 25 unit, which is a fully microprocessor controlled 2.5 kW CW CO_2 laser source of transverse flow type.

SL 25 combines the innovative optical design with higher levels of automatic control, assuring a simple and reliable operation in industrial environment.

From the industrial user point of view, the required high laser reliability is essentially consisting of the long term output power stability, the possibility of programming automatic working cycles in conjunction with CNC equipment or processor controllers, the availability of automated auxiliary laser procedures (start-up, shutdown, conditioning) and the reduction of the warm-up and maintenance times.

SL 25 incorporates all the features mentioned above, allowing this laser source to be used also by not highly skilled operators. This is a very important point, because the availability of highly skilled operators may be the bottle-neck for the penetration and the growing of the laser technology in industrial environment.

Therefore SL 25 is well introduced in the development trend of industrial lasers in the near future, which must be directed towards the improvement of the laser performances and, at the same time, towards the extremely simplification of all the operations regarding laser working and source maintenance. In these terms the industrial laser must be considered as an ordinary machine-tool.

DESIGN CONSIDERATIONS

Basically SL 25 unit consists of the laser head, the beam diagnostic unit, the power and control unit and the cooling unit (Fig. 1).

Fig. 1. SOITAAB SL 25 laser unit.

Laser head

The laser head is designed with a modular mechanical structure which allows to double the output power, connecting two 2.5 kW modules together. In such a way it is possible to obtain a 5 kW laser from two SL 25 units, whose hardware is unchanged, except the length of the cavity rods.

A 3000 rpm axial blower makes to flow the gas mixture through the discharge region at a velocity of 50 m/s. The gas pressure is 40 mbar and the gas consumption is 100 Nl/h.

The discharge module consists of a water cooled cathode and an uncooled anode, separated by a 5 cm gap. The cathode is a special pipe electrode, while anode is segmented in a set of resistively ballasted pads. Approximately 10% of the supplied electrical power is dissipated by ballast resistors located inside the vacuum vessel. Pyrex plates are used as a duct of the gas flow across the discharge volume.

The optical cavity is designed with innovative concepts. The mechanical structure of the cavity is entirely put inside the vacuum vessel. This solution avoids the misalignment of the optics when the vacuum vessel is evacuated, after having carried out the cavity alignment by He-Ne laser at atmospheric pressure.

The optical cavity is a 5 passes resonator, folded by means of copper mirrors. All the cavity optics are mounted in two identical compact mechanical blocks. The block of rear optics contains a power sensor to perform the laser power feedback. This sensor makes possible to monitor and control on-line the laser power, avoiding the need of using a chopper blade inserted in the output beam path. Chopper blade causes in fact disturbances on the output power, which affect the quality of the processed workpieces. Due to the particular scheme of the used optical cavity, the welding resonator may be changed into the heat-treatment resonator only replacing one mirror. Both the type and the position of all the other cavity optics remain unchanged. Therefore the prealignment procedure of the cavity by He-Ne laser is extremely simplified and time-saving. The welding cavity is an unstable resonator with magnification ratio of 2 and output beam diameter of about 45 mm; alternatively a low order mode gaussian beam from a stable resonator can be used for welding processing. The output beam for heat-treatments is a multimode Laguerre-Gauss beam with a flat top intensity profile and diameter of 40 mm.

SL 25 laser source is equipped with a very compact beam diagnostic unit (Fig. 2), mounted on the laser head. The unit dimensions are 450 mm x 500 mm x 350 mm. The diagnostic unit contains the beam shutter, with opening and closing time of 0.1 s, the He-Ne laser, providing both the prealignment of the cavity and the external optics up to the workpiece, and the output beam mode viewing system. The unit allows the user to make an easy on-line inspection of the output beam characteristics and to operate the drives of the cavity alignment, which is accomplished moving one motor-driven mirror. Due to the beam diagnostic unit, also the on-line alignment of the cavity is greatly simplified. In fact the operator can see the image of the output beam, by means of

Fig. 2. SL 25 laser beam diagnostic unit.

an IR sensitive plate, and, at the same time, act on the alignment drives, avoiding the need of using several perspex plates placed in the beam path.

Power and control unit

SL 25 laser unit is microprocessor controlled. The microprocessor is caring of all the internal functions within the laser, such as the control of gas flows and pressures, water flows and temperatures and safety devices.

Due to the microprocessor control, the user can program from the console keyboard all the working cycles required by the industrial needs (ramp-up, dwell time, power levels, ramp-down, waiting time, cycles number). The working cycle is performed by the output power controlled operation, which is obtained using a power feedback. The minimim ramp-up/down time of the laser output power is about 0.1 s and ± 2 % of long term power stability is achieved, Moreover the power feedback reduces to few minutes the time to reach the laser steady-state from cold start. All the operations concerning laser start-up, shutdown and source conditioning after maintenance are fully automated in order to make easier and more reliable the use of the laser in industrial environment and to reduce most of the time consuming procedures. Finally the laser can also be remotely operated in conjunction with CNC machines and process controllers.

LASER PERFORMANCES

The results of some preliminary welding tests are shown in Fig. 3. The figure shows the penetration depth on stainless steel AISI 304 vs. the workpiece velocity. The 2.5 kW laser beam has been focused on the test specimens with a meniscus ZnSe lens having the focal length of 5 inches and a coaxial He shielding gas flow at the rate of 15 l/min has been used.

CONCLUSIONS

In order to extremely simplify all the operations regarding laser working and source maintenance, a novel highly reliable 2.5 kW CO_2 industrial laser has been built.

The microprocessor control and the incorporated laser beam diagnostic unit allow the user to accomplish in a very simple and efficient way all the operations required in industrial laser processes.

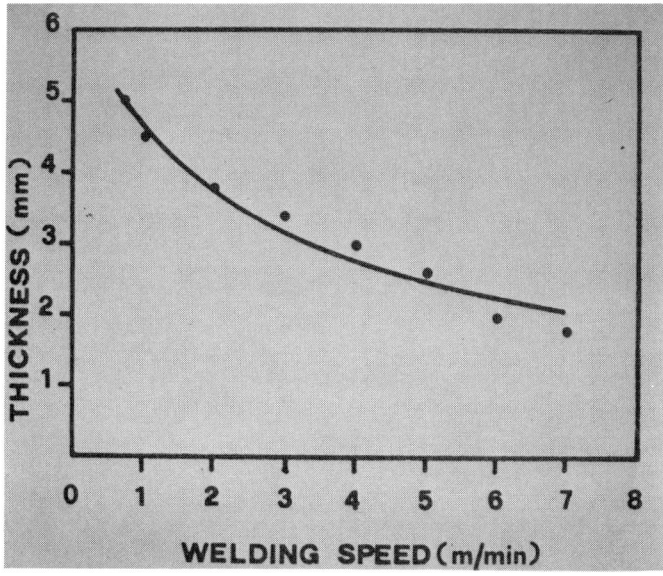

Fig. 3. Penetration depth on AISI 304 vs. the welding speed. Focal length 5"; He flow rate 15 l/min.

Chapter 5

RESEARCH AND DEVELOPMENT IN LASER TECHNOLOGY AND APPLICATION

MEASURING OF SURFACE ROUGHNESS BY HOLOGRAPHIC PHASE SHIFTING INTERFEROMETRY

Ming Chang and Ching-Piao Hu*

Department of Mechanical Engineering, Chung-Yuan Christian University, Chung Li, Taiwan
*Department of Mechanical Engineering, National Taiwan University, Taipei, Taiwan

ABSTRACT

An optical surface roughness measuring instrument has been developed which provides a noncontact method of obtaining surface characteristics of engineering surfaces. The system consists of a standard Twyman-Green interferometer controlled by a piezoelectric transducer, a linear array of photodiode detectors, and a microcomputer. The combination yields a system that measures the height variations of surfaces to a high degree of precision. Theoretical analysis and experimental systems are described. Surface roughness measurements are made in mirror, pre-coating surface, epoxy-painting surface, and fine-machined aluminium surface. Experimental results show that the measuring range of the sensitivity is from below 0.001 μm up to 2 μm rms. Due to the ability of high precision and fast measurement speed, an on-line metrological tool for roughness measurement of engineering surfaces can be developed.

Key words: surface roughness measurement, Twyman-Green interferometer, phase-shifting interferometry, detector array, computer control.

INTRODUCTION

The measurement of the surface roughness of engineering surfaces is becoming increasingly important in industry. Until now, the usual method to determine the surface roughness is the stylus method [1], that amplifies and records the vertical motions of a stylus as it is moved at a constant velocity across the surface to be measured. The diameter of the stylus is very small, typically 2 μm, to detect finer asperities. A load of about 25 mg is applied during tracing. This technique is quite adequate for hard materials. However, particularly in the area of fine machining, the process has fairly large disadvantages as: (1) The stylus exerts a pressure on the surface to be tested so that the surface is damaged. (2) Point-by-point measurement take a longer profile measurement time.

For fine-machined surfaces and soft surfaces, noncontacting techniques such as optical and interference methods are preferable. The state-of-the-art techniques in optical roughness measurement [2-5] including specular reflectance, total integrated scatter, diffuseness, angular scattering distributions, speckle, ellipsometry, and interferometry. Overall, no currently available technique combines accuracy and speed needs for profile measurement of engineering surface. The need for a better technique to measure quantitatively surface microtopography of fine-machined surface and soft surface is obvious.

This paper describes a noncontact optical measurement instrument for measuring surface roughness that uses the principle of interferometry, electronic phase measurement techniques, and computer analysis. The main advantages of this instrument over other optical methods including high accuracy, simple design, short measurement time, and its good sensitivity in the measuring range from 0.001 μm to 2 μm rms surface roughness. This instrument is available commercially and has been used in many areas of application [6-8].

OPTICAL AND DIGITAL INSTRUMENT

Figure 1 is a schematic of the optical and digital instrument.

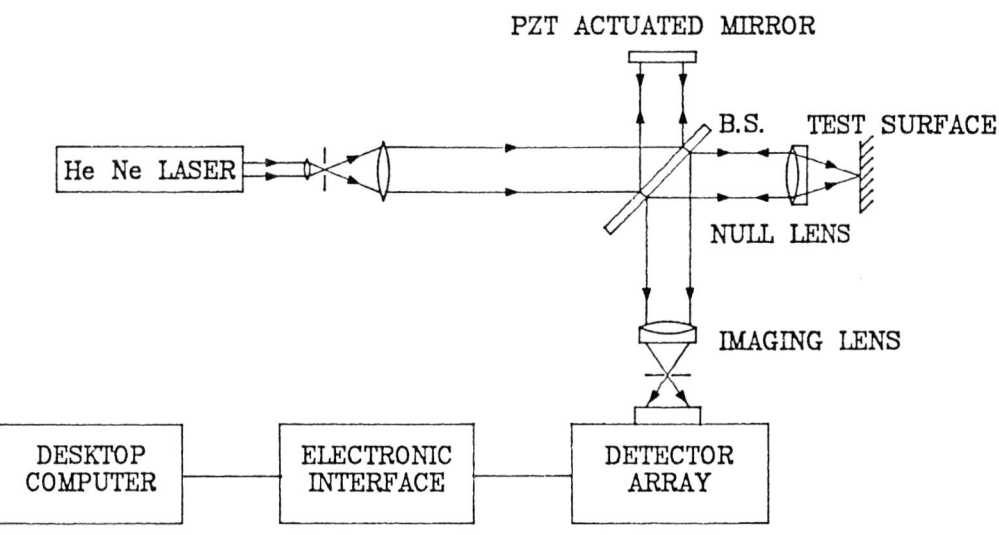

Fig. 1. Schematic of the optical and digital instrument for surface roughness measurement.

A Twyman-Green interferometer is modified to incorporate an optical phase-shifting device and a solid-state, linear array of 1728 photodiode detectors, which provides the capability of accurate surface height measurements. Using the linear detector array, surface profiles over a small region of a sample can be obtained, and surface height measurements made available at each detector location. For random surface roughness measurements, the surface height data form a basis of rms surface roughness, is defined by

$$\text{rms} = \left(L^{-1} \int_0^L Z^2(x) dx \right)^{\frac{1}{2}}, \qquad (1)$$

where the quantitative surface height data $Z(x)$ can be obtained using phase shifting interferometry [9,10] which is a technique that can determine the shape of a surface or wavefront by calculating a phase map from measured intensities, and rms is equal to one standard deviation of the profile about the mean line.

The Twyman-Green interferometer is built by mounting the reference mirror on a piezoelectric transducer (PZT). While voltage is applied to the PZT transducer, providing a phase modulation by modulating the optical phase difference between the test and the reference arms of the interferometer. This causes the interference fringes to shift in position. The PZT transducer is either stepped or ramped four times, each corresponding a constant phase shift. After each phase shift, the interference pattern from the interference of the wavefronts caused by the test surface and the reference surface is recorded by the photodiode detector array and stored in a microcomputer.

The detector output is digitized by a 10-bit A/D converter, and the results are put into computer storage using a direct memory access interface. The microcomputer then solves for the phase values (which are proportional to the surface heights) after the four measurements by using the algorithm described below. The microcomputer also controls the PZT through a D/A converter. The calculated surface heights are then sent, via a GPIO parallel interface, to a desktop computer.

For this phase-shifting technique to work, the phase difference between adjacent pixels in the measured wavefront must be less than π. This restriction is simply a consequence of the sampling theorem. It will limit the measurable imaging area of the test surface (which is adjusted by the null lens), and thus limits the test sensitivity.

SURFACE ROUGHNESS MEASUREMENT ALGORITHM

For a two-beam interferometer, the detected intensity due to the interference of the two beams is

$$I(x,y) = I_0[1 + \gamma \cos\phi(x,y)], \qquad (2)$$

where $I(x,y)$ is the intensity distribution in the interference plane, I_0 is the average intensity, γ is a constant representing fringe contrast, and $\phi(x,y)$ is the phase difference between the test and the reference arms of the interferometer that we want to measure.

When a phase shift is induced into the reference wave, the resulting interference fringes is also shifted the same amount. If four frames of intensity data are recorded with the phase of the reference beam changed by an amount 2α between each readout, we have

$$A(x,y) = I_0\{1 + \gamma\cos[\phi(x,y) - 3\alpha]\},$$
$$B(x,y) = I_0\{1 + \gamma\cos[\phi(x,y) - \alpha]\},$$
$$C(x,y) = I_0\{1 + \gamma\cos[\phi(x,y) + \alpha]\}, \qquad (3)$$
$$D(x,y) = I_0\{1 + \gamma\cos[\phi(x,y) + 3\alpha]\}.$$

The phase distributions $\phi(x,y)$ at each detection point (x,y) can be determined by

$$\phi(x,y) = \arctan\left[\frac{\sqrt{[(A-D) + (B-C)][3(B-C) - (A-D)]}}{(B+C) - (A+D)}\right]. \qquad (4)$$

A similar method for measuring the phase is the ramping technique. Ramping means to move the reference mirror at a rate linearly with time. The computer reads the intensity readings out of the integrating detector array as the reference mirror moves. While the phase of the reference wave is shifted linearly with time over a range of -2δ to $+2\delta$, each integrated intensity frame corresponding phase shift δ, then the measuring phase $\phi(x,y)$ can also be obtained from Eq. (4).

Since the surface height distribution is relative to a mean value, the decision of the initial phase data is arbitrary. Also, the calculation of Eq. (4) is independent of the amount of the constant phase shift between each frame of intensity data, this technique eliminates the need for calibration of the phase shifter PZT.

Once the phase, $\phi(x,y)$, is determined across the interference field, the optical path difference (OPD) for the object wavefront relative to the reference wavefront is easily determined from the phase

$$OPD(x,y) = \phi(x,y)\lambda/2\pi. \qquad (5)$$

It is related to the surface height distribution $Z(x,y)$ by a multiplicative factor

$$Z(x,y) = \frac{OPD(x,y)}{\cos\theta + \cos\theta'} = \frac{\phi(x,y)\lambda}{2\pi(\cos\theta + \cos\theta')}, \qquad (6)$$

where θ is the object illumination angle, and θ' is the view angle measured relative to the surface normal. This factor is one-half for a double-pass interferometer like a Twyman-Green.

Once the height distribution across the sample is measured, the data can be fitted in a least-square method to determine the average height and the tilt across the sample. The average height is subtracted from the height distribution. The tilt is also subtracted because it is arbitrary and depends upon the adjustment of the reference surface in the interferometer.

RESULTS AND DISCUSSION

As mentioned previously, the interference fringes is read out by a solid state detector array. Figure 2 shows the typical output for looking at a finished surface. Figures 3 ~ 6 consist of a series of plots we have measured in studying the tribology of different surface media including aluminium-coating surface (mirror and pre-coating surface), epoxy-painting surface, and fine-machined aluminium surface. These plots can be made for a single data set or for the average of many data sets.

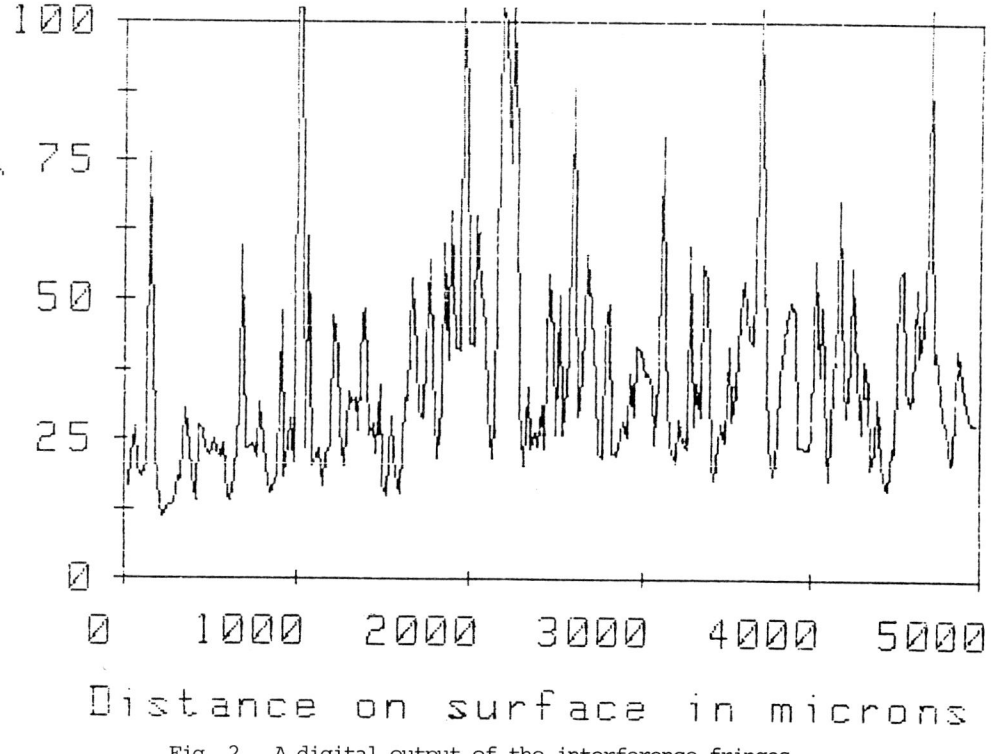

Fig. 2. A digital output of the interference fringes formed by the reference wave and the test surface.

In obtaining the surface profile, tilt is always subtracted, and the rms and peak-valley are calculated. In the histogram of the surface height distribution, the surface height values are displayed normalized to the maximum. A bias of the distribution of surface heights away from zero shows more holes or peaks in the surface.

The experimental results show that the measuring range of the sensitivity for now is from below 0.001 μm up to 2 μm rms value. For the roughness measurement with surface roughness is over 2 μm, this technique is still available by means of decreasing the test area and adding the number of the pixels of the detector array. The repeatability depends upon how rough the surface is

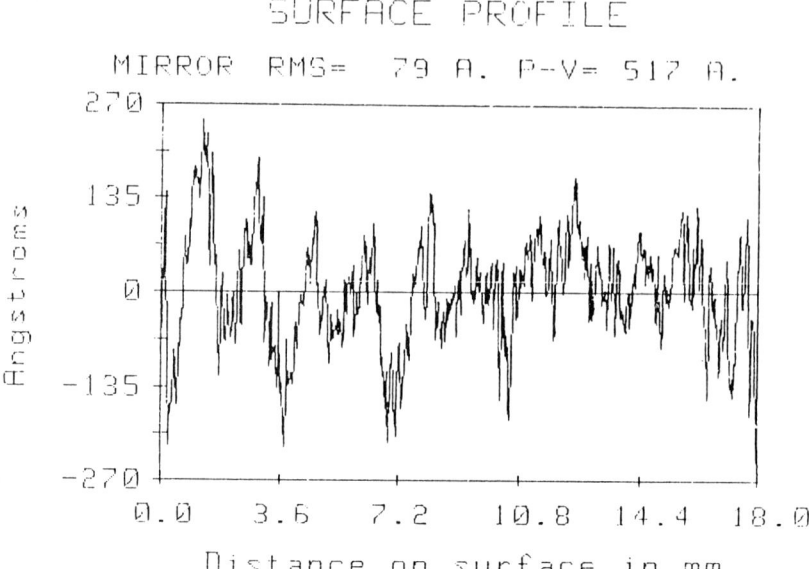

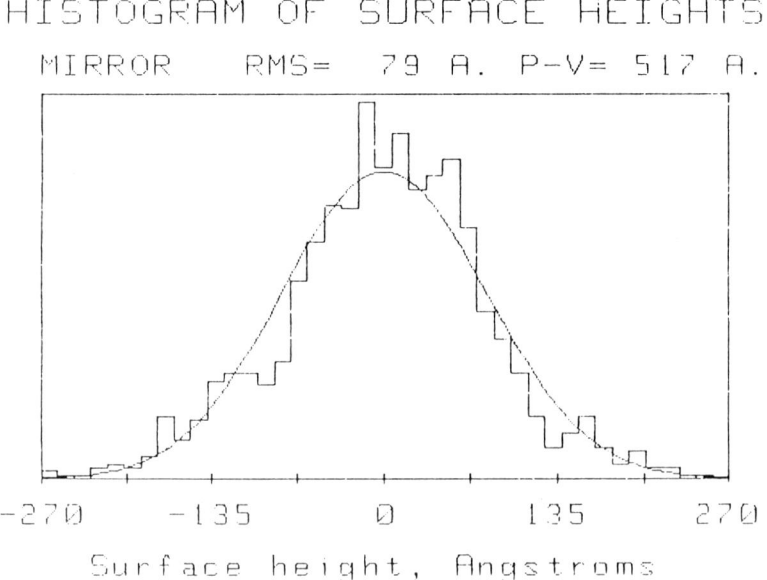

Fig. 3. (a) Surface profile for a test mirror.
(b) Histogram of surface heights for Fig. 3(a).

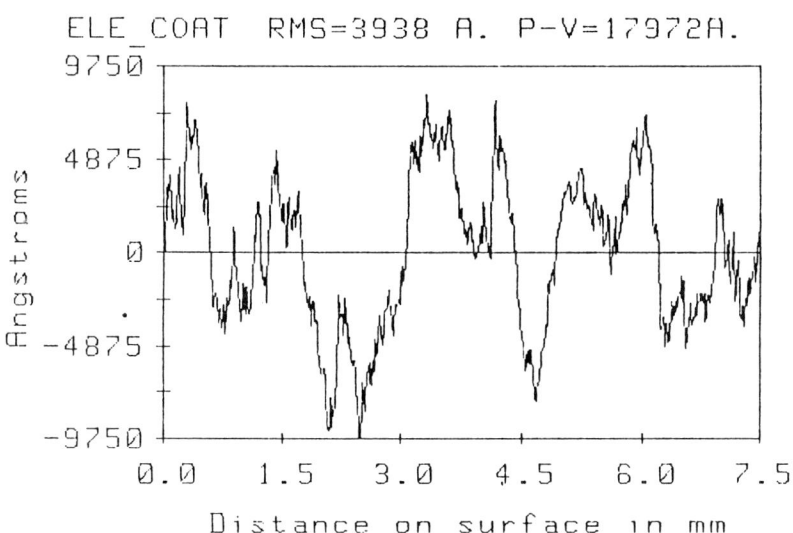

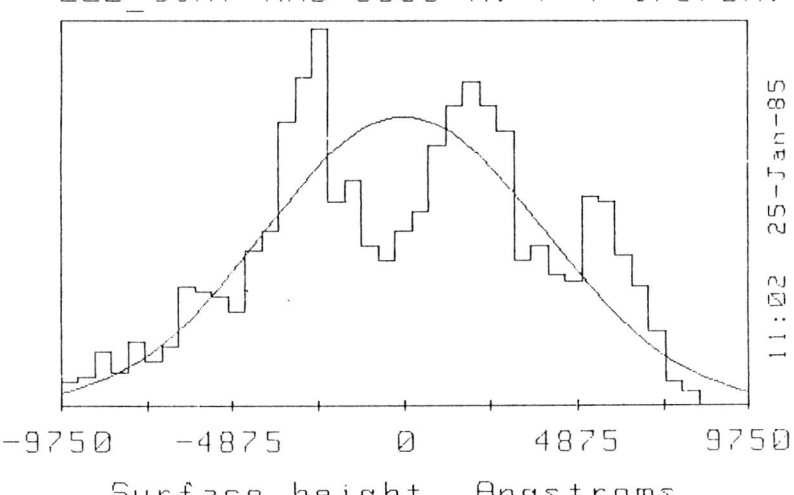

Fig. 4. (a) Surface profile for a pre-coating surface.
(b) Histogram of surface heights for Fig. 4(a).

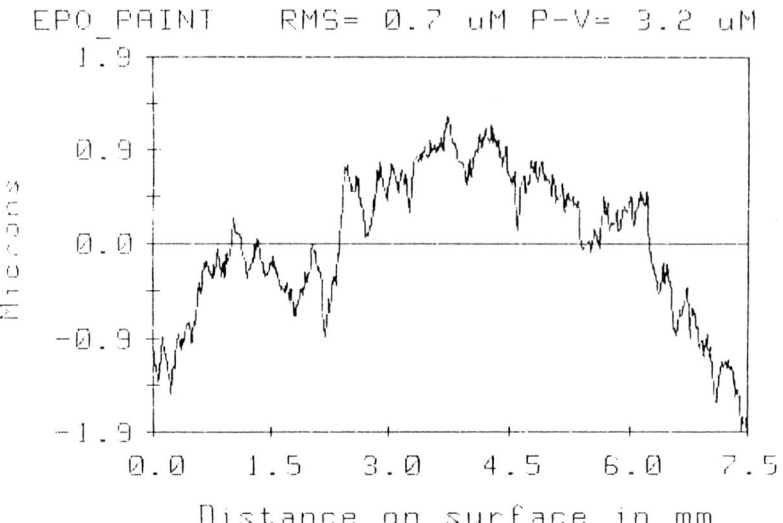

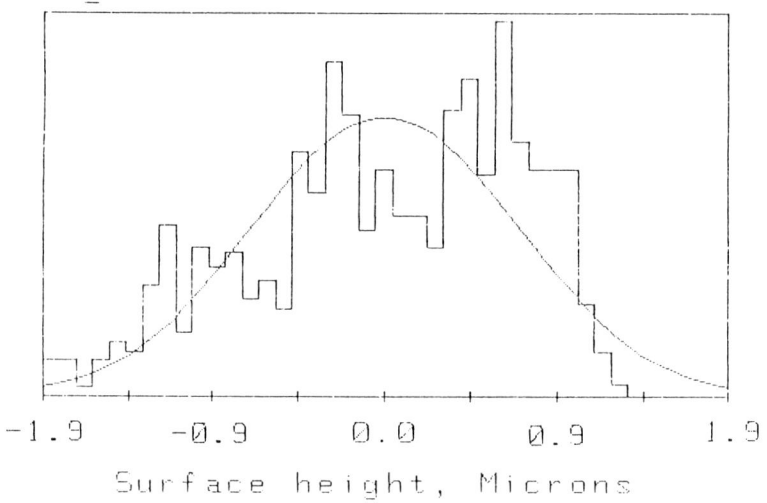

Fig. 5. (a) Surface profile for an epoxy-painting surface.
(b) Histogram of surface heights for Fig. 5(a).

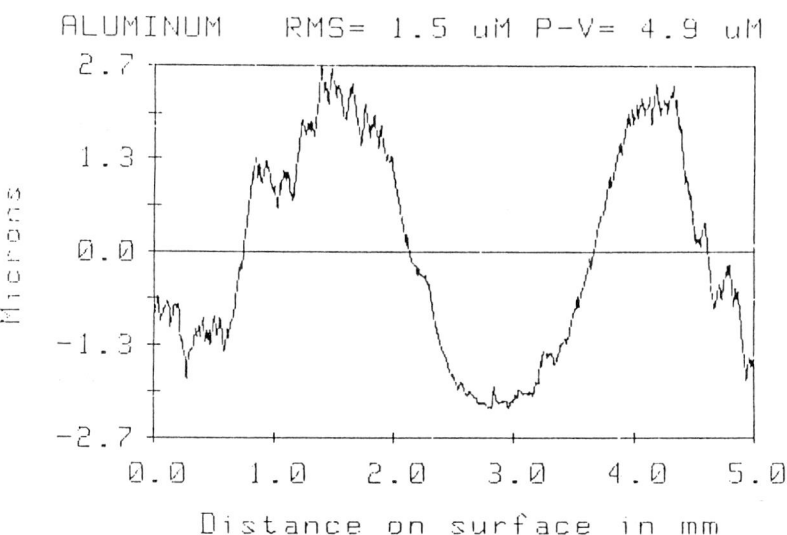

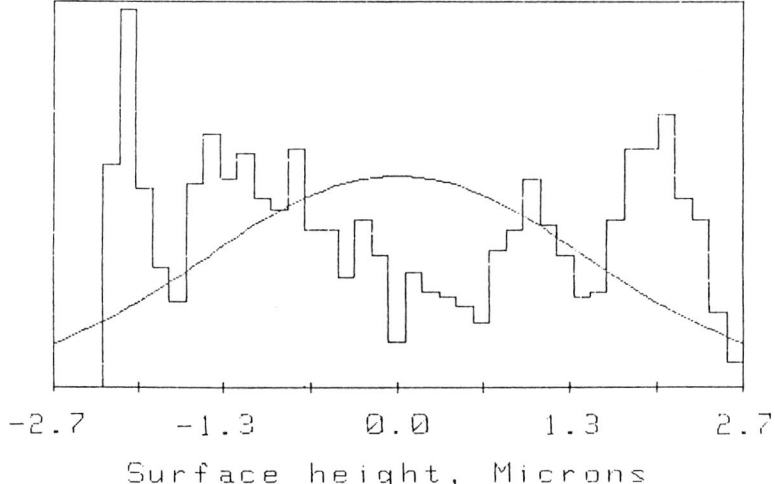

Fig. 6. (a) Surface profile for a fine-machined aluminium surface.
(b) Histogram of surface heights for Fig. 6(a).

and how good the surface reflectivity is. A typical repeatability for aluminium is 5% the rms value.

This technique uses computer control to take data and calculate surface height distribution. A single surface profile measurement can be taken in as short a time as 0.1 seconds with a graphics display of the surface profile in about 3 seconds. Owing to the on-line surface roughness measurement of engineering surfaces is becoming increasingly important, the future work will strive to improve the overall system and software to effectively execute a high speed on-line observation.

To obtain the full-field surface characteristics, scans at different locations on the surface should be made. The rms value of each scan or profile should be averaged to obtain a better indication to the overall height distribution rather than the local behaviour.

As a final discuss, the optical roughness measurement techniques developed here are not dependent on any single interferometric implementation. A variety of interferometers can be modified to incorporate a phase-shifting device and an integrating detector array. The combination of these techniques will be a tremendous aid to precise machining such as the overall flatness of optical disks, the surface of the diamond-turned head wheel of video recorders, and many other things.

CONCLUSIONS

A noncontact surface roughness measuring instrument has been successfully developed by using Twyman-Green interferometer. From the digital outputs of four successive intensity data frames of the interference pattern, the surface height distribution of test surface are calculated using phase-shifting interferometry technique. The instrument can be applied to any kind of surface media and is especially helpful for the roughness measurement of fine machined surfaces and soft surfaces. The accuracy decreases with larger roughness value and lower surface reflectivity. The typical error for aluminium material is 5% the rms value. Due to the ability of high precision and fast measurement speed, the on-line roughness measurement of engineering surfaces may be executed.

REFERENCES

1. Thomas, T. R. (1978). Surface Roughness Measurement: Alternative to the Stylus. No. 4.2, NELEX 78, National Engineering Laboratory, East Kilbride, Glasgow.
2. Vorburger, T. V. and Teague, E. C. (1981). Optical techniques for on-line Measurement of Surface Topography. Prec. Eng., 3, 61-83.
3. Teague, E.C., Vorburger, T. V. and Marstre, D. (1981). Light Scattering from Manufactured Surfaces. CIRP Annals, 30, 563-569.
4. Warneke H. J. and Ahlers R. J. (1984). Proc. IMEKO - Symposium on Measurement and Estimation, Bressone, Italy.

5. Brodmann, R., Gerstorfer, O. and Thurn, G. (1985). Optical Roughness Measuring Instrument for Fine-Machined Surfaces. Opt. Eng., 24, 408-413.
6. Wyant, J. C., Koliopulos, C. L., Bhushan, B. and George, O. E. (1984). An Optical Profilometer for Surface Characterization of Magnetic Media. ASLE Trans., 27, 101-113.
7. Chang, M., Hu, C. P., Lam, P. and Wyant, J. C. (1985). High Precision Deformation Measurement by Digital Phase Shifting Holographic Interferometry. Appl. Opt., 24, 3780-3783.
8. Creath, K. (1985). Phase-Shifting Speckle Interferometry. Appl. Opt., 24, 3053-3058.
9. Carre, P. (1966). Installation et utilsation du comparateur photoelectrique et interferential du Bureau International des Poids et Mesure. Metrologia, 2, 13-23.
10. Koliopoulos, C. L. (1981). Interferometric Optical Phase Measurement Techniques. Ph. D. Dissertation, U. Arizona.

SUPERRESOLUTION IN MICROSCOPY THROUGH HOLOGRAPHIC SYNTHETIC APERTURE

P. De Santis, F. Gori, G. Guattari and C. Palma

Dipartimento di Fisica Universita' La Sapienza — P. le A. Moro, 2–I–00185 Roma, Italy

ABSTRACT

A synthesis principle, implemented through holography, has been applied to microscopy at large space-bandwidth product. Indicatively, the method succeeds in doubling the aperture of a good microscope objective. The experimental test shows the feasibility of the method.

INTRODUCTION

Ernst Abbe stated the limit of resolution of optical systems for the case of coherent illumination (1). In up-to-date terms, every optical system transfers a limited band of spatial frequencies that depends on the used wavelength and on the numerical aperture of the system itself (2).

Attempts to overpass the classical limit of resolution are known as superresolution techniques and have been experimented since the very times of Rayleigh. The research in this field has followed many ways: superresolving pupils have been invented (3) and extensively studied (4), although practical devices seem difficult to realize. Many a posteriori techniques, i.e. techniques for obtaining supersolution by processing the ordinary image, have been studied and implemented; among them, we may quote eigenfunction technique (5), iterative procedures (6) and singular value decomposition (7). Another class of attempts have been performed to artificially increase the bandpass of the optical system. This can be obtained by shifting to the passing region, with a proper coding technique, those frequencies which would be otherwise stopped by the system. The correct final image is then obtained through subsequent decoding and suitable frequency re-shifting (8,11). This procedure can be thought of as a synthetic aperture technique (2). Finally, the general problem of optical superresolution has been recently reviewed and the influence of some general parameters (like the space-bandwidth product) on the information capacity has been evidentiated (9).

In this paper, we intend to show how a synthesis technique can be used for obtaining superresolution in microscopy. Superresolution is, of course, very important in microscopy, where the imaging system is often characterized by a high space-bandwidth product. This prevents from using several techniques that give good results only for low space-bandwidth products (4-7). The method we propose is based on a holographic technique. As is well-known, the merge of microscopy and holography dates from the very birth of the holography (2). A holographic method has been already applied to cases of low space-bandwidth product (11) and, here, we extend it to a case of microscopy with a high space-bandwidth product. Superresolution is achieved (up to a factor of two for a good microscope objective) through the synthetic realization of an effective aperture larger than the actual aperture of the optical system.

THEORY

We consider a one-dimensional plane transmitting object (for the sake of simplicity, we limit ourselves to the one-dimensional case, the extension to two dimensions being straightforward). Making reference to an axis ξ we associate to the object an amplitude transmission function $\tau(\xi)$ (see Fig. 1) and we denote by $\tilde{\tau}(\nu)$ its Fourier transform (F.T.). We shall assume that the spatial frequency spectrum of the object is contained in the interval $(-\nu_M, \nu_M)$.

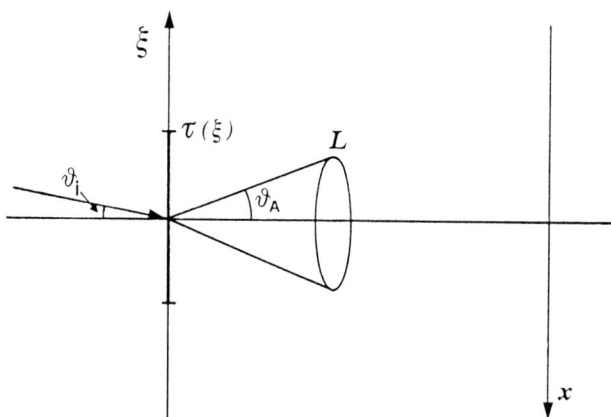

Fig. 1. Imaging with oblique illumination.

Let a plane wave (with wavelength λ)

$$V_i(\xi) = A \exp(2\pi i \xi \sin\theta_i/\lambda) \qquad (1)$$

impinge on the object plane with an angle θ_i with respect to the optical axis and let θ_A be the angular aperture of a microscope objective L, imaging the object on the x-axis. The amplitude distribution just behind the object plane is

$$V(\xi) = V_i(\xi)\tau(\xi) = \tau(\xi) \; A \; \exp(2\pi i \xi \sin\theta_i/\lambda). \qquad (2)$$

Because of its finite angular aperture, the microscope objective cuts off the spatial frequencies at a maximum value $\nu_A = \sin\theta_A/\lambda$. If $\nu_A < \nu_M$, object details corresponding to spatial frequencies greater than ν_A are lost and a superresolution technique is necessary in order to recover them.

Let us assume for the objective a rectangular transfer function rect $(\nu/2\nu_A)$, giving rise to an impulse response $2\nu_A\mathrm{sinc}(2\nu_A x)$. Taking into account the linear magnifying power M of the microscope objective, the image amplitude distribution is

$$U(x) = V(x/M) * \mathrm{sinc}(2\nu_A \cdot x/M), \qquad (3)$$

where * stands for convolution and, as usual, $\mathrm{sinc}(x)=\sin(\pi x)/\pi x$. Here and in the following unessential proportionality factors are omitted. From eqs. (2) and (3) it is immediately seen that the F.T. of $U(x)$ is

$$\tilde{U}(\nu) = \tilde{\tau}\left[M(\nu - \frac{\sin\theta_i}{M\cdot\lambda})\right] \cdot \mathrm{rect}(M\nu/2\cdot\nu_A). \qquad (4)$$

Because of the frequency shifting $\nu_r = \frac{\sin\theta_i}{M\lambda}$, the image spectrum contains object frequencies belonging to the interval $(-\nu_A-\nu_T, \nu_A-\nu_T)$ (see Fig. 2). By suitably changing θ_i, we can allow all the bandwidth of the object to pass through the objective aperture in sequential times.

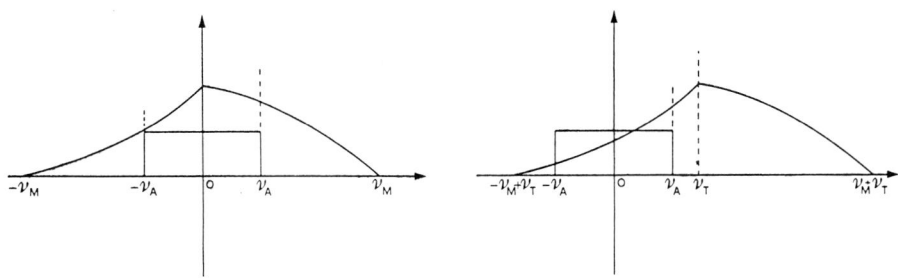

Fig. 2. Frequency shifting of the object spectrum.

Holography provides a mean to record different "images" corresponding to different θ_i on a same holographic plate and to reconstruct a correct superresolved image. To explain the method let us refer to Fig. 3. Suppose the object is successively illuminated with plane waves at angles θ_{ik} ($k = 0,1,2,\ldots n$) with respect to the optical axis, in such a way that different and contiguous portions of the object bandwidth pass through the pupil of the optical system. This happens if the frequency shifting ν_T is just equal to the bandwidth $2\nu_A$ of the objective pupil. Of course, the process must be repeated at the symmetrical angles $-\theta_{ik}$ in order to transfer the opposite part of the object band. Therefore, we have

$$\sin\theta_{ik}/M = \lambda k \nu_T = 2\lambda k \cdot \nu_A, \qquad (k=0,1,2,\ldots n) \qquad (5)$$

We now take a hologram of the k-th image with a reference plane wave R_k making an angle ϕ_k with respect to the optical axis and record multiple exposures on the same plate P. For every exposure, interference between the image amplitude distribution and the reference wave takes place. The holographic virtual image term has the F.T.

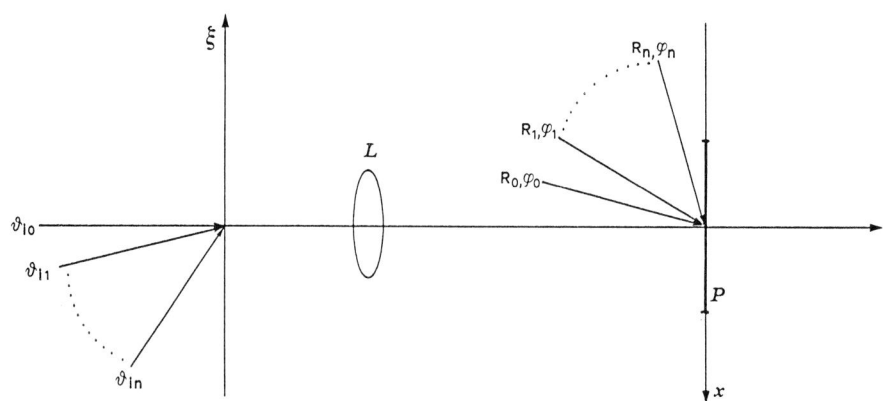

Fig. 3. Multiple-exposure hologram recording.

$$\tilde{F}_k(\nu) = \tilde{\tau}\left[M(\nu - \frac{\sin\theta_{ik}}{M\cdot\lambda} + \frac{\sin\phi_k}{\lambda})\right]$$
$$\cdot \operatorname{rect}\left[\frac{M(\nu+\sin\phi_k/\lambda)}{2\nu_A}\right], \qquad (k=0,1,2,\ldots n) \qquad (6)$$

If the angles ϕ_k satisfy the conditions

$$M(\sin\phi_k - \sin\phi_0) = \sin\theta_{ik} = M2\lambda k\nu_A, \qquad (k=0,1,2,\ldots n) \qquad (7)$$

then eq. (6) becomes

$$\tilde{F}_k(\nu) = \tilde{\tau}\left[M(\nu + + \frac{\sin\phi_o}{\lambda})\right] \cdot \text{rect}\left[M(\nu + \frac{\sin\phi_o}{\lambda} + 2k\nu_A)/2\nu_A\right], \quad (k=0,1,2,\ldots n) \quad (8)$$

Reconstructing with an illuminating plane wave making an angle ϕ_O with respect to the optical axis the virtual image term gives rise to an image contribution whose F.T. is

$$\tilde{G}_k(\nu) = \tilde{\tau}(M \cdot \nu) \cdot \text{rect}\left[M(\nu + 2k\nu_A)/2\nu_A\right], \quad (k=0,1,2,\ldots n) \quad (9)$$

In eq. (9) the shifted rect functions are contiguous, so that, if their sum amounts to the entire object spectrum (see Fig. 4), viewing the hologram illuminated at an angle ϕ_O is equivalent to viewing the object through a normally illuminated optical system having a synthetic angular aperture able to transfer the entire bandwidth $2\nu_M$.

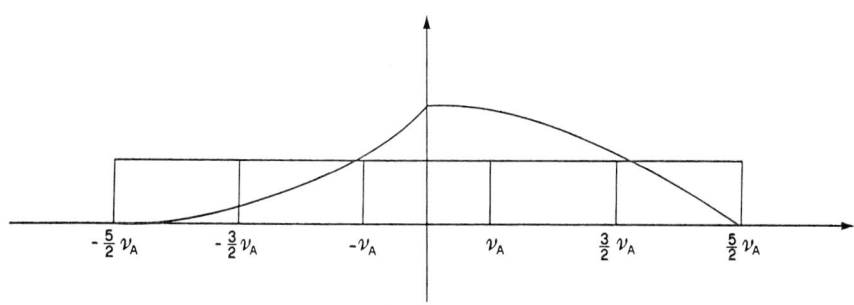

Fig. 4. Holographic reconstruction of the object spectrum.

EXPERIMENT

The experimental setup is represented in Fig. 5. The object G is a sinusoidal amplitude grating previously realized through interference of two simmetrical plane waves making angles $\theta' = \pm 62°$ with respect to the normal axis. With this configuration, the spacing P of the grating equals a portion 1/1.77 of the wavelength of the light we used (He-Ne laser)

$$p = \lambda/|2\sin\theta'| = \lambda/1.77 = 0.36\mu m \quad (10)$$

The imaging lens L is an achromatic dry microscope objective having a magnifying power 80X and a numerical aperture (N.A.) 0.9; the corresponding angular aperture is $\theta_A = 64°12'$. If

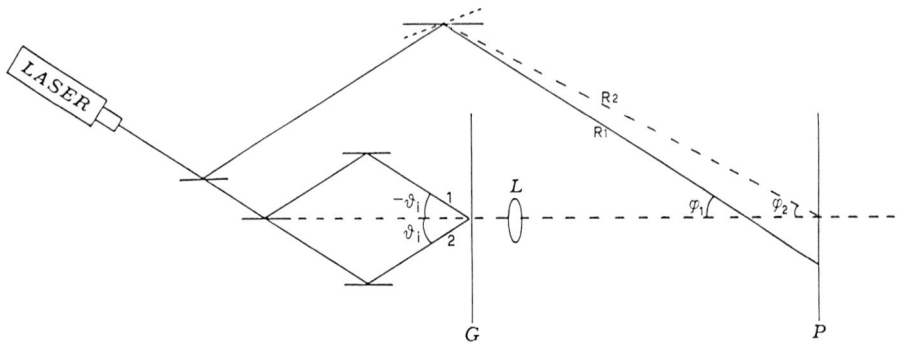

Fig. 5. Experimental setup.

normally illuminated, the grating would diffract its first orders at angles θ_D satisfying the conditions $\sin\theta_D = \pm\lambda/p\ (|\theta_D|<90°)$. In our case, this would give imaginary diffracted orders and no image of the grating could be obtained through this objective.

In our experiment, two oblique plane waves impinged on the grating at symmetrical angles with respect to the optical axis (illuminations 1 and 2 in Fig. 5). In correspondence to them, two holograms were recorded on the same plate P with reference waves R_1 and R_2, respectively. For the sake of symmetry, we choose $\theta_i = \theta' = 62°$. In this condition, the grating G acts as a hologram, reconstructing the beam 2 when illuminated with the beam 1 and vice-versa. The values chosen for θ_i assure that the diffracted waves can enter the angular aperture of the objective and an image of the grating can be formed on the plate P. In the same time, the best diffraction efficiency is obtained (2). The angles ϕ_1 and ϕ_2 of the reference waves in Fig. 5 have been chosen according to eq. (7), that is

$$\sin\phi_1 = \sin\phi_o + \sin\theta_i/M, \quad \sin\phi_2 = \sin\phi_o - \sin\theta_i/M, \quad (11)$$

where M is the linear magnifying power corresponding to the experimental distance D and ϕ_0 is the angle of the reference wave that would be used for a normal illumination of the object. We choose $\phi_0 = 14°$ and used the experimental values D=800 mm, M=56. From eqs. (12), it results ϕ_1 = 14 15'12", ϕ_2 = 13 04'48".

The double exposure hologram, realized in this way, has been illuminated with a reference wave at the angle ϕ_0. The image is formed through a lens with a magnification M', so that the total magnification M" is M"=M·M'. The final image is shown in Fig. 6, together with the (central) spot of the impulse response of the objective (taken at a distance D' such that the experimental magnification is M").

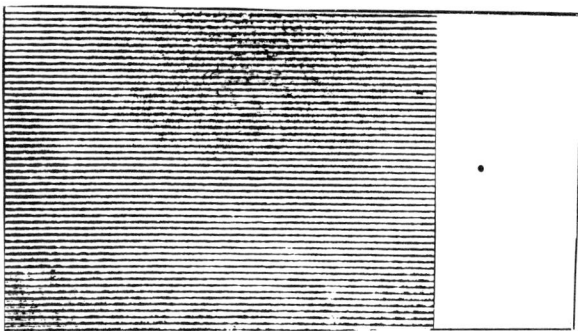

Fig. 6. Reconstructed grating image and optical system spread function.

Visual comparison gives a qualitative idea of the superresolution obtained with this technique. Quantitatively, the angular aperture synthesized is 124°, while the angular aperture of the objective is 64°12'. Roughly speaking, a doubling of the aperture has been obtained.

CONCLUSIONS

We proposed the application to microscopy of a holographic method for obtaining superresolution through a synthesis technique. Differently from other superresolution methods, this one works also with optical systems having large space-bandwidth products. As a matter of fact, at the maximum of its potentiality (grazing illumination, i.e. $\theta_i = \pm 90°$), the holographic synthetic method we proposed is able to double the bandpass of the best dry objective (N.A.=1), evidentiating details of the order of $\lambda/2$.

We performed an experimental check of the method, by holographically synthesizing an angular aperture nearly double of the real angular aperture of the microscope objective used for the experiment.

We think that an effort should be made toward the implementation of the method in the framework of the real-time holography. Such an extension should greatly enhance the potentiality of the method in the practical use.

REFERENCES

1. Die Lehre von der Bildenstehung in Mikroskop von E. Abbe (Vieweg, Braunschweig, 1910).
2. J. W. Goodman - Introduction to Fourier Optics - MacGraw Hill, 1968.
3. G. Toraldo di Francia, Nuovo Cimento, Suppl. 9, 426-438 (1952).
4. B. R. Frieden, Opt. Acta 16, 795-807 (1969); G. R. Boyer, Appl. Opt. 15, 3089-3093 (1976); R. Boivin and A. Boivin, Opt. Acta 27, 587-610 (1980); I. J. Cox, C. J. R. Sheppard and T. Wilson, J. Opt. Soc. Am. 72, 1287-1291 (1982).
5. C. K. Rushforth and R. W. Harris, J. Opt. Soc. Am. 58, 539-545 (1968); B. R. Frieden, in Progress in Optics, E. Wolf ed. (Pergamon, New York, 1971), vol. 9, pp. 311-407.
6. R. W. Gerchberg, Opt. Acta 21, 709-720 (1974); A. Papoulis, IEEE Trans. Circ. Syst. CAS-22, 735-742 (1975); P. De Santis and F. Gori, Opt. Acta 22, 691-695 (1975); C. K. Rushforth and R. L. Frost, J. Opt. Soc. Am. 70, 1539-1544 (1980).
7. M. Bertero, C. De Mol and G. A. Viano, in Inverse Scattering Problems in Optics, H. P. Baltes, ed. (Springer-Verlag, Berlin, 1980).
8. W. Lukosz, J. Opt. Soc. Am. 56, 1463 (1966); A. Bachl and M. Lukosz, J. Opt. Soc. Am. 57, 163 (1967); M. A. Grimm and A. W. Lohmann, J. Opt. Soc. Am. 56, 1151 (1966).
9. W. T. Cathey, B. R. Frieden, W. T. Rhodes and C. K. Rushforth, J. Opt. Soc. Am. A1, 241 (1984); J. Cox and C. J. R. Sheppard, J. Opt. Soc. Am. A3, 1152 (1986).
10. C. W. McCutchen, J. Opt. Soc. Am. 57, 1190 (1967).
11. M. Ueda, T. Sato and M. Kondo, Opt. Acta, 20, 403 (1973).

HOLOGRAPHIC OPTICAL ELEMENTS ON DICHROMATED GELATIN, FROM REFRACTIVE TO DIFFRACTIVE OPTICS

P. Meyrueis* and R. Piel*

*ENSPS, Strasbourg, France
**X-IAL, Strasbourg, France

1. INTRODUCTION

It has been known for centuries that a piece of glass cut in a spheric or a near spheric way on one or two faces can focus light and then can transform the image of the surrounding perceived by eye for, correction of vision, microscopes, telescopes, etc.

The basic phenomenon that allows this is the refraction of light when light passes from one transparent material to another with a different index.

The diffraction of light, while discovered later, gave something that is similar to refraction, but requires a grating or a hole.

Up to now refraction of light has been used to design and realise lens systems used for many applications.

But the limits are well known in aperture, aberration, dimension, weight and manufacturing cost. The lens systems were greatly improved by coatings, gradient index substrate, aspheric forms. But a theoretican limit exists, and effective technology is now very close to the theoretical limit of the possibility of lens systems based on refraction.

Among an increasing number of authors we explore diffractive optics based on holographic optical elements (HOEs) made on dichromated gelatin with our optrigelac process that gave them advantages including high diffraction efficiency, large size, and off axis asymmetric optics (Fig. 1).

2. METHODS OF PRODUCTION OF HOE

The holographic grating is a microfringe pattern produced in precise coherent conditions by the interference of several wavefronts. The holographic grating can be computer calculated directly or indirectly. Directly by plotting the grating with an electronic

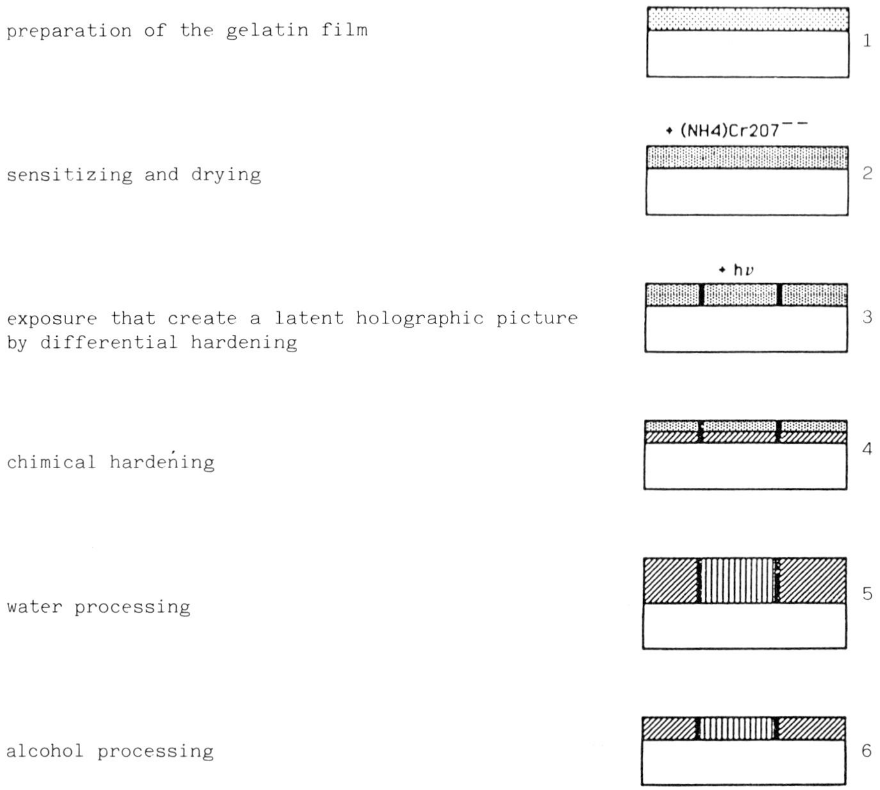

preparation of the gelatin film	1
sensitizing and drying	♦ (NH4)Cr2O7⁻⁻ 2
exposure that create a latent holographic picture by differential hardening	♦ hν 3
chimical hardening	4
water processing	5
alcohol processing	6

Fig. 1. Optrigelac method of production of HOE.

gun on a recording plate, or indirectly by using a computer made hologram that will generate with a coherent light wavefront that will interfere at a given location to give the desired holographic pattern. Point sources can also be used for generating spheric wavefronts for simple lenses. The set up for the recording is organized by computer programs.

An hybrid method both analogic and digital was also developed by Armtai from the Weizmann Institute to produce complex HOE from interfering wavefronts produced by simple holograms by changing the geometrics and the wavelength between recording and read out of holograms.

We are developing computer software for the realisation of an extended range of HOE for: head up displays, light multiplexing on optical fiber network, camera lenses, communication antenna for coherent light, solar captor, scanners for printers or laser T.V., laser disc lens etc.

HOE are lighter, more compact, have large aperture potential, can be manufactured easily and quickly. Their main inconvenience is in the aberrations that exist when the wavelength and angle of use differs from the recording parameters. A great part of these problems can be corrected by the use of a HOE system composed of several holograms with a diaphragm located between them that can limit the aberration.

Wemgartner in PTB produced holographic microscope objectives and photographic camera lenses that give good results with this method the centre of the image field being completely free of aberrations even for large aperture. One surface produced for one wavelength, so that this kind of imaging lens has no spherical aberration. It is possible to go further by introducing a certain spherical aberration into the hologram to cancel or balance the spherical aberration caused by a wavelength difference between the one used for production and the one used to image through the HOE.

It is necessary to use a multi-holograms lens because a simple holographic lens has a rather large coma. This coma degrades the imaged picture for a large aperture.

This side effect of HOE can be compared with the effect of parabolic mirror used in telescopes that are also aberration free on the optical axis but whose usable image field is small. One can imagine a large aperture multi-holographic lens telescope that will give with some counterpart very interesting results for space telescope for instance. But some software problems in the design of such systems have to be solved. Because the aberration of a simple glass lens increases rapidly with higher apertures, by an adapted glass and holographic lens combination putting diaphragms at the right place, the coma of the system can be made zero in the Seidel approximation. The wavefront aberrations are calculated in the classical case as a polynomial expression of the pupil coordinate and in this sense the Seidel approximation of the coma is of the third order in the pupil coordinate. The coma of such a glass lens can be aberration balanced by holography The Seidel coma is then not zero but very small and of the opposite sign to the coma for the 5th and 7th order. It is practically possible to balance the coma of the glass lens by the design of the holographic lens.

We can now examine the aberration of a diffractive and refractive hybrid system. The holographic lens corrects the spherical aberration of the glass lens and partially its chromatic aberration. It's sufficient to use a spherical wavefront aberration of the 4th and 5th order in the hologram in order to obtain a rather good correction of the spherical aberration of the system and diffraction limited behaviour. The data to take into account are the glass index, the radius of curvature of the lens, the two aberration coefficients of the holographic lens representing a spherical aberration of the 4th and the 6th order and its focal length. The index is chosen as high as possible to have as large a radius of curvature as possible that will give smaller geometrical aberration. There is a simple relation between the radius of curvature and the focal length. The focal length of the holographic lens can be used for the chromatic aberration of the system. Field curvature cannot be corrected as perfectly as astigmatism.

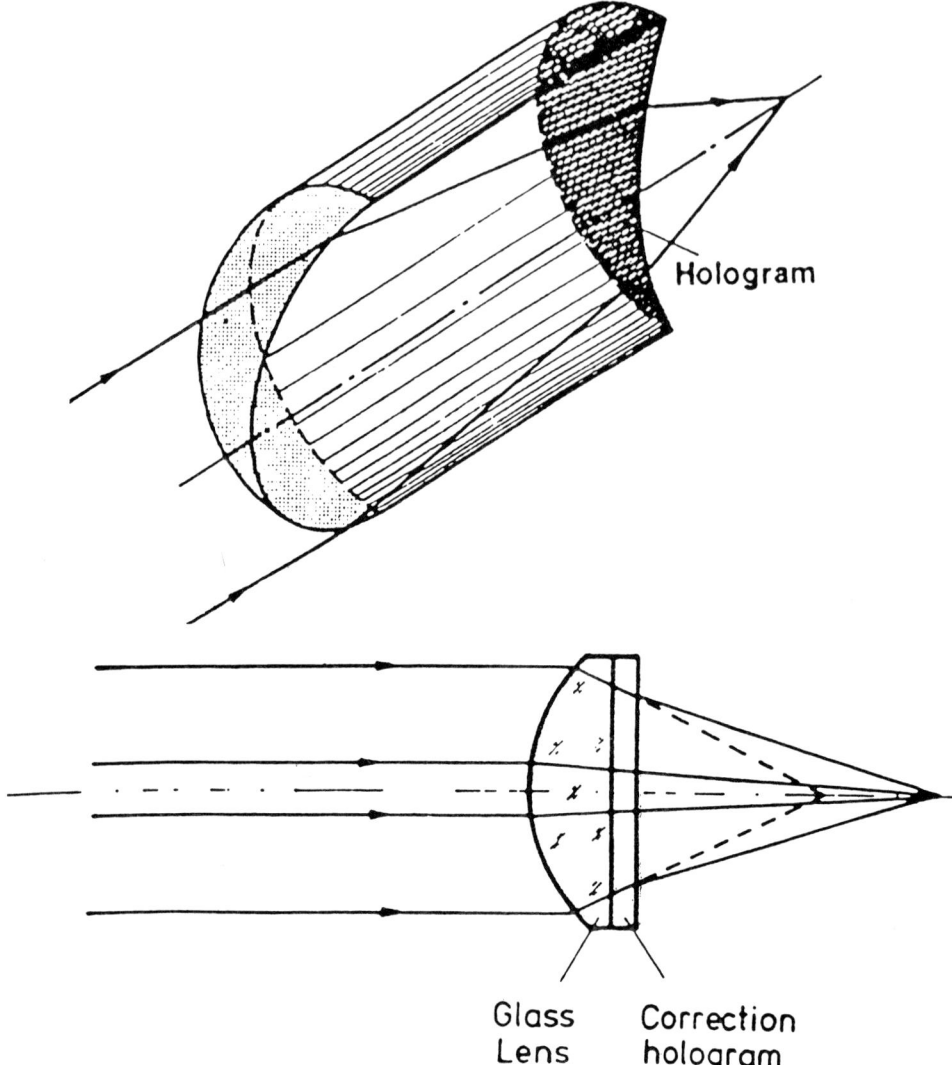

Fig. 2. Weingartner holographic optics for microscopy (2A) and videodisc (2B).

Holographic lenses also act as "intelligent mirrors" or be combined with mirrors. The main advantage of classical mirrors are they are free from chromatic aberration, they have small geometrical aberrations due to the smallness of the curvature of the spherical surface, and they can be used in a large spectrum range. But they have a blind zone around the centre of the aperture. It is possible only to realise long focal length and moderate image field and the centre of the optics, the best for aberration, cannot be used. Holography allows off axis use in good conditions with small focal length and with holography correcting

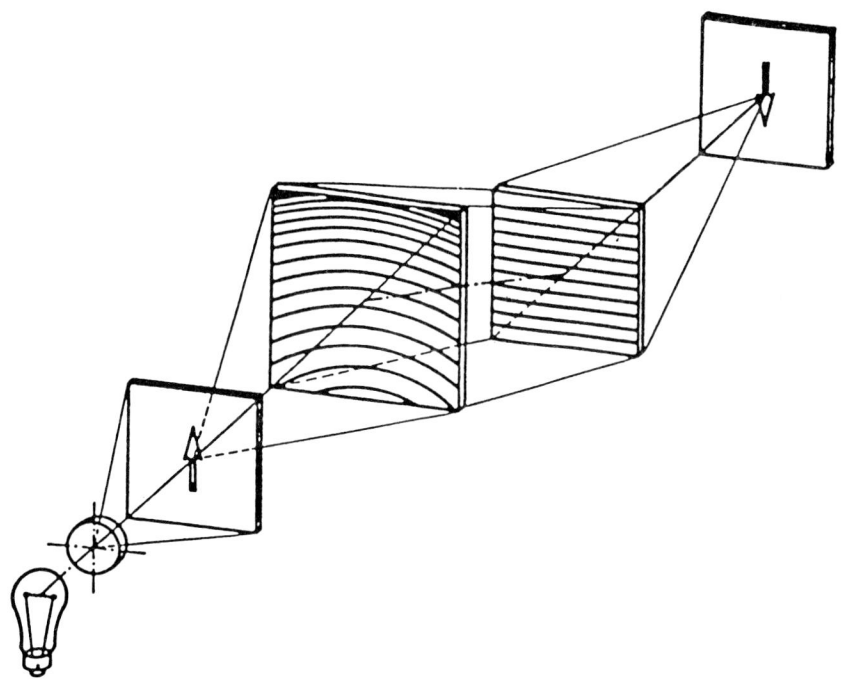

Fig. 3. White light projection of an image with diffractive optics.

the aberration. Reflection holograms done with the OPTRIGELAC process are good for that. For the best result, the hologram must be of a curved surface in order to correct longitudinal chromatic aberration while being almost achromatic. The use off axis of a second hologram for correcting the resulting chromatic aberration will correct geometrically and chromatically correct the system for good field and good colour performance in daylight.

3. THE MATERIAL FOR HOE

The possible materials for making HOE are silver halide emulsion, thermoplastic, film, photopolymers, BSO and similar crystals, and the dichromated gelatin family. We have developed a method for producing HOE called OPTRIGELAC which is described below.

Silver halide and other materials have a limited diffraction efficiency. OPTRIGELAC allows a high diffraction efficiency and a good angular discrimination. The basic material is gelatin doped with ammonium dichromate. The wavelength and angular selectivity is adjustable with low noise. But these results are obtained only with an appropriate method for the preparation of the material, the coating, sensitizing and development procedure.

We are going to describe the model and practical working procedure to have an excellent result.

The advantages of the OPTRIGELAC product are:

- low diffusivity
- high diffraction efficiency (98 %)
- high resolution (more than 5.000 line/mn)
- a reconstruction wavelength that can be selected
- a bandpass wide or narrow that can be precisely controlled
- an adjustable angular selectivity
- a possible reprocessing for optimizing the parameter in several steps
- an adjustable thickness.

The essential conditions for having a good OPTRIGELAC hologram are:

- the initial thickness of the gelatin
- the choice of the gelatin (quality)
- the initial hardness of the gelatin
- the bichromate concentration of the sensitizing bath
- drying conditions : temperature, moisture, time
- the recording wavelength
- the delay between exposure and processing
- the PH of the bath
- the recording set up (geometry)
- the energy of the recording
- the temperature of the bath.

The characteristics of the HOE produced on this basis are:

- diffraction efficiency
- diffusivity
- wavelength of use
- spectral selectivity
- angular selectivity.

The processing method while being more complex can be summarized with the following steps:

1) preparation of the material
2) coating a substrate (glass or plastic)
3) sensitizing of the material
4) exposure
5) processing with water
6) processing with alcohol
7) drying
8) protection

The order of these steps is well defined but some "inter steps" can exist depending on the result wanted.

4. HOW TO MAKE AN OPTRIGELAC HOLOGRAM FOR A HOE

a) Preparation of the material

One method is not suited for "professional use". This method consists of using the gelatin from photographic plate by removing the silver halide. The thickness is not regular enough and the quality of the gelatin is not correct, but it is possible to have in this way 3D pictures for entertainment.

The second method consists of preparing and coating the material. The most important point is the hardness of the gelatin, which derives from the number of links between a chemical material and gelatin molecules. By controlling the hardness it is possible to obtain either holograms produced by surface modulation or by phase modulation (modulation of the index of the material).

For having index modulation holograms you need a hardness more important than for having hologram with surface modulation. But if the hardness is too high we are going to reduce index modulation. If it is too low the signal to noise ratio will be low. It is very important to have a stable moisture control in surrounding of the HOE manufacturing system (35 %) and a temperature at 18°C. If it is too warm we are going to have an amorphous material and if it is too cold a crystal state. A too low moisture level will limit the molecular movement before the structuration of the fibers, too much moisture will show the reaction, and lengthen the fiber that will affect the micromechanics characteristics of the film. The range of moisture and temperature allowed will make possible a good compromise through a gel state that is similar to the crystallisation of polymer.

In water, gelatin molecules are organized in spiral. A good drying will give a uniform transparent film the main characteristic of which will depend on the drying temperature. 10 °C is the right temperature to have a good "tri molecule arrangement" on a spiral.

b) How to sensitize a HOE OPTRIGELAC

The natural gelatin does not sense light. Ammonium bichromate is used as "sensitizer" by dipping the HOE into a 1 to 25 % solution. A better sensitivity will come from a higher concentration that will also modify the spectral sensitivity.

Best results are in U.V. decreasing to the green but dyes allow to use OPTRIGELAC in any wavelength recording energy is between 100 and 400 $mJ:cm^2$ depending on the dye.

c) Exposure of the OPTRIGELAC HOE

Reflection and transmission holograms are to be exposed in a different way. Transmission holograms are recorded by projecting light on one side of the gelatin layer, reflection ones are recorded by using both sides. Arriving on a side the light will have a maximum energy level that will decrease progressively in the layer by absorption. This phenomenon will give a rather uniform energy distribution inside the material in the reflection case and a gradient in the transmission. The energy gradient will depend on the beams ratio and their incidence angle. The curve will not be the same when the thickness varies.

d) Final liquid processing of HOE OPTRIGELAC

2 baths, water and ALCOHOL.
After exposure extra ammonium bichromate is removed by a mixture of water and chemical fixer. It is at this moment always possible to modify the hardness, pH, the temperature and the time by reprocessing to give the desired diffraction efficiency, signal-to-nose ratio, wavelength of use, color selectivity etc.

The second bath is alcohol. It will eliminate the water between the gelatin molecule and materialise the modulation index that is recorded.

Several baths with a programmed concentration and temperature are used, the last one is at 20°C and 100 % concentration.

Concentration and temperature act on diffraction efficiency colour selectivity and signal-noise ratio.

Ethanol is used for short wavelength but with lower efficiency, isopropanol is good if used with adapted concentration and temperature for long wavelength that can be precisely adjusted.

The basic recording phenomenon is a molecular reorganization and deformation of the molecule in the effect of light.

The spectrum bandpass is the inverse function of the thickness and a linear function of the index modulation. For instance to have a high diffraction efficiency and a narrow band the thickness must increase and the modulation must lower. The thickness can be modified by the pH of the bath and by the concentration of alcohol.

After the last alcohol bath the gelatin is dried. The gelatin spatial area that has absorbed energy will be harder and will thus absorb less water that will be removed by alcohol faster. At a given step exposed area will be dry, non exposed will be humid, in these area fiber will be reorganized up to the last moment, density and index will be high in the hard area and low in the "soft" area. Molecule density is lower in the non-exposed area than in the exposed area.

A very good quality can be obtained by a serial of progressive reprocessing.

A differential spatial characteristic in volume and in surface is possible with the OPTRIGELAC process that open new way of making HOE.

5. CONCLUSION

We can say that diffractive optics is still in infancy. It is at the state of development at which was refractive optics in Galileos time. But the way is opened.

Concepts, methods and materials exist that have already permitted practical and commercial success in head up displays, laser disc lenses, glass lens testing, optical fiber multiplexing, spectroscopy and protection goggles. It is just a beginning, the first black and white picture was shot with an all holographic lens a few months ago. Some companies are starting a new department, or small companies are created for diffractive optics like X-IAL in Strasbourg that propose a catalogue of products and services in HOE. We don't think that diffractive optics will replace refractive optics for all its uses but it will improve many and give new openings as in photonics computing.

REFERENCES

* P. Meyrueis et al. Les enjeux industriels de la photonique, Tome I et II - La documentation francaise - 1983.
* P. Meyrueis and C. Liegeois. Un nouveau matériau pour multiplexage holographique sur réseau de fibres optiques - annales des télécommunications - février 1986.
* P. Meyrueis, C. Liegeois and R. Piel. Brevets n° 85 1874 France - Procédé de production d'un support holographique à base de gelatine bichromatée.
* P. Meyrueis and B. Hill. Méthode et système de réalisation de matrice holographique sur photoresist pour duplication d'hologramme en grande série. Conférence européenne d'holographie professionnelle (HOLOPRO 85) Belfort juin 1985 - Proceedings european photonics association n° 23.
* P. Meyrueis and C. Liegeois. Holographic optical elements for optical computing with MUX and DEMUX and spatial light modulator optical society of america meeting 31 april 1986 Hawaii USA.
* P. Meyrueis. Synthèse d'image par holographie et ordinateur - INA Publication Forum Nouvelles images INA - Fevrier 1984.
* Y. Amitai - A. A. Friesem. Recursive techniques for designing fourier transform holographic lenses - SPIE Vol 700 1986 - International optical computing conference.
* J. Weingartner. An innovative holographic microscope objective - OPTIK 68-2 - 1984 - 185-190.
* J. Weingartner. An holographic mirror objective - OPTIK 65-1 - 1983 - 49-61.
* R. C. Fairchild and J. R. Fienup. "Computer originated hologram lenses". Opt. Eng. 21. 133 (1982).
* K. A. Winick and J. R. Fienup. "Optimum holographic elements with nonspherical wave front". J.O.S.A. 73, 208 (1983).
* J. Kedmi and A. A. Friesem. "Optimal holographic fourier transform lens". Appl. Opt. 23. 4015 (1984).
* J. N. Latta. "Computer based analysis of hologram imagery and aberration". Appl. Opt. 10. 599-608 (1971).
* E. B..Champagne. "Nonparaxial imagining, magnification and aberration properties in holography". J.O.S.A. 57. 51 (1967).
* M. Akagi. Spectral sensitization of dichromated gelatin. Society of photographic scientists and engineers (1973). Tokyo Meeting Proceedings (1974). pp. 248-250.
* T. Kubota, T. Ose, M. Sashu and K. Honda. Hologram formation with red light in methylene blue sensitized dichromated gelatin. Applied Optics, USA (1976), 15, n° 2, pp. 556-558.
* A. Graube. Dye sensitized dichromated gelatin for holographic optical element fabrication. Photographic science and engineering, USA (1978), 22, n° 1, pp. 37-41.
* A. Graube. Holograms recorded with red light in dye sensitized dichromated gelatin. Optics Communic., USA (1973), 8, n° 3,

pp. 251-253.
* M. Sasaki, K. J. Honda and S. I. Kikuchi. Studies of photosensitive dichromated materials. Report of the Institute of Industrial Science, University of Tokyo, n° UDC 541-547.
* C. Pearce. Spectral sensitization of dichromated gelatin for an improved holographic material. US Army Electronics R et D Command. Report DELNV-TR-0020 (avr. 1983).
* R. Curran and T. Shankoff. The mechanism of hologram formation in dichromated gelatin. App. Optics USA (1970), 9, n° 7, pp. 1651-1657.
* D. Meyerhoffer. Dichromated gelatin. RCA Review, USA (1976), 33, n° 110, pp. 75-99.
* S. Case and R. Alferness. Index modulation and spatial harmonic generation in dichromated gelatin films. Appl. Phys., USA (1976), 10, pp. 41-51.
* D. M. Samoilovich, A. Zeichner and A. A. Friesem. The mechanism of volume hologram formation in dichromated gelatin. Photographic Science and Engineering, USA (1980), 24, n° 3, pp. 161-166.
* S. Sjolinder. Dichromated gelatin and the mechnism of hologram formation. Photographic Science and Engineering, USA (1981), 25, n° 3, pp. 112-118.

3rd International Conference on Lasers (iitt)
Kongresshaus Zurich

March 31 - April 1, 1987

A FAST EXPERIMENTAL METHOD OF MEASURING LASER BEAM ABSORPTION AS A FUNCTION OF TEMPERATURE IN SOLIDS

R. Dekumbis, H. Mayer and Ph. Fernandez

Centre de Traitement des Matériaux par Laser (CTML)
Federal Institute of Technology, Lausanne, Switzerland

ABSTRACT

The presented method is based on the measurement of the heat flow in a thin disk-sample heated by a laser beam and cooled in a gas jet.

From the slopes of the heating and cooling curves the absorption may be obtained as a function of temperature. Care must be taken in using reliable data for the specific heat of the material investigated.

Results from measurements of CO_2-Laserlight-absorption performed on steel X20CrMoV 12 1 and aluminium Al 99.99 are presented, with emphasis on the influence of surface roughness.

INTRODUCTION

In laser surface hardening the absorption can vary considerably depending on the surface conditions (roughness and/or coating used) and surface temperature. In order to modelize the heat flow and temperature field during laser hardening with an acceptable degree of precision, reliable data on absorption are necessary as a boundary condition.

The present paper describes a simple and fast method for measuring absorption as a function of temperature in solids, when the corresponding specific heat is known. The method is based on the temperature measurement during laser-heating of a small cylindrical specimen cooled by a constant shielding-gas flux.

2. THEORETICAL BACKGROUND

Under the assumption of a homogeneous temperature distribution in the sample, the total heat flux during <u>heating</u> Q_h at the temperature T can be written as:

$$Q_h(T) = m \cdot c_p(T) \cdot \dot{T}_h(T) = A(T) \cdot P_o - P_{vh}(T) \quad [W] \quad (1)$$

where

- m : mass of sample [kg]
- c_p : true specific heat [J/kg/K]
- $\dot{T}_h$: heating rate [K/s]
- A : Absorption [1]
- P_o : Laser power on the sample surface [W]
- P_{vh} : Power loss during heating [W]

On <u>cooling</u>, the total heat flux Q_c becomes:

$$Q_c(T) = m \cdot c_p(T) \cdot \dot{T}_c(T) = -P_{vc}(T) \quad [W] \quad (2)$$

where

- $\dot{T}_c$: cooling rate [K/s]
- P_{vc} : Power loss on cooling [W]

subtraction of equation (2) from (1) gives

$$\begin{aligned} Q_h(T) - Q_{hc}(T) &= m \cdot c_p(T) \cdot (\dot{T}_h(T) - \dot{T}_c(T)) \\ &= A(T) \cdot P_o - P_{vh}(T) + P_{vc}(T) \end{aligned} \quad (3)$$

We assume the power loss at the same temperature on heating and cooling to be equal:

$$P_{vh}(T) = P_{vc}(T) \quad (4)$$

$$A(T) = m \cdot c_p(T) \cdot (\dot{T}_h(T) - \dot{T}_c(T)) / P_o \quad (5)$$

$\dot{T}_h$ and $\dot{T}_c$ can be calculated from the time-temperature-curve; m and P_o are known while c_p data can be found in the literature [1,2].

Note that equation (5) is only valid over a temperature range where no phase changes take place.

3. EQUIPMENT AND EXPERIMENTAL SETUP

As a power source we used a 1.5 kW CW-CO_2-Laser from ROFIN SINAR featuring a stable near top hat mode. A quarterwave beambender produced a circular-polarized laser beam, thus eliminating any orientation-dependent effects. The beam was then focalized by a 150 mm off-axis parabolic copper mirror.

The disk shaped sample (diameter 19.2 mm, height 2 mm) was placed on a thermally isolating ceramic seat at the bottom of a shielding gas chamber (Fig. 1). To avoid excessive oxidation, Helium was blown at 25 l/min from two sides onto the sample by two separate

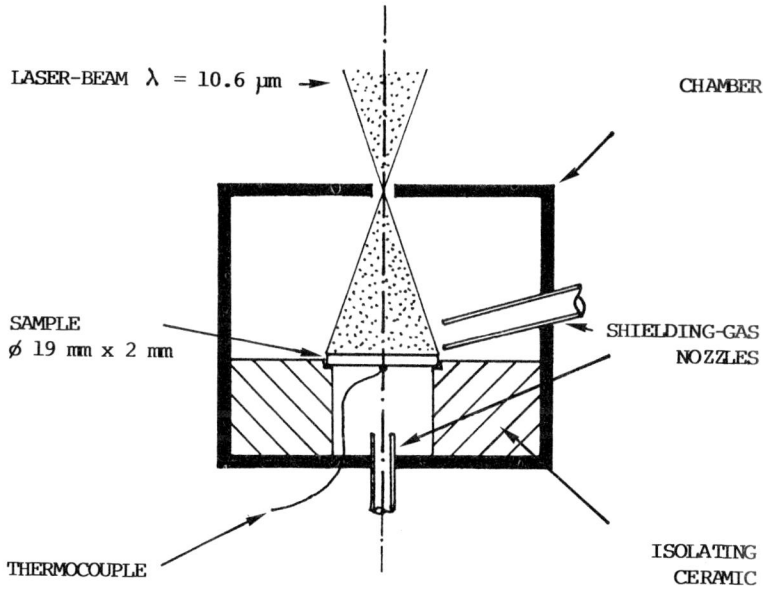

Fig. 1. Experimental setup.

nozzles. The focal point of the laser beam was positioned in the center of an aperture at the top of the chamber. In this way the sample was irradiated by a defocalized beam with a diameter of 16,3 mm ($1/e^2$). The mean intensity was 420 and 630 W/cm².

Type K-thermocouples (⌀ 0,1 mm) were fixed on the sample's bottom side to measure the temperature. A data-acquisition, reading at 5 to 10 Hz, was employed. Heating and cooling rates were then determined by stepwise linear regression over the range of temperature to be investigated. Absorption was obtained for each temperature interval using equation (5) and the c_p-formula given in Table 1.

Measurements were made on a commercial steel X20 CrMoV 12 1 and pure aluminium Al 99.99 (Table 1). Samples were ground or polished by different abrasives in order to vary the surface roughness (Table 2).

4. RESULTS

Figure 2 shows the thermal cycle of samples S2 and A2 (Table 2) with the corresponding heating and cooling rate and the calculated absorption as a function of temperature.

When the laser is **switched** on or off, there are transient stages where the temperature distribution in the sample changes until a stationary state is attained (Fig. 2b).

Table 1. Materials Data

Material	Steel X20CrMo V 12 1	Pure aluminium
Chemical composition [weight %]	C 0,24 Cr 12,04 Si 0,49 Mo 1,08 Mn 0,68 Ni 0,72 P 0,017 V 0,29 S 0,005	Al 99,99
Condition	1050°C/1h/Oil quench 700°C/2h	as cast
True c_p [J/kg/k] T in [°C] range: 100-650°C	$432 + 0,769 \cdot T +$ $-1,78 \cdot 10^{-3} \cdot T^2 +$ $+2,68 \cdot 10^{-6} \cdot T^3$ [1]	$\overset{*}{3}33 + 2,772 \cdot t +$ $-4,064 \cdot 10^{-3} \cdot t^2 +$ $+2,344 \cdot 10^{-6} \cdot t^3$ $t = T + 273$

*least square fit from data points of reference [2].

Table 2. Surface conditions for samples on which absorption was measured. Roughness measurements were made after laser absorption measurements, following DIN 4768/1 (Ra), DIN 4768 (R_z) and DIN 4762/1 (R_t).

Sample no.	Abrasive Medium	Roughness μm R_a	R_z	Mean absorption at 250°C [%]
S1 (steel)	Diamond : 1 μm (alc)	0,04	0,3	10,2
S2 "	SiC paper P1000 (aq)	0,05	0,4	10,5
S3 "	" " P 500	0,07	0,7	10,5
S4 "	" " P 220	0,32	2,8	13,5
S5 "	" " P 80	0,43	3,1	15
A1 (Al)	Diamond : 1 μm (alc)	<< 1*	< 1*	2,4
A2 "	SiC paper P1000 (aq)	0,67	4,6	7,5
A3 "	" " P 500	0,89	6,6	10,9
A4 "	" " P 80	1,97	12,5	10,1
A5 "	" " P 220 (dry)	1,58	12,5	10,0

*estimate

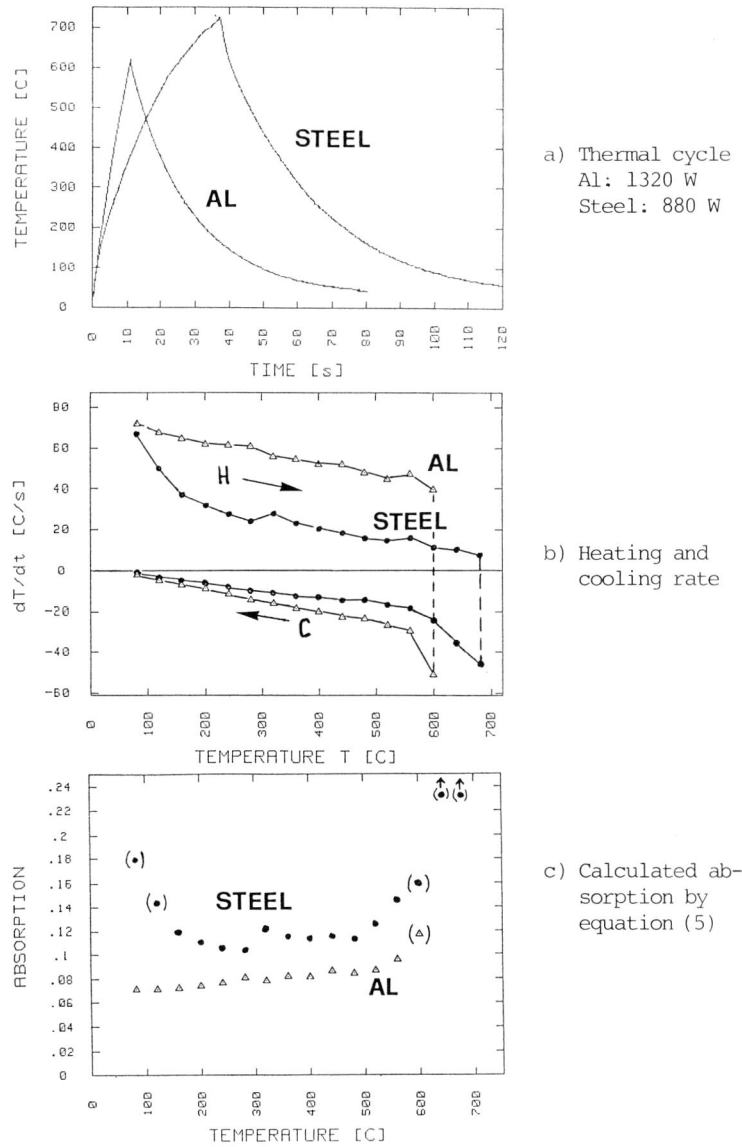

Fig. 2. Absorption measurement for aluminium (sample A2) and steel (sample S2). Values between parentheses are not valid due to non-stationary-state conditions.

This behaviour is much more pronounced with steel because of its lower thermal conductivity. Calculation of the absorption is only valid outside the transitory regions (Fig. 2c).

Results for steel and aluminium are shown in Figs. 3 and 4. Absorption changes only very little with temperature, but strongly depends on surface roughness. All temperature-absorption curves have about the same slope. At least for aluminium, maximum absorption is obtained when the surface roughness R_z is close to the wavelength of the laser source (Fig. 5).

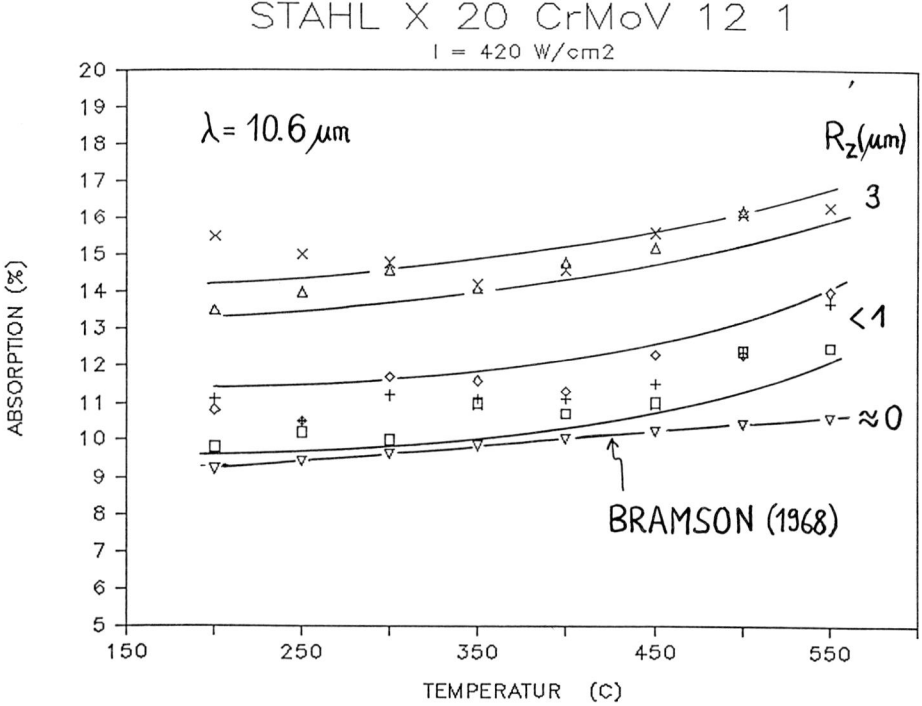

Fig. 3. Results for steel and comparison with calculated values using BRAMSON's formula [3] and electrical resistivity data given by RICHTER [1].
■S1   +S2   ◆S3   △S4   ▽Bramson   ×S5

There is a remarkable correlation between surface roughness (R_z), absorption and the laser-wavelength (Fig. 5). However, this observation is purely empirical, and it is beyond the scope of the paper to give a physical explanation for this.

ACKNOWLEDGMENTS

This work was sponsored by the Swiss Government and the Kommission zur Forderung der Wissenschaftlichen Forschung (KWF). We are also grateful to Mr. Christopher Brown (LMM/EPFL) for his help in surface roughness measurements.

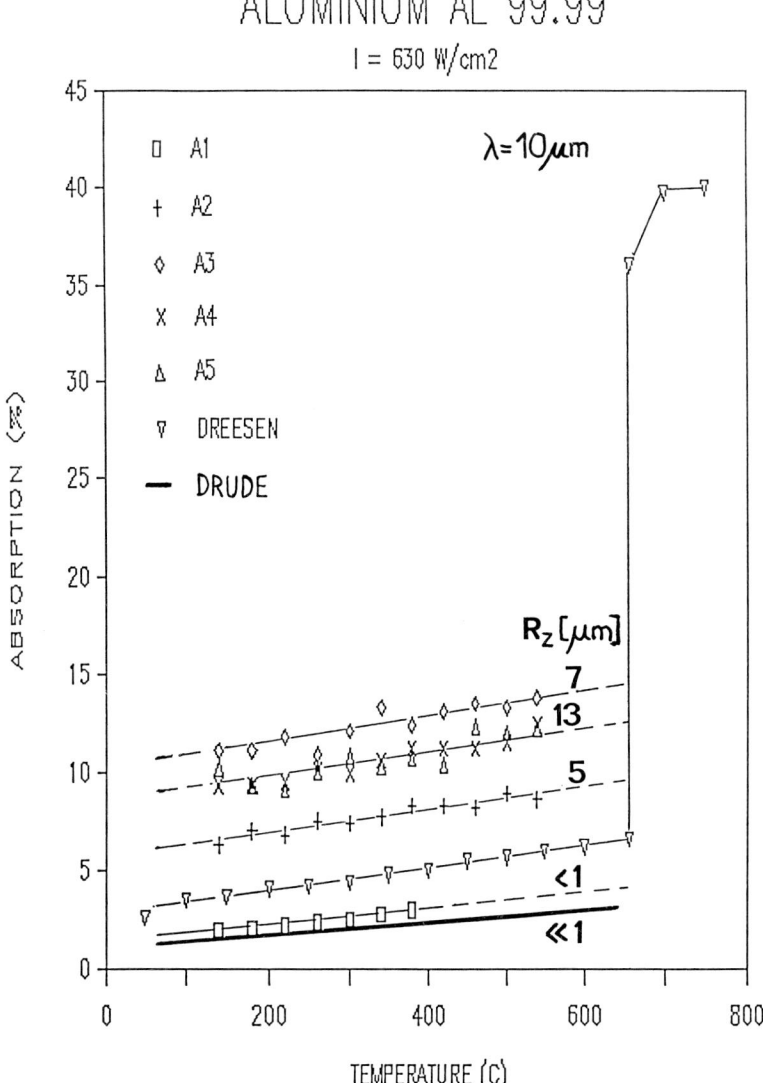

Fig. 4. Results for pure aluminium and comparison with data from DREESEN et al (1984) and calculated values from DRUDE theory [4].

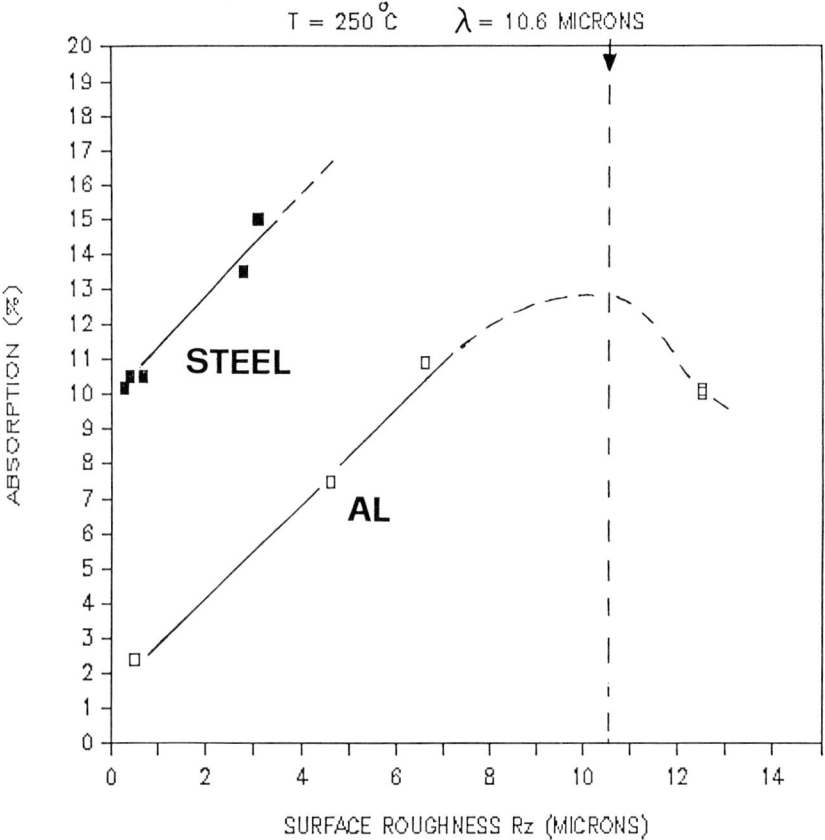

Fig. 5. Absorption at 250°C as a function of roughness R_Z (mean value of max. height between peaks and valleys).

REFERENCES

1. F. Richter. "Physikalische Eigenschaften von Stählen und ihre Temperaturabhängigkeit", Stahleisen-Sonderberichte, Heft 10 (1983).
2. Y. S. Touloukian and E. H. Buyco. "Specific Heat", Thermophysical Properties of Matter, Vol. 4, Plenum (1970).
3. M. Bramson. "Infrared Radiation: A handbook for applications" Plenum Press, New York (1968), p. 127.
4. H. G. Dreesen, C. Hartwich, J. H. Schaefer and J. Uhlenbusch, J. Appl. Phys., 56 (1), July (1984).

OPTICAL FIBER TEMPERATURE SENSING BY MODE FILTERING

A. Cahkari* and P. Meyrueis**

*Societe Cordon Eurelectric Strasbourg
**E.N.S.P.S., Strasbourg

1. INTRODUCTION

It is now well known that in the future many measurements of data in the field of physics will be made using optical fiber sensors. This has been recognized by analysts such as Frost and Sullivan and Kessler Intelligence.

The first optical fiber systems were for military use in sonar and gyroscope systems. It was only by exploiting the military potential of optical fibers that the high cost of the development of such systems was met. These sensors generally used phase effects. Some other optical fiber sensors appeared later using modification of light parameters reflected or transmitted by a meterial at one end of the fiber, for instance liquid crystal or bubble detection. Other fibers use the polarization effect for measuring electromagnetic fields. In this paper we propose a new way of using the fiber itself as a sensor by exploiting a mode filtering phenomenon which is described below.

2. COUPLING OPTICAL FIBERS AND EMITTERS BY BENDING

Light power is usually injected into an optical fiber through an end which has been cut and polished. This method is expensive, slow, and not very reliable. We propose to effect the same light power injection process by bending of the fiber. This process can be applied to any fiber, including plastic ones. With rather low efficiency two fibers can be coupled this way.

This method also allows the injection of white light into a multi-mode fiber, which is unlike any other method. Such a device is not affected by any electromagnetic noise, and it can work in an industrial environment, for example with robots.

The bending of the fiber induces a local modification of the numerical aperture, core diameter and local length of the fiber. The enlarged numerical aperture modified by bending allows the

injection of light through a lateral surface of the fiber. We
tested this method on several fibers, including a PCS200 with a
core diameter of 200 μm and a numerical aperture of 0.4. The
bending radius is limited by the mechanical characteristics of
the fiber; for the PCS200 above this limit it is 3 mm. We studied
the influence of the bending radius on the amount of light injected into the fiber.

The coupling of one emitter by bending the fiber gives 50% transmission between the two ends of the fiber. We have built several
prototypes that consist of a hollow plastic cylindrical device on
which a fiber is bent. A low cost LED is then used to illuminate
the system. The exact percentage transmission of the two branches
can lead to the use of one branch as a reference if they both have
the same length.

3. TEMPERATURE SENSING BY MODE FILTERING

A reverse effect of the bending is to lose light travelling in
the fiber at the bending section. This loss can be modulated by
temperature, but the same result can be obtained by bending the
fiber or by changing a given length of the cladding of the fiber
for a material that will have a refractive index which varies
with temperature. We will describe these two possibilities.

a) Bent fiber sensing by mode filtering

Suppose that we have a bent fiber system inside a hollow plastic
holder. This fiber system is composed of fibers with several
core diameters. Strains vary with the core diameter and the
radius of curvature of the fiber (see Fig. 1).

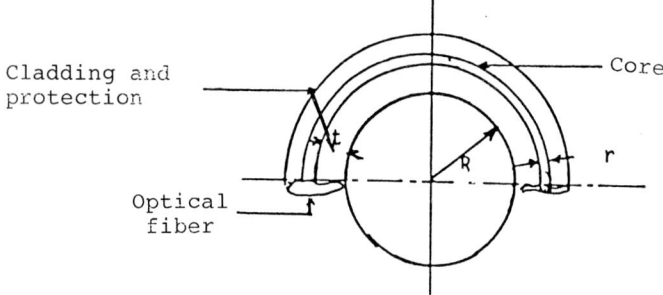

Fig. 1. Strain system.

Let R be the radius of curvature, r the core diameter and t the
thickness of the cladding and the protection of the fiber. It
is possible to calculate the length modification in the fiber:

$$[L - l = Lr/R + r + t]$$

where
L is the length of the curved fiber,
l is the length of the fiber without bending and
L - l is the length modification of the fiber by curving.

Strain ε and force f can be modelled by:

$$[\varepsilon = L - l/L]$$
$$[f = \varepsilon E]$$

where
E is Young's modulus.

Figure 2 shows the length modification of the fiber on bending against the radius curvature. The knowledge of the mechanical properties of the materials used in the fiber gives us values for establishing a mechanical critical radius R_{CM}. So it is very possible to inject light into the fiber through a lateral surface.

In the bent portion of the fiber, some of the guided energy is lost as partially refracted light. The relation between radius of curvature and energy loss is as follows:

$$(a(l) = 10\log(1 - D(n^2)/R(ON)^2)$$
$$= 10\log \frac{P_2}{P_1}$$

where
a is an energy loss factor,
D is the core diameter in the fiber,
n is the refractive index of the core,
ON is the numerical aperture,
R is the radius of curvature of l,
P_1 is the input power and
P_2 is the output power.

The total loss expected for optical critical radius gives a value for the critical radius R_{CO} of

$$[R_{CO} = Dn^2/(ON)^2]$$

In a gradient index fiber, light is propagating between two cylinders with r_1 and r_2. When the fiber is bent the propagation constant changes, for a radius r_0, propagation conditions are not satisfied if $r_0 = r_2$ and the energy carried by the modes is lost. To control the phenomenon it is merely necessary to determine all the modes entering when $r_0 < r_2$.

An experiment has been devised to validate this theory with a two-branch "energy bend injected" sensor. One branch is used for reference, the other is used as a sensor. The measurement of the loss related to the bending makes possible the use of the system as a sensor. Having found a modification in the propagation condition, we can model the following. Maxwell's equations govern the guided propagation in an optical fiber with β being the propagation constant. We can determine propagation

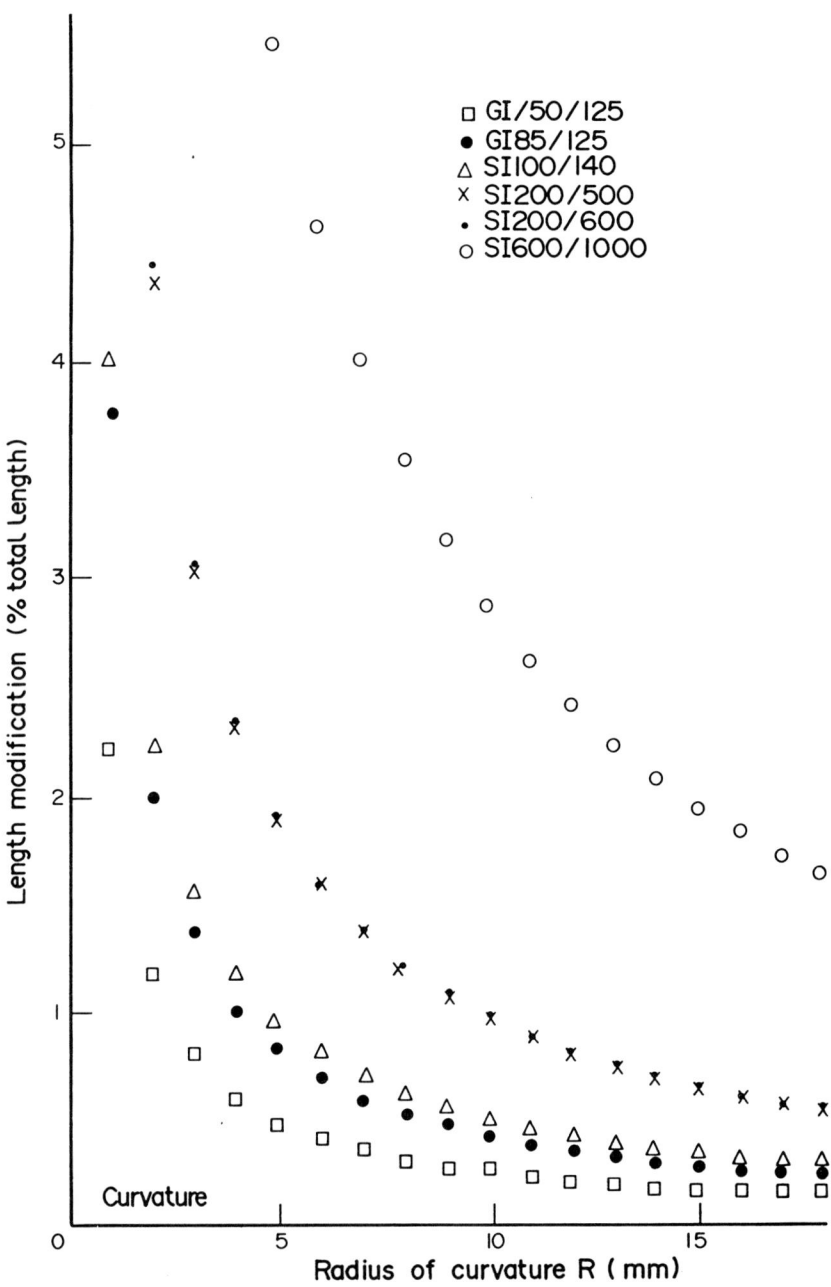

Fig. 2. Length modification of the optical fiber against ratio of curvature.

properties by using a normalized frequency or a wavelength.
Propagation conditions can be expressed by:

$$[K^2 n_1^2 - \beta^2 - (\frac{V}{r})^2 \geq 0]$$

The maximum value that β can take is obtained for $V = 0$, that is,

$$\beta_{max} = Kn_1,$$

and the minimal value is

$$\beta_{min} = Kn_2;$$

so we have the following propagation conditions:

$$[Kn_2 \leq \beta \leq Kn_1].$$

We can either use the transmission or the reflection method. The sensing effect through bending works with any fiber but it is much easier to use a fiber made of different materials. If, for instance, the fiber possesses physical properties which are different, then the effect is amplified.

b) Temperature sensing by substituting an adapted material for cladding

We can obtain the same mode filtering effect by removing or by etching (using a laser) the cladding of a fiber and replacing it by a liquid or solid material which will change its refractive index with temperature (see Fig. 3), or by doping the original cladding of the fiber locally with a material which will give it an index modulation potential. We will, in this way, have a transducer which will filter modes as a modulator related to the temperature of the fiber.

Thus we will have the possibility of changing the numerical aperture locally. This can be repeated several times with several substitute materials around the optical fiber in order to obtain the required measurement.

Paraffin has been used as a substitute with a modified length of 10mm. The paraffin was held by a cylindrical metallic tank with an inside diameter of 600µm.

c) Experimental results

We have experimentally tested the loss related to the core refractive index, the core diameter, the numerical aperture and the bending radius. This is shown in Fig. 4 where we used a 50% transmission bending coupling. When comparing between theory and experimental values concerning step index the fiber GI150/125, SA100/140, PCS 200, gives good results (as shown in Fig. 4). We have determined with the help of a model the range of curvature for a given fiber and the R_{CO} (total loss for radius of curvature),

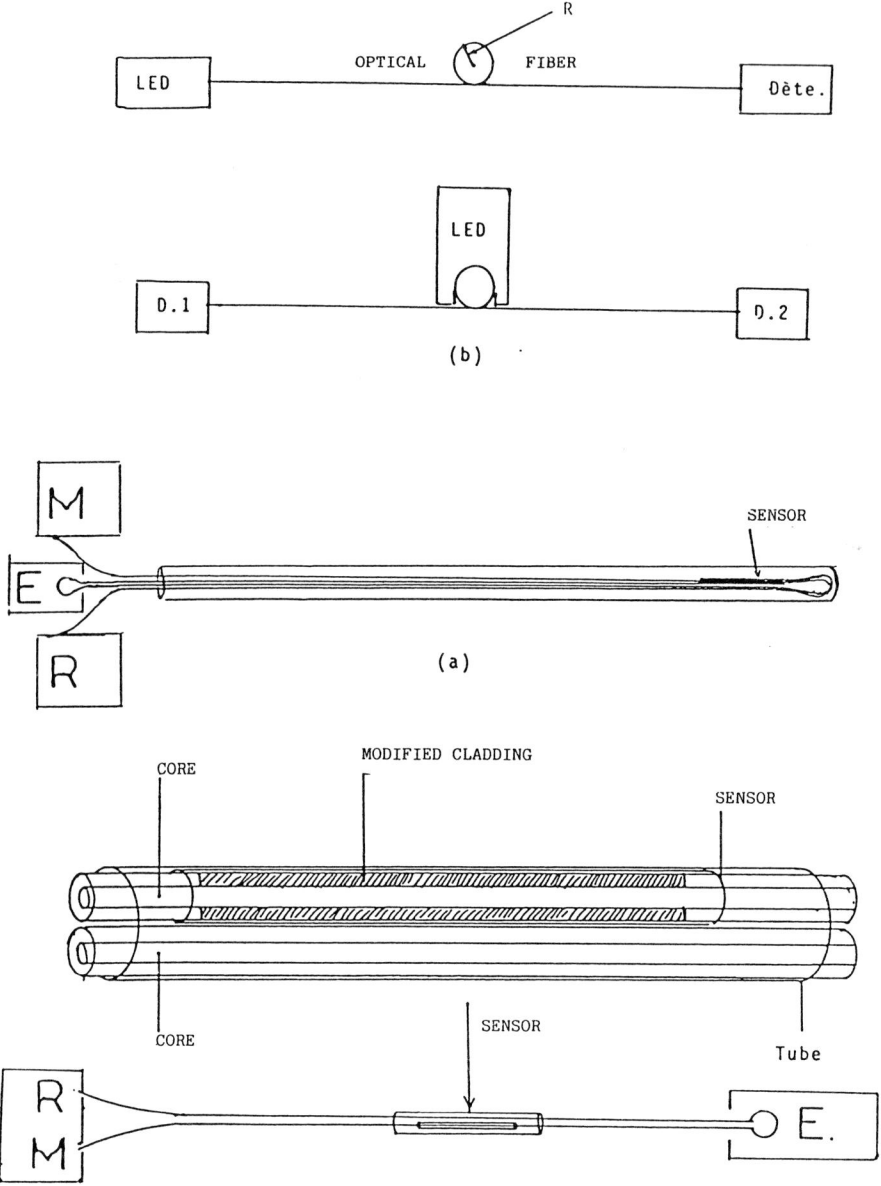

Fig. 3. Temperature optical fiber sensor with injection by bending.

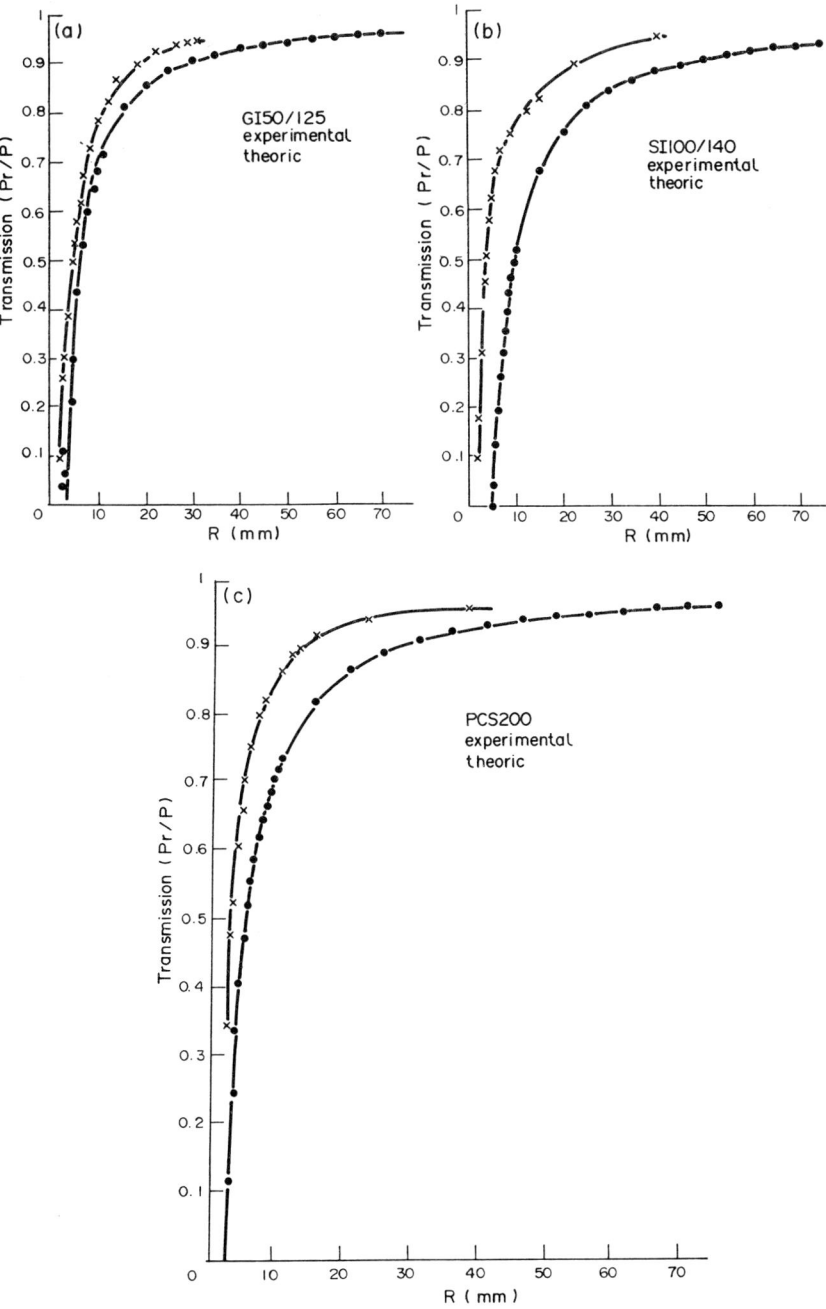

Fig. 4. Loss with the curvature for GI150/125 (a); SA100/140 (b); PCS200 (c).

the limit being the maximum mechanical resistance radius R_{CM}. An increase of the length of the sensing fiber increases the sensitivity.

The sensitivity of the sensor is related to the sensor length and the cladding refractive index, which was tested by replacing one part of the cladding by a material different from the original one. We can obtain even better results by cladding substitution and bending in order to achieve the best results in terms of sensitivity and effect differentiation (Fig. 5).

4. CONCLUSION

We have designed and built an apparatus to measure temperatures by using mode filtering in an optical fiber. This system achieves a high degree of accuracy and can be produced at low cost due to the coupling to LED by bending. The signal can be analyzed by means of simple electronics. We think that this system has an important future. It will be produced commercially be Eurelectric Cordon Company (FRG). This work was carried out with the assistance of Eurelectric Company and the French-German Institute for Applied Research.

REFERENCES

1. A. Chakari and P. Meyrueis; Council of Europe; AS/Science/Phot/(38)3; P.33; (1986).
2. Charles K. Kao, McGraw-Hill Book Company, "Optical fiber systems, Technology, Desing and Applications".
3. A. Chakari, P. Meyrueis and M. Grosmann, 1st International Conference and Exhibition of fibers optics applications to developing, Loubljana, Yugoslavia 14-17 Oct. 1985 published FODC, proceeding P.1 1985.

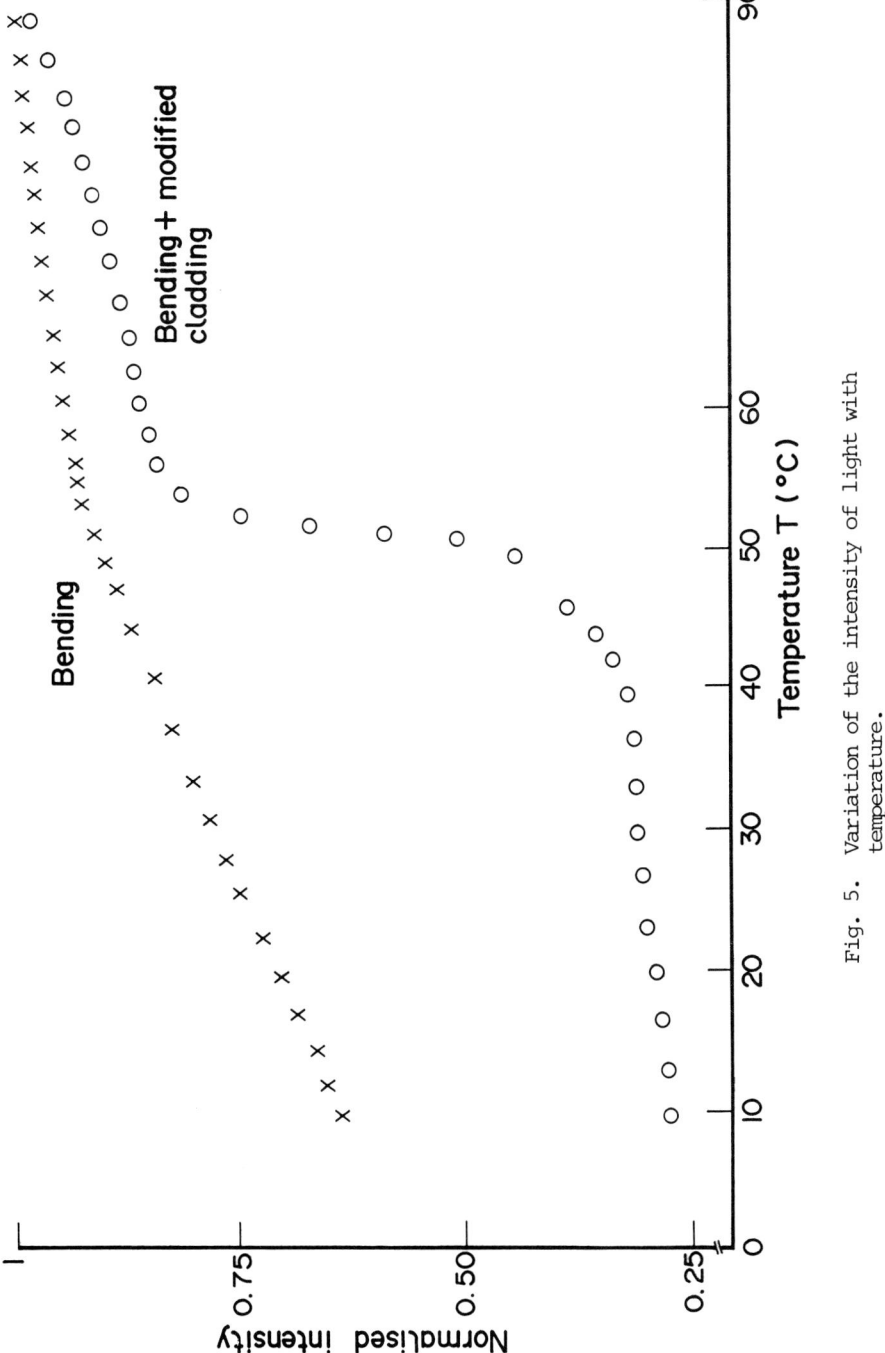

Fig. 5. Variation of the intensity of light with temperature.

SURFACE MICRO ANALYSIS WITH LASER FOURIER TRANSFORM AND SYNTHETIC HOLOGRAM PROCESSING

E. Soubari, P. Meyrueis and M. Torzynski

Ecole Nationale Superieure de Physique de Strasbourg, France

1. INTRODUCTION

The analysis of surfaces for mechanical applications is becoming increasingly important with the development of sophisticated structures.

Measurement of the topology and topogarphy of mechanically finished surfaces was first performed by contactual methods. As time progressed, transducers were improved by electronics or optics. Eventually, light was used as the sensing medium, to be followed by laser. Most recently, speckle and optical Fourier transformation of coherent light surface data have been used. This last method, developed for the most part at ENSPS, is the most efficient, perhaps because the amount of surface data sensed was very important (Fig. 1).

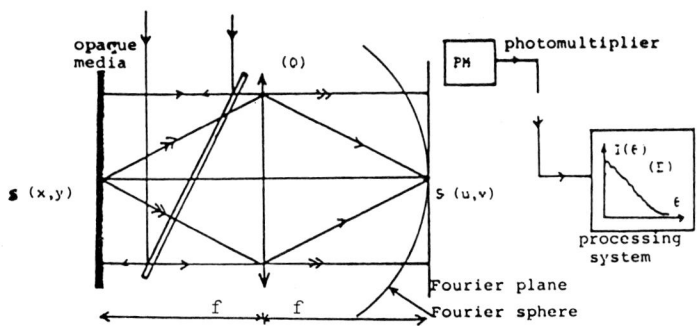

Fig. 1. Diffraction of a coherent plane wave by a reflecting opaque object.

This importance induces a large computer time for extracting the

more significant data through a CCD interface. It was necessary
to use this method at its full potential with the effective possibilities of present computers, to have a pre-processing at very
high speed and a large band. We have developed a method using
computer-made holograms that solves this problem to a large extent.
The optical processing of the data by the holographic processor
is fast and reliable. We have in this way conceived and realised
a hybrid processor (analogic - digital), mixing electronic and
photonic processing. It can be improved in many ways, opening
the possibility to define new surface standards. The roughness
methods commonly used were set from the results of contactual
surface measuring systems. The photonics methods give much more
complete results that can be the base of new surface standards
(Figs. 2 & 3).

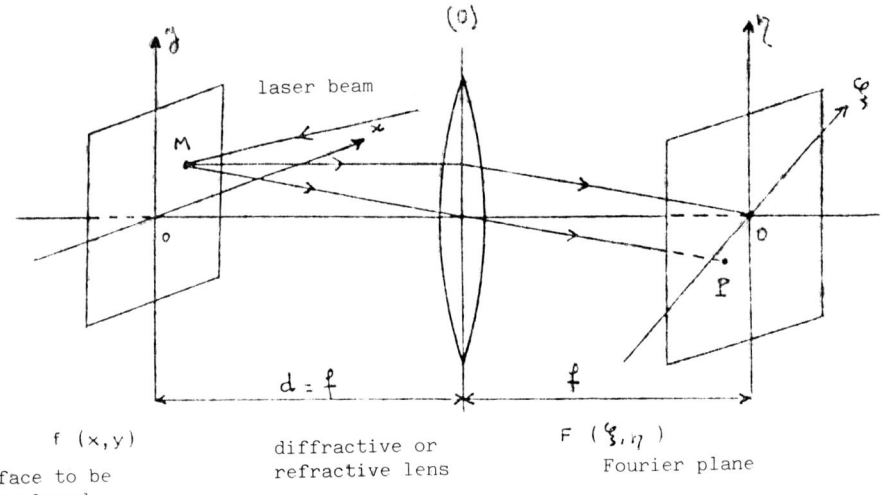

f (x,y)
Surface to be
analysed

diffractive or
refractive lens

F (ξ, η)
Fourier plane

Fig. 2. Basic set-up for roughness measurement.

2. PRINCIPLE OF THE METHOD

a) Basic principle

For describing the structure of the electromagnetic field of light
it is possible to say that a complex E.M. is the superposition
of elementary plane waves of sinusoidal form. This Fourier method
can be applied to the light structure and to a surface structure,
a surface being defined as an interface between a fluid and a
solid. A complex surface can be described by a model superposition of sinusoidal surface. Every sinusoidal surface will be
described by two parameters: its period and its amplitude. If
a plane coherent wave interacts with this surface the resulting
wavefront (after the interaction) will carry the tridimensional
data concerning the area illuminated, in a phase modulated form.
These encoded OATA can be read in the Fourier plane of a lens.
The lens will do a decoding of phase in a similar process of the
encoding-decoding of holography. By putting an adapted hologram

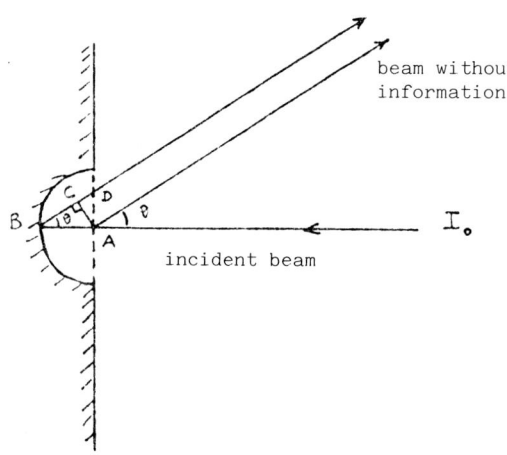

Fig. 3. Relation between roughness of a surface and phase difference of laser beam.

in the Fourier plane it will be possible to select and process data before a CCD sensor and a classical image **processor**. This digital image processing will be very simplified. A microcompute will be used and the measurement will take a short time (Figs. 3 - 6).

b) Photonics modelisation of surface roughness

A consider a diffuse surface illuminated by a coherent plane wave This surface is located at a distance d from a converging lens. The delay caused by the surface is $f(x,y)$. The amplitude in the focal plane of the lens is given by the Fresnel Kirchoff formula.

$$F(\xi,\eta) = \frac{1}{j\lambda f} \text{Exp } j(1-\frac{d}{f})^k(\xi^2 + \eta^2) \iint_{(D)} f(x,y) \text{ Exp}[-j\frac{k}{f}(\xi x + \eta y)] dx\, dy \quad (1)$$

- λ is the light wavelength
- K is the wavenumber
- x,y are the spatial coordinates
- D is the surface to be analysed
- ξ,η are the spatial coordinates of P in the Fourier plane.

A roughly polished surface has a $\Delta\phi$ phase difference between a diffracted and a reflected beam on the surface defects.

It can be written:

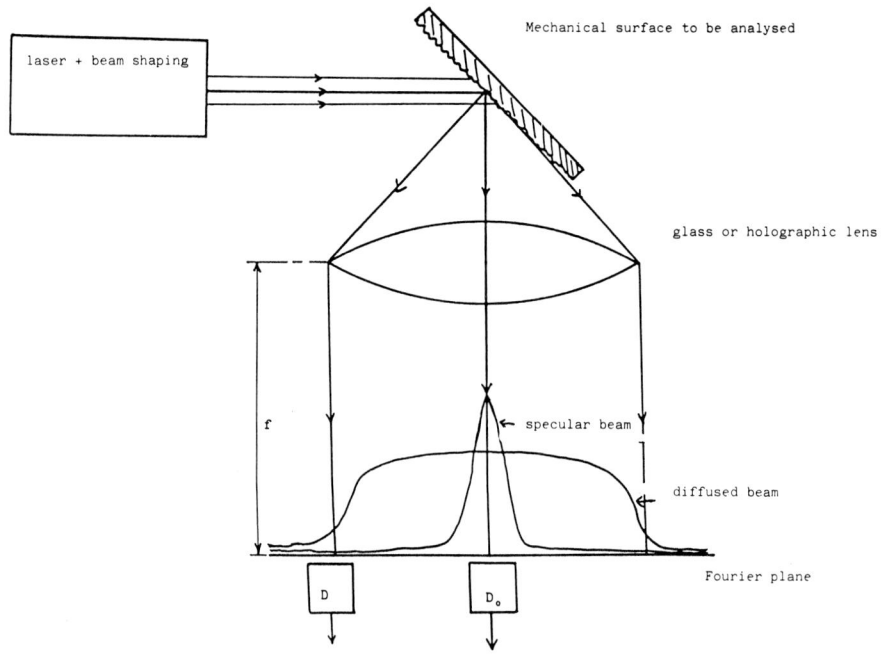

Fig. 4. Set-up for roughness measurement with T.F.

$$\Delta\phi = k(ABC) = k\ z(x,y)\ (1 + \cos\theta)\ k\ z(x,y)\ \frac{(2+\theta^2 + \ldots)}{2!} \quad (2)$$

If θ is small, $\theta^2 \cong 0$ $\quad \Delta\phi = 2kz(x,y)\ \frac{4\pi}{\lambda}\ z(x,y)$

We can deduce the distribution of the amplitude reflected by the surface:

$f(x,y) = R(x,y)\ \text{Exp}[2jkz(x,y)]$

$R(x,y)$ is the distribution of reflected light.

$z(x,y)$ is the distribution of surface roughness.

Relation between the roughness distribution and the diffraction pattern: the diffraction pattern can be approximated by the Fourier transform.

$$\begin{array}{l} f(x,y) = R(x,y)\ \text{Exp}\ 2jkz(x,y) \\ \text{F.T.}\ \downarrow \\ F(\xi,\eta) = \text{T.F.}[R(x,y)\ \text{Exp}\ 2jkz(x,y)] \end{array} \quad (3)$$

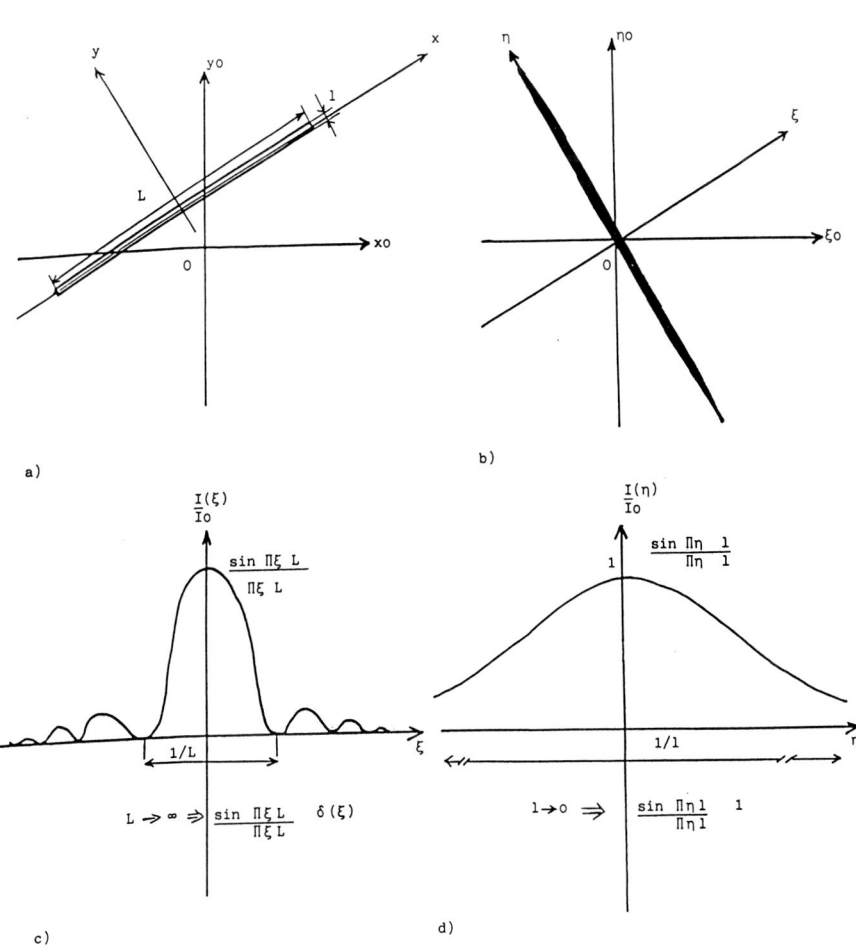

Fig. 5. Influence of the dimension L and l of a linear defect on the optical F.T.

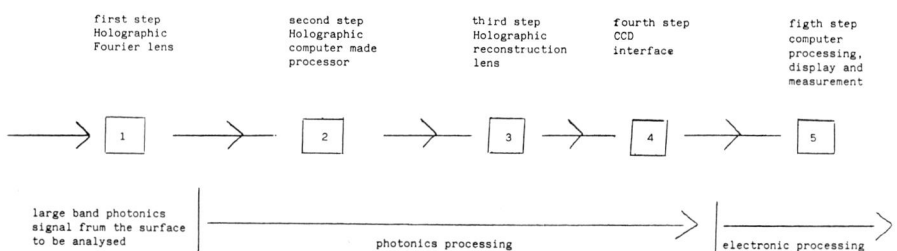

Fig. 6. Basic set-up for the implantation of the holographic processor for a better efficiency this processor can be made on dictramated guatin.

The energy spectrum is fiven by the intensity:

$$I(\xi,\eta) = |F(\xi,\eta)|^2 \qquad (4)$$

If $R(x,y)$ is a constant and $z(x,y) < \frac{\lambda}{10}$, the energy spectrum approximates to the roughness spectrum

$$I(\xi,\eta) = \left| T.F. \; c \; 1+2jkz(x,y) + \ldots \right| \\
= \left| c \; \delta(\xi,\eta) + 2jkZ(\xi,\eta) + \ldots \right| \qquad (5)$$

where $Z(\xi,\eta) = F.T.z(x,y)$
$\delta(\xi,\eta)$ is a Dirac distribution
c is a constant
F.T. stands for Fourier transform.

We can now write

$$I(\xi,\eta) \propto |Z(\xi,\eta)|^2 = T.F.*\overline{z}(x,y)*z(x,y) \\
= T.F. \; Cz\overline{z}(x,y) \qquad (6)$$

Czz is the auto-correlation function of $z(x,y)$
* : Convolution Symbol $\overline{z}$: Conjugate of z

By detecting the intensity $I(\xi,\eta)$ that corresponds to the energy spectrum of the amplitude of light reflected by the surface we can deduce the statistical distribution of roughness.

c) Analytical shape of the energy spectrum

If we approximate the minimum defect by a circular hole of 2 Å diameter with a large number N we have for N identical defects:

$$I_N(\xi,\eta) = N\, I(\xi,\eta)$$

where $I_N(r) = N\, I(r)$ (7)

with $r^2 = \xi^2 + \eta^2$

r is the radial distance in the Fourier plane.

For a pattern of circular defects with variable radius with M class of defects we have:

$$I(r_j) = I_i(r_j)\, P(a_i) \tag{8}$$

$I_i(r_j) = G_{ij}$ is a characteristic function that we can write:

$$G_{ij} = \frac{\sum_{i,j} \pi^2 a_i^4}{\lambda^2 f^2} \left[\frac{2J_1(X)}{X}\right]^2 \quad \text{with } X = \frac{2\pi a_i r_j}{\lambda f}$$

$P(a_i)$ is the statistical weight
$J_1(x)$ is a Bessel function of the first order and type we can now write the matrix equation :

$$[I] = [G] \cdot [P]$$

This formula allows the determination of the size and statistical distribution of the surface defects from the optical energy spectrum. For every rj there corresponds an intensity $I_j = I(r_j)$ measured in the Fourier plane. That is only the addition of the effects of defects with the same size within an angle j. By doing several measurements of I_j and the associated r_j we can solve this problem for any defects on a rough surface (Figs. 7 & 8).

There are two cases:
- the number of measurements of I_j is equal to the number of classes chosen for ai. The G_{ij} matrix is a square matrix and the resolution of the linear system is done by inversion of G_{ij}

$$P(a_{ij}) = G^{-1} \cdot I(r_j)$$

- the number of measurements is very large compared to the number of classes of ai.

We use the mean square method that allows the normalisation of the linear system of equation

$$P(a_i) = (G^*.G)^{-1}.G^*.I(r_j) \quad \text{with } G^* = {}^tG \text{ transposed matrix}$$

Some other numerical methods exist. They approximate the optical diffraction pattern by mathematical functions or by polynomial approximation or by a continuous decomposition of Fourier transformation.

314

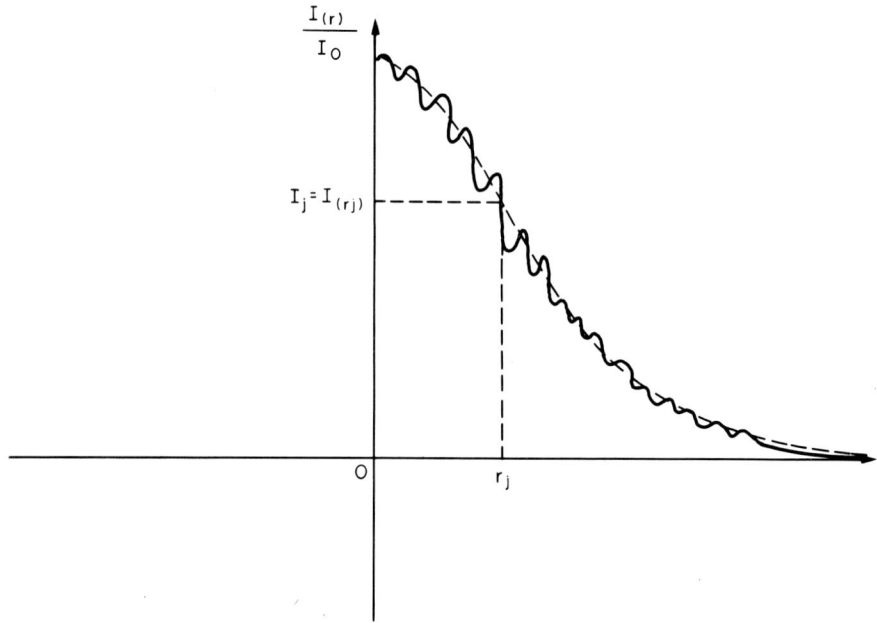

Fig. 7. Radial profile of the diffraction pattern.

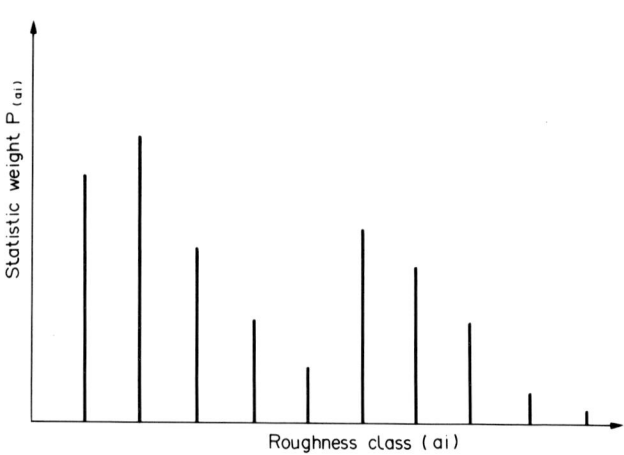

Fig. 8. Repartition of the statistic weight in relation with roughness.

The same results can be obtained for transparent materials.

It is possible to use holographic lenses for instance made with the OPTRIGELAC method (high efficiency material and process developed by E.N.S.P.S.) for the Fourier effect. These lenses can be perfectly corrected and can have a large aperture with any focal length. The system can be very compact, very stable and resistant to surroundings perturbation (Fig. 9).

d) The holographic pre-processor

As we said previously, the computer made holographic processor can be used to accelerate the output data sensed by the previous method, before computer processing.

We have for this purpose studied and realised a set of holograms allowing almost all the classic known functions of digital image processing. One hologram will be necessary for every function but the hologram can be positioned on a disc that will give the right function very quickly. The mechanical positioning of the hologram is possible with an accuracy of one micron. We have developed a system for testing the holograms and we check their quality with pictures processed by computer and by our method (Figs 10 & 11).

We have realized these operators by the method of synthetic rectangular openings and the phase determination by a computer Univac 1110 of 320 Kwords (36 bits each) and incremental tracer Benson (1/10 mm increment). The drawings are realized off line; the writing instructions are recorded on a magnetic tape which is then taken by the tracer. The size of the binary drawing, 30 x 30 cm^2, is reduced to the scale of 1/300 to obtain a hologram of 1 x 1 mm^2 on a holographic Kodak SO-253 film of 1250 lines per mm of resolution power. The camera used is a Canon of 50 mm and f 1.8. The lighting was assured by four lamps totaling 2000 W for a shutter speed of 0.25 s and a diaphragm of f8.

The calculation time-1mn drawing time was 1 hour (off-line setting).

Omnidirectional nondirectional spatial differentiation reveals sharp changes in the luminous intensity levels in images.

Numerical Fourier holography suffers from the method of transfer of information by photoreduction. For this reason, it is more convenient to process the images in an optico-electronic way. Systems in real time, based on noncoherent-coherent converters are very promising. They will allow direct display of the computer-calculated holograms on light modulating devices that can be reused many times.

In our hybrid system, the coherent optical device realizes the bidimensional Fourier transform of an image, the convolution and the correlation.

The electronic hardware is reserved for the controls and for the precise calculations that optics cannot handle.

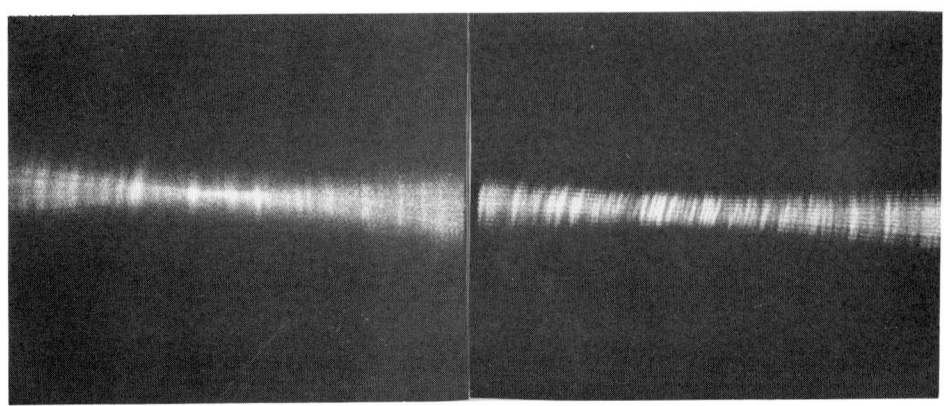

A B

C D

Fig. 9. Examples of Fourier transform of surfaces (stainless steel) with Fourier lens
A : grinded Ra = 1,6 µm
B : grinded Ra = 3,2 µm
C : circular polished Ra = 0,05 µm
D : circular polished Ra = 0,2 µm.

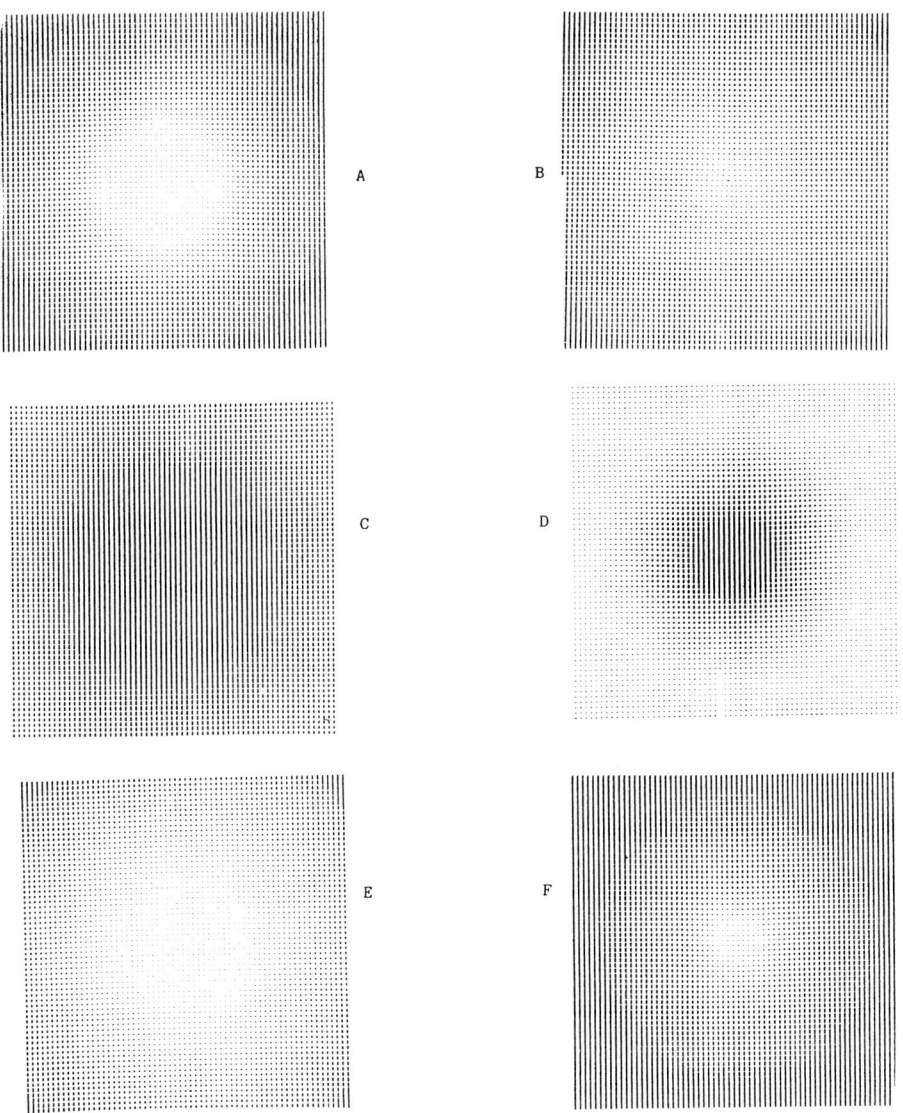

Fig. 10. Examples of holographic processors. A: Laplacian operator; B: gradient operator; C: gaussian operator low pass filter; D: Maxwellian (low pass); E: gaussian operator high pass filter; F: exponential operator high pass filter.

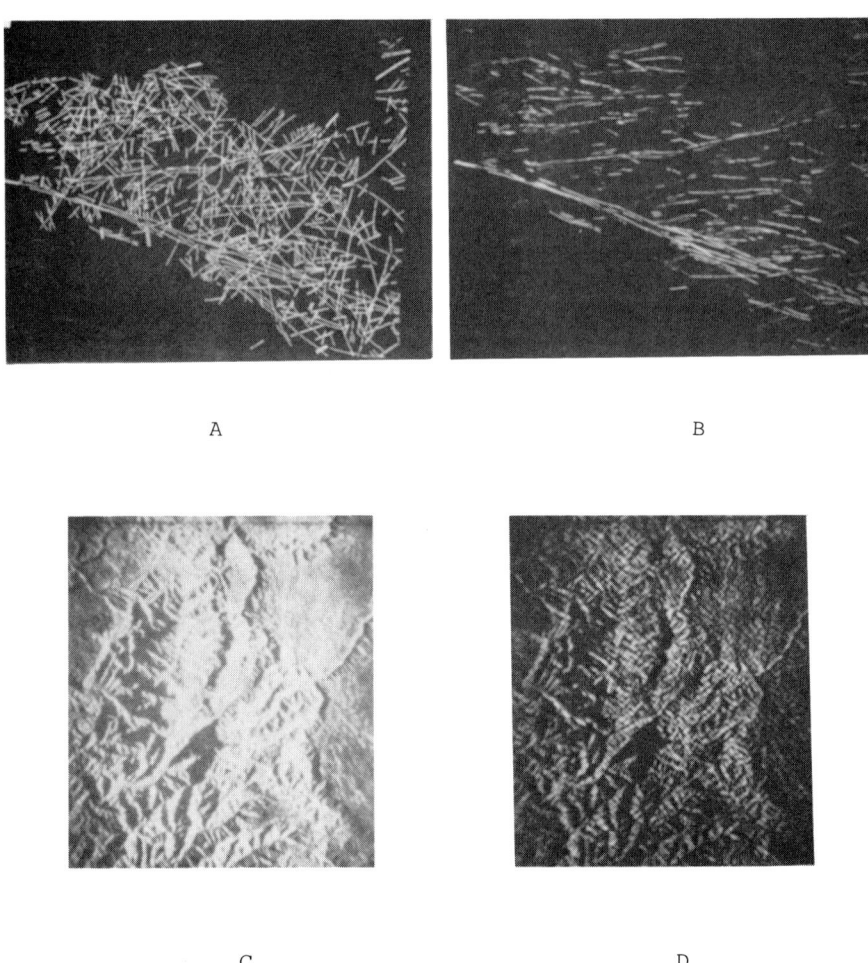

Fig. 11. Examples of image processed by Holographic processor (case: macrosurface)
A: Initial image of cracks
B: Differential selection of cracks by directional filter
C: Initial image of a surface received by a side looking radar
D: Image processed by optical differential laplacian operator.

In such a hybrid system, each domain finds its possibilities. Electronics permits the automatic spatial spectral analysis of Fraunhofer diffraction patterns. The interest of such a hybrid system was demonstrated during the processing of a great number of photographs (real time processing).

The result obtained is satisfactory in spite of the technical difficulties coming from the great number of holographic element manufacturing steps: limitation by the plotting table (resolution of the plotting, photoreduction, chemical processing resulting from the transfer of information on holographic film of resolution power), and the mechanical positioning of the hologram in a plane (6° liberty).

We have for instance generated holographic spatial high-pass filters that find applications in imagery, for the restitution of minute details and contours (Laplacian and gradient operators mathematical exponential) (Maxvellian operator + Gaussian operators whose transfer functions are given on the figures).

Example of method of realisation of holographic processor: gradient and Laplacian.

Let $s(x,y)$ = the bi-dimensional entry signal to the optical system of linear filtering, which has the transfer function:

$$H(u,v) = F.T.[h(x,y)]$$

$h(x,y)$ is the impulsive response
F.T. = Fourier transform
u,v = spatial frequencies
x,y = spatial variables

The response $r(x,y)$ of the optical system is given by the convolution $s(x,y)$ in $h(x,y)$.

$$r(x,y) = s(x,y) * X\ h(x,y) \xrightarrow{F.T.} R(u,v) = S(u,v).H(u,v)$$

If we want to apply a differentiation operator to the entry signal, for example, the gradient which is mathematically defined by:

$$\vec{\nabla} = \frac{\partial}{\partial x} + j \frac{\partial}{\partial y} \text{ with } j = \sqrt{-1}$$

$$H(u,v) = TF\ (\vec{\nabla}) = 2\pi(-v + ju)$$

$$H(u,v) = 2\pi\sqrt{u^2 + v^2}\ \text{Exp}\ [j\ \text{Arctg}\ (\frac{u}{v})]$$

$$|H(u,v)| = \text{amplitude} = 2\pi\sqrt{u^2 + v^2}\ \text{presents circular symmetry}$$

The response to the gradient operator is:

$$r\vec{\nabla}\ (x,y) = \frac{\partial}{\partial x} s(x,y) + j \frac{\partial}{\partial y} s(x,y) = \vec{\nabla}\ s(x,y)$$

The intensity is:

$$I(u,v) = \left| \frac{\partial}{\partial x} \times (s(x,y)) \right|^2 + \left| \frac{\partial}{\partial y} (s(x,y)) \right|^2.$$

The Laplacian which is scalar can be represented by only one parameter, the transmittance.

$$r \, \Delta(x,y) = s(x,y) \text{ is a scalar}$$

$$I \, \Delta(u,v) = \left| \frac{\partial^2}{\partial^2} s(x,y) + \frac{\partial^2}{\partial^2} s(x,y) \right|^2$$

3. EXPERIMENTAL SYSTEM

We have validated the above concept on several experiments.

The surface acquisition system is composed of an optical system for sending and processing the beam and a CCD sensor connected to a computer working with specialised software, which we have developed.

The laser used is HeNe in the visible spectrum (red = 633 nm) of 5 mW. The holographic lenses is made on DCG with the OPTRIGELAC method and has a focal length of 300 mm and a diameter of 80 mm. The beam used is parallel after collimation and has an effective diameter of 10 mm. We use a matrox PIP card an IBM PC and a Fairchild CCD camera to complete the system.

We have tested several samples already measured by classical ways provided by the French research center on machine tools. We found a good correlation on the 0.5 to 10 micron defect range, between the classical and the method we have developed, but our method gives much more information that for the moment cannot be related to any data base. This information is very useful in advanced mechanics, for instance the statistical repetition of defects on a given surface sensed at very high speed. We give the statistical weight of roughness repartition on a 10 points sampling. Figure 12.

4. CONCLUSION

We have validated through experiments and systems the efficiency of Fourier transform in the analysis of surfaces anested by computer holography.

The advantages compared to other methods are: noncontactual, nondestructive, fast, global and complete, low cost. It can be extended with the same advantages to screw analysis, transparent material control, powder and particle control. With new roughness standards and perhaps with new units it can be possible to have absolute high quality measurements. The main originality of our work lies in the introduction of optical computing methods in the measurement of surfaces.

We thank John Caulfield, adjunct professor at ENSPS and Director of the Applied Optics Science Center, University of Alabama, Huntsville, for rewarding discussions on the subject of his paper

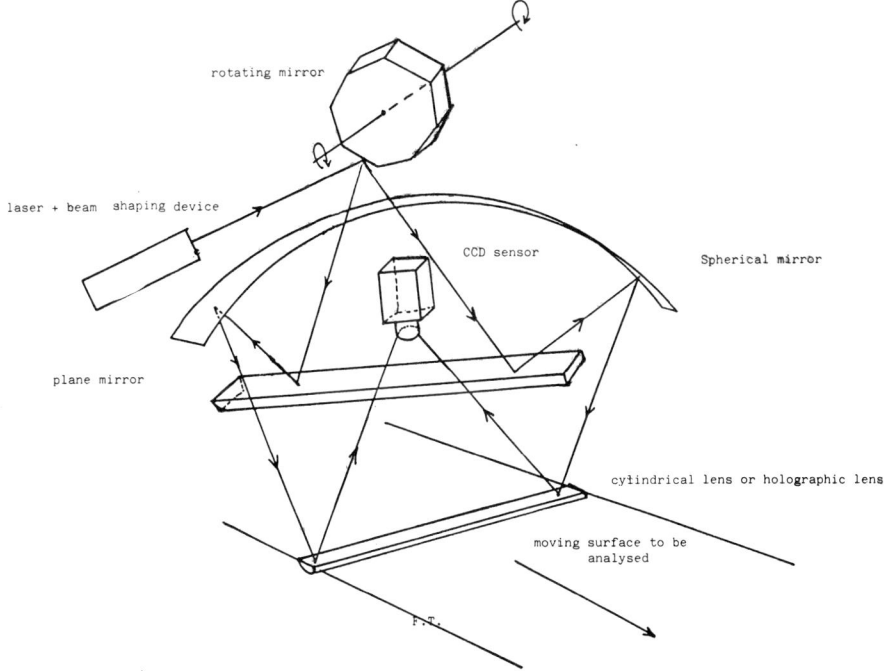

Fig. 12. Project of a moving surface analyser with F.T.

REFERENCES

1) P. Meyrueis et al. Les enjeux industriels de la photonique tome 1 et 2 - la documentation Francaise 1983.
2) Goodman, J. W. (1972). "Introduction à l'optique de Fourier et à l'Holographie", Masson et Cie.
3) Cathey, W. T. (1974). "Optical Information Processing and Holography", John Wiley and Sons, New York.
4) Sawatari, T. and Mueller (1977). "Surface Flaw Detection Using Optical Filters" International Advances in Non-destructive Testing. Vol. 5, pp. 1-15, Copyright 1977 Cordon and Breach Science Publishers, Inc. Primited on the United States of America.
5) Inari, T. (1980). "Inspection and Monitoring of Surface Roughness by Laser", Mitsubishi Electric Corporation Products and Development Laboratory, Amagasaki, Japan.
6) P. Meyrueis, C. Liegeois and M. Grosmann. Automatic sciences measuring system applied to biostereometrics, Springer Verlag Optical Sciences, Vol. 18 (1979).
7) P. Meyrueis, C. Draman and P. L. Wendel. Digital image processing with coherent light a method and some applications, Optical Engineering, Vol. 22, no. 3, Mai-June 1983.
8) P. Meyrueis. Photonics International, Mars 1984.
9) P. Meyrueis, M. Grosmann and C. Liegeois. Mesure controle et essais en lumière cohérente Mesure, régularisation, automatisme Août Septembre 1979.

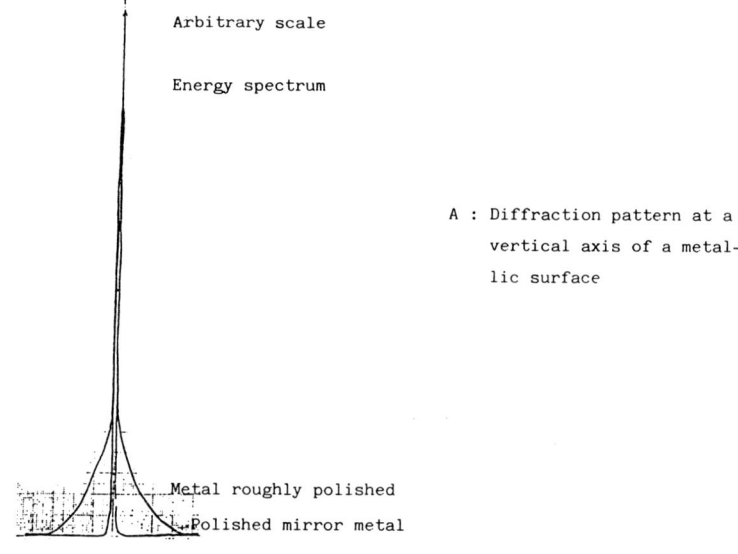

Pattern distances diffraction of a metallic surface

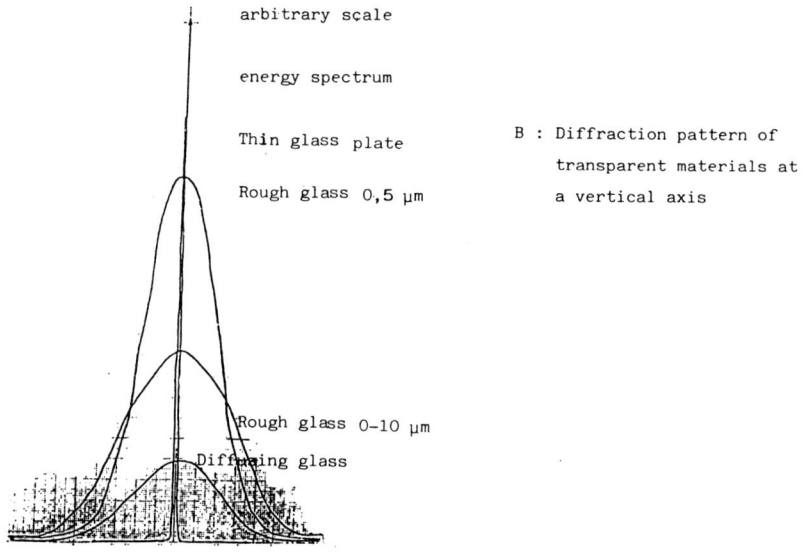

Diffraction pattern of transparent materials

Fig. 13. Example of diffraction pattern of metallic and transparent materials.

10) P. Meyrueis and C. Liegeois. Methode de réalisation et d'utilisation d'éléments optiques Holographiques à haute performance. Spectra No 105 - Vol. 13 Septembre 1985.
11) P. Meyrueis, P. Smiegielski and A. Dubourg. Effective practical use of Holography and related technologies, SPIE Vol. 210 - 1980.
12) E. Soubari, C. Liegeois and P. Meyrueis. Periodic structure defects in mechanical surfaces diagnosed through diffraction phenomena analysis. SPIE Vol. 240 - 1980.

This work was supported by the French-German Institute for the Application of Research and the Conseil Regional d'Alsace

FLEXIBLE METAL WAVEGUIDE FOR CO_2 LASER PROCESSING

M. Torjinsky*, P. Meyrueis* and C. Liegeois**

*Ecole Nationale Superieure de Physique de Strasbourg
**XIAL Strasbourg

SUMMARY

In this paper we describe a method and a system for carrying CO_2 laser light for laser processing, particularly with a robot arm.

1) Brief history of hollow CO_2 laser waveguides

It is not possible to use an optical fiber to bring CO_2 laser light at 10.6 micron wavelength from one point to another, because there is no existing material able to be transformed into flexible optical fibers that is transparent at this wavelength. Generally, laser light at 10.6 microns is carried through multimirror systems using a complex mechanical design allowing the mirror to stay parallel in a limited range of displacement. But these systems are fragile, expensive and deteriorate the beam rather rapidly with time, especially with the small defects that appear on the surface of the mirrors, or their positioning.

We have therefore explored other possibilities. We have selected for miscellaneous reasons the hollow waveguide. After some complementary experiments, we selected the metallic structure. For the geometry we have tested both circular and helicoidal waveguides.

For the circular waveguide, the losses are around 80% with polished copper with a diameter of 5 mm with a bending radius of 1 m. Results are better is we work in the TEM_{OO} mode. A variant is an "open circular waveguide". Fig. A.

This guide allows lower losses, but prevents the position of the beam at any point in a given volume. Besides of that, it is dangerous for leaks with misuse.

The helicoidal waveguide mainly uses a "one face effect", that is to say that the beam is only reflected on one face. This phenomenon induces a simplification of the geometry (no parallelism is necessary), but there is a very serious security problem. The device that behaves like a phone cord can leak if it is squeezed too heavily or if it is wrongly handled.

There is a possible semi-rectangular configuration that can be very good when it is straight, but much less so when it is bent (Fig. B). In addition it is almost impossible to manufacture.

2) Rectangular waveguide

The design of the waveguide is displayed in Fig. 2. The complete system will include a CO_2 laser, polarizing mirrors and cylindrical lenses that will couple the laser beam with the guide and that will handle the output of the guide itself.

The guide is composed of two metal sheets (spring steel) with a thickness of 1 mm and 10 mm wide. The width can be modified depending on the power to be carried up to 5 cm. The inside layer is in gold or palladium. The lateral walls are 0.5 mm thick. They are made of another sheet glued to the two other parts. The lateral walls are 1 cm wide.

Loss with the guide on a straight position:

If we consider the nth incident made, it will have the losses that can be modeled by:

$$N(0).A(0)$$

where N(0) is the number of reflections, and, A(0) is the loss with every reflection. It is possible to quantify the loss in function of the thickness of the guide and its width. We observe a rapid increase of the loss with the increase of these parameters.

Loss with the guide bent:

Under a given curvature ray Ro, only one wall is supporting the reflected beam.

Modeling of the loss uses Maxwell equations. The loss is then independent of the wavelength, of the curvature ray and of the thickness of the guide. The straight guide being dependent on these parameters, there is a transition point between these two configurations from Ro = 10 cm. From experimental work, we can notice that surface defects for the inside part of the guide is the most important parameter.

Loss of the twisted guide:

This configuration can be considered as a mixing of the previous ones. But some specific phenomena occur. The plane polarisation parallel to the wall induces the increase of the percentage of low loss made. The twisting of the guide transforms the propagation mode. It is shown and validated experimentally that the loss from twisting is independent from the mode. Meanwhile, we can have an important mode degradation if the input angle is not accurate enough. This accuracy is around 0.6°. By tests, we have evaluated the accuracy of the positioning of the wall; it is very close to the theoretical value, i.e. around 0.3 mm.

We can say roughly that the losses of the straight guide are proportional to 1/a (a being the active thickness of the guide), the loss of the bent guide being proportional to a. For a given application and a given way of operating the guide, an optimum exists.

Practical realisation of the guide: (Figs. 1-6)

The lateral and the upper and lower walls are glued or laser welded. The mounting is described in FIG. with several possibilities. The XIAL product GOLAS manufactured with this method has the following parameters in the case, for example, of low laser power:
- power: 10 to 100 W.
- material: high elasticity steel gold plated.
- thickness: (intern) of the guide 0.5 mm.
- width: 8 mm.
- length: 0.5-1.5 m.
- minimal curvature ray: 15 cm.

CONCLUSION

We think that the system that we describe above with its advantages and its inconvenients is well suited for laser processing. We do not have enough experience is real workshop use to be sure that the life duration is what we expect it to be as we have simualted (around 2000 hours). We do not see any other possibility of waveguiding 10.6 micron laser light for CO_2.

We hope that our solution will be useful in robotics applications of CO_2 laser and in surgery applications of the lasers.

XIAL company expects to manufacture the CO_2 laser light waveguide from the end of 1987 under the name GOLAS.

REFERENCES

R. Ohlmann, P. Richards and M. Tinkhamm, 1958, JOSA, 46, 6, 531-3.

T. Poehler and R. Turner, 1970, App. Opt. 9, 4, 971-3.

P. Smith, P. Maloney and O. Wood, 1973, App. Phy. Lett. 23, 9, 524-

M. Nishihara, T. Inoue and J. Koyama, 1974, App. Phy. Lett. 25, 7, 391-

E. Garmire, T. McMahon and M. Bass, 1976, App. Opt. 15, 1, 145-50.

K. Larkmann and U. Steier, 1976, App. Opt. 15, 5, 1334-40.

E. Garmire, T. MacMahon and M. Bass, 1976, App. Phy. Lett. 29, 4, 254-

H. Nishihara, T. Mukai and T. Inoue, 1976, App. Phy. Lett. 29, 9, 577-

E. Garmire, 1976, App. Opt. 15, 12, 3037-9.

E. Garmire, T. MacMahon and M. Bass, 1977, App. Phy. Lett. 31, 2, 92-4.

H. Krammer, 1978, App. Opt. 17, 2, 316-9.

M. Marhic, L. Kuan and M. Epstein, 1978, App. Phy. Lett. 33, 7, 609-.

M. Marhic, L. Kuan and M. Epstein, 1978, App. Phy. Lett. 33, 10, 874.

E. Garmire, T. MacMahon and M. Bass, 1979, App. Phy. Lett. 34, 1, 35-7.

E. Garmire, T. MacMahon and M. Bass, 1980, JOE, 16, 1, 23-36.

Y. Mizushima, T. Sugeta and H. Nishihar, 1980, App. Opt. 19, 19, 3259-60.

M. Miyagi, 1981, App. Opt. 20, 7, 1221-9.

M. E. Marhic and E. Garmire, 1981, Appl. Phy. Lett. 38, 10, 74.

M. E. Marhic, 1981, App. Opt. 20, 19, 3436-41.

G. F. Vander Voort, 1984, Handbook of Metals, 35.

M. F. Von Allmen, 1984, Phy. Proc. Mat. 86, 49-74.

V. Bartimord, A Cutolo and S. Solimeno, 1985, Phy. Proc. Mat. 86, 231-61.

G. Fritzsche, 1985, Phy. Proc. Mat. 86, 469-74.

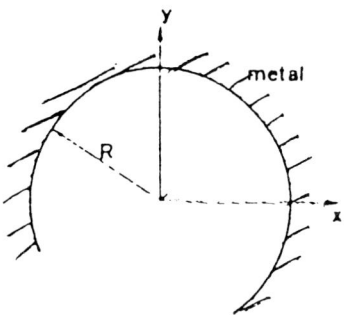

Fig. A.

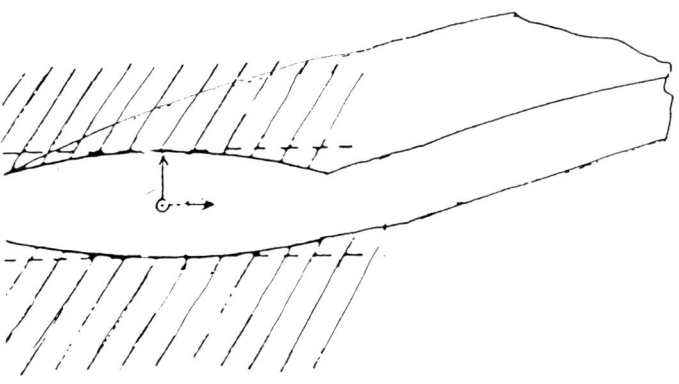

Fig. B.

LEGEND OF THE INDEX
Figs. 1 to 6

- 1- Rectangular waveguide in a twisted position
- 2- Laser source
- 3- Target
- 4- Beam expander
- 5- Cylindrical lens
- 6- Holding of the input system
- 7- Input of auxiliary gaz
- 8- Axis of input
- 9- Output of the waveguide
- 10- Output holding
- 11- Output cylindrical lens
- 12- Focusing lens
- 13- Robot holder
- 14- Clutch
- 15- Auxiliary gaz output
- 16- Protection cladding
- 17- Upper wall of the waveguide
- 18- Lower wall of the waveguide
- 19- Lateral wall of the waveguide
- 20- Idem
- 21- Grinded lateral component
- 22- Idem
- 23- Grinded upper and lower walls
- 24-
- 25- Working surface of the lateral wall
- 26-
- 27- Simplified wall
- 28-
- 29- Glue or welding line
- 30-

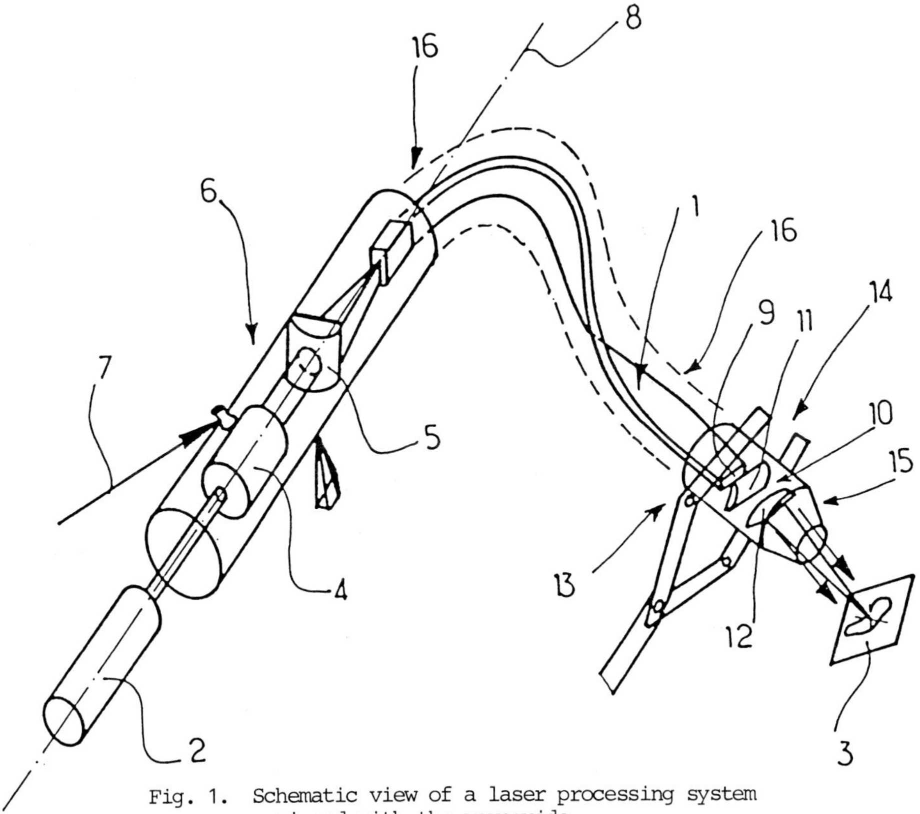

Fig. 1. Schematic view of a laser processing system equipped with the waveguide.

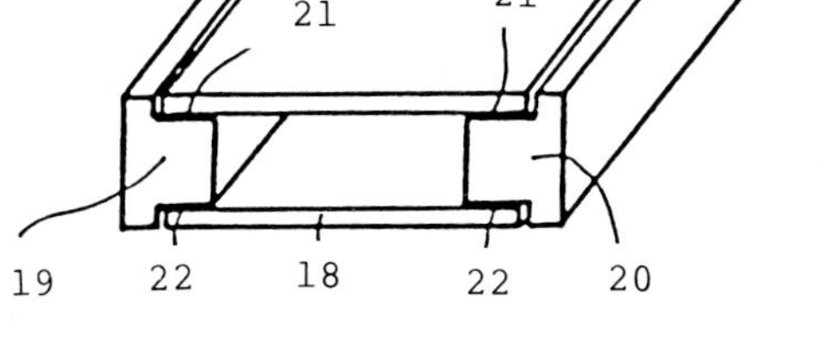

Fig. 2. Description of the architecture of the waveguide for low power.

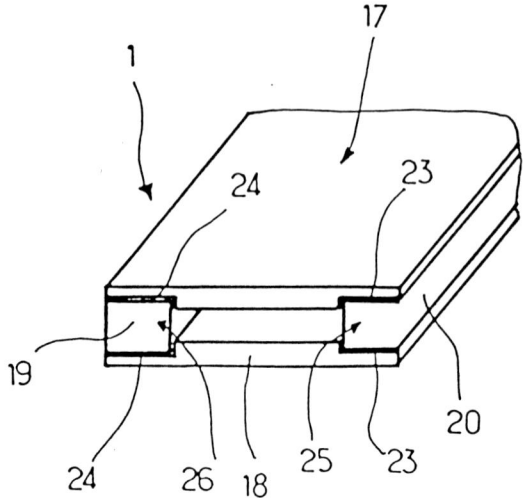

Fig. 3. Description of the architecture of the waveguide for high power.

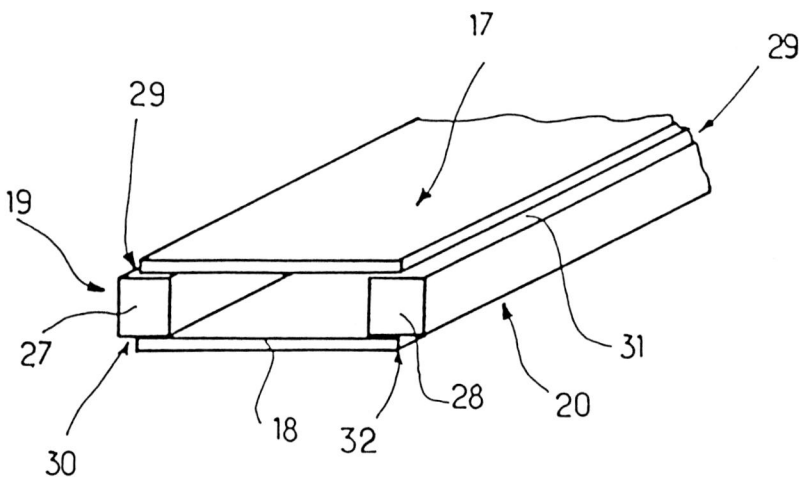

Fig. 4. Architecture for economical welded waveguide.

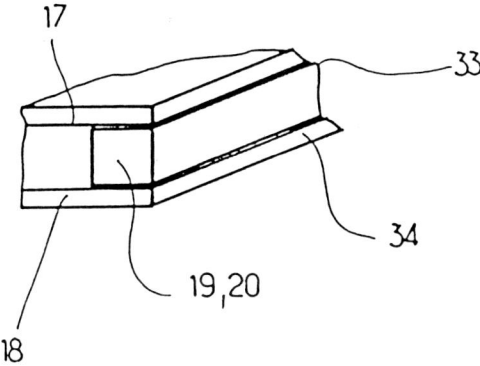

Fig. 5. Architecture for economical glued waveguide.

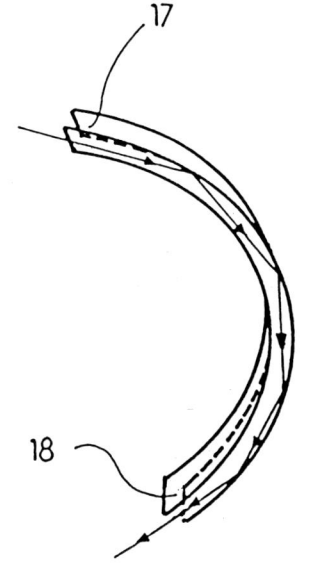

Fig. 6. Propagation of light in a bended waveguide.

AUTHOR INDEX

Antona, P.L., 143
Appiano, S., 143

Beitialarrangoitia, J.C., 213, 227
Bouzid, A., 307

Cahkari, A., 297
Ching-Piao, Hu, 259

de Beurs, H., 27
de Hosson, J.T.M., 27
Dekumbis, R., 289
De Santis, P., 271

Fantini, V., 251
Fernandez, P., 289

Galantucci, L.M., 57
Garcia de Vicuna, G.E., 213, 227
Ghosh, S.K., 213, 227
Giordano, L., 49
Godijk, J., 131
Gori, F., 271
Guattari, G., 271

Hornbogen, E., 117

Incerti, G., 251

Kahrmann, W.N., 41
Kreutz, E.W., 159

Liegeois, C., 325

Magnanelli, S., 57
Mandziej, S., 131
Mayer, H., 289
Meyrueis, P., 279, 297, 307, 325
Ming, Chang, 259
Mordike, B.L., 3, 41
Moschini, R., 143

Palma, C., 271
Piel, R., 279

Querry, M., 195

Ramous, E., 15, 49
Riabkina-Fishman, M., 101
Ruta, G., 57

Seegers, M.C., 131
Shao, Huimeng, 89
Soubari, E., 307
Staniek, S., 117

Takemoto, M., 75
Tiziani, A., 49
Torzynski, M., 307, 325

Wissenbach, K., 159

Zahavi, J., 101
Zhang, Luting, 89
Zhu, Jingpu, 89

JAN 0 4 1990